Evolution in Health and Disease

Evolution in Health and Disease

EDITED BY

Stephen C. Stearns

Basle, Switzerland

OXFORD • NEW YORK • TOKYO
OXFORD UNIVERSITY PRESS
1999

Oxford University Press, Great Clarendon Street, Oxford OX2 6DP

Oxford   New York
Athens   Auckland   Bangkok   Bogota   Bombay
Buenos Aires   Calcutta   Cape Town   Dar es Salaam
Delhi   Florence   Hong Kong   Istanbul   Karachi
Kuala Lumpur   Madras   Madrid   Melbourne
Mexico City   Nairobi   Paris   Singapore
Taipei   Tokyo   Toronto   Warsaw

and associated companies in
Berlin   Ibadan

Oxford is a trade mark of Oxford University Press

Published in the United States
by Oxford University Press Inc., New York

A catalogue record for this book is available from the British Library

Library of Congress Cataloging-in-Publication Data
Evolution in health and disease / edited by Stephen C. Stearns.
Based on a conference held in Sion, Switzerland, 6–12 Apr. 1997.
Includes bibliographical references and index.
1. Medical genetics—Congresses. 2. Human evolution—Congresses.
3. Diseases—Causes and theories of causation—Congresses.
I. Stearns, S. C. (Stephen C.), 1946– .
[DNLM: 1. Evolution congresses   2. Adaptation, Physiological
congresses.   3. Health congresses.   4. Disease congresses.   QH
366.2 E926   1999]   RB155.E96   1999   616'.042—dc21   98–21522

ISBN 0 19 850110 2 (Hbk)
ISBN 0 19 850445 4 (Pbk)

Typeset by Footnote Graphics, Warminster, Wilts
Printed in Great Britain by
The Bath Press, Avon

To Silke Bernhard, MD, Dr hc

1928–1993

who designed the Dahlem Conference format
that was essential
for the success of our meeting.

She dedicated her life to improving communication
among scientists.

These scientists dedicate their book to her.

Her contribution lives on.

ACKNOWLEDGEMENTS

This book is based on a conference held in Sion, Switzerland, 6–12 April 1997 and made possible by the generous support of the following institutions:

European Science Foundation, Strasbourg

Institut Kurt Bösch, Sion

Kontaktgruppe-Forschung, Basle Chemical Industry

Swiss Nationalfonds, Berne

The Wellcome Trust, London

SmithKline Beecham, London

Swiss Academy of Medical Sciences, Basle

Swiss Academy of Natural Sciences, Berne

Francine Crettenand at Institut Kurt Bösch helped to organize the conference. Werner Arber made the right phone call at the right time, and Michel Delaloye provided essential support. All the authors thank them all.

CONTENTS

CONTENTS

CONTRIBUTORS

PETER AABY
Epidemiology Research Unit, Statens Seruminstitut, Artillerivej 5, 2300 Copenhagen, Danemark

ROY M. ANDERSON
Department of Zoology, South Parks Road, Oxford OX1 3PS, UK

RUSTOM ANTIA
Department of Biology, Emory University, 1510 Clifton Road, Atlanta, Georgia 30322, USA

CHARLES BANGHAM
Department of Immunology, Imperial College School of Medicine, St Mary's, London W2 1PG, UK

FERNANDO BAQUERO
Department of Microbiology, Ramon Y. Cajal Hospital, C. Colmenar KM9 100, 28034 - Madrid, Spain

GUIDO BARBUJANI
Dipartimento di Biologia, Università di Ferrara, via L. Borsari 46, I-44100 Ferrara, Italia

DAVID J.P. BARKER
MRC Environmental Epidemiology Unit, Southampton General Hospital, Southampton S016 6YD, UK

RICHARD BAX
SmithKline and Beecham Pharmaceuticals, New Frontiers of Science Park, Third Avenue, Harlow, Essex CM19 5AW, UK

ROBIN I.M. DUNBAR
Population Biology Research Group, School of Biological Sciences Nicholson Building, University of Liverpool, Liverpool L69 3BX, UK

GEORGIA DUNSTON
Department of Microbiology, Howard University College of Medicine, 520 West Street, Room 3012, Washington DC 20059, USA

S. BOYD EATON
Departments of Radiology and Anthropology, Emory University, Atlanta, Georgia 30322, USA

STANLEY B. EATON III
829 Harold Ave SE, Atlanta, GA 30316-1219, USA

DIETER EBERT
Zoology Institute, Rheinsprung 9, CH-4051 Basle, Switzerland

PAUL EWALD
Department of Biology, Amherst College, Amherst, Massachusetts 01002-5000, USA

LAURENT EXCOFFIER
Department of Anthropology, University of Geneva, 12 G. Revillod, CH-1227 Geneva, Switzerland

CALEB E. FINCH
Department of Gerontology, Gerontology 324A, University of Southern California, University Park, mc-0191, Los Angeles, CA 90089-0191, USA

ROGER GOSDEN
University of Leeds, School of Medicine, Division of Obstetrics and Gynaecology, D Floor, Clarendon Wing (LG1), Belmont Grove, Leeds LS2 9NS, UK

SUNETRA GUPTA
Department of Zoology, South Parks Road, Oxford OX1 3PS, UK

DAVID HAIG
The Botanical Museum, Harvard University, Cambridge, MA 02138, USA

JOHN M. HANCOCK
Gene and Genome Evolution Group, MRC Clinical Sciences Centre, Royal Postgraduate Medical School, Hammersmith Hospital, Du Cane Road, London W12 0NN, UK

IAN HASTINGS
Institute of Cell, Animal and Population Biology, University of Edinburgh, West Mains Road, Edinburgh EH3 9JT, UK

EVELYNE HEYER
Laboratoire d'Anthropologie Biologique, Musée de l'Homme 17 pl du Trocadéro, F-75016 Paris, France

ADRIAN V.S. HILL
Molecular Immunology Group, Room 375, Institute of Molecular Medicine, John Radcliffe Hospital, Headington, Oxford OX3 9DU, UK

EDWARD C. HOLMES
Department of Zoology, South Parks Road, Oxford OX1 3PS, UK

SHARON L.R. KARDIA
Department of Human Genetics, University of Michigan, Ann Arbor, Michigan 48109-0618, USA

THOMAS B.L. KIRKWOOD
Department of Geriatric Medicine, 3.239 Stopford Building, Oxford Road, Manchester M13 9PT, UK, and Collegium Budapest (Institute for Advanced Study), Szentharomsag utca 2, H-1014 Budapest, Hungary

JAN KLEIN
Max-Planck-Institut für Biologie, Abteilung Immungenetik, Corrensstrasse 42, 7400 Tubingen, Germany

JACOB KOELLA
Department of Zoology, University of Aarhus, DK-8000 Aarhus C, Denmark

ANDRÉ LANGANEY
Department of Anthropology, University of Geneva, 12 G. Revillod, CH-1227 Geneva, Switzerland

ED LEGRAND
R.W. Johnson Pharmaceutical Research Institute, 1000 Route 202, Raritan, NJ 08869, USA

BRUCE LEVIN
Department of Biology, Emory University, 1510 Clifton Road, Atlanta, Georgia 30322, USA

MARC LIPSITCH
Department of Biology, Emory University, 1510 Clifton Road, Atlanta, Georgia 30322, USA

RUTH MACE
Department of Anthropology, University College London, Gower St, London WC1E 6BT, UK

ANGELA MCLEAN
Institute for Animal Health, Compton, Nr Newbury, Berkshire RG20 7NN, UK

GEORGE M. MARTIN
University of Washington, Department of Pathology, Box 357470, Seattle, WA 98195, USA

JOHN MAYNARD SMITH
School of Biological Sciences, University of Sussex, Falmer, Brighton BN1 9QG, UK

URS A. MEYER
Biozentrum, Klingelbergstr. 70, CH-4056 Basle, Switzerland

MANFRED MILINSKI
Universität Bern, Zoologisches Institut, Abteilung Verhaltensökologie, CH-3032 Hinterkappelen, Switzerland

ARNO G. MOTULSKY
Departments of Medicine and Genetics, Box 356423, University of Washington, Seattle, WA 98195-6423, USA

RICHARD MOXON
University Department of Paediatrics, Level 4, John Radcliffe Hospital, Headington, Oxford OX3 9DU, UK

RANDOLPH M. NESSE
Department of Psychiatry, University of Michigan Medical School, 5057 ISR, Box 1248, Ann Arbor, MI 48106

LINDA PARTRIDGE
Department of Genetics and Biometry, University College London, 4 Stephenson Way, London NW1 2HE, UK

GILLES PISON
Laboratoire d'Anthropologie biologique, UMR 152 CNRS, Musée de l'Homme, 17 place du Trocadéro, 75116 Paris, France

SILVANO PRESCIUTTINI
Dip. Sc. Amb. Terr- Genetica, Via San Giuseppe 22, I-56100 Pisa, Italy

ANDREW READ
Cell, Animal and Population Biology, University of Edinburgh, Edinburgh EH9 3JT, UK

HEINZ RICHNER
Universität Bern, Zoologisches Institut, Abteilung Verhaltensökologie, CH-3032 Hinterkappelen, Switzerland

MICHAEL R. ROSE
Department of Ecology and Evolutionary Biology, School of Biological Sciences, University of California, Irvine, CA 92717, USA

ALICIA SANCHEZ MAZAS
Department of Anthropology and Ecology, University of Geneva 12, rue Gustave-Revilliod, CH-1227 CAROUGE, Geneva, Switzerland

AKIRA SASAKI
Department of Biology, Faculty of Science, Kyushu University, Fukuoka 812, Japan

DENIS SHIELDS
Genetics Department, Trinity College, Dublin 2, Ireland

NOEL SMITH
School of Biological Sciences, University of Sussex, Falmer, Brighton BN1 9QG, UK

TOM SMITH
Swiss Tropical Institute, Socinstrasse 57, CH-4051 Basle, Switzerland

STEPHEN C. STEARNS
Zoology Institute, Rheinsprung 9, CH-4051 Basle, Switzerland

JARI STENGÅRD
Department of Human Genetics, University of Michigan, Ann Arbor, Michigan 48109-0618, USA

BEVERLY I. STRASSMANN
Department of Anthropology, University of Michigan, Ann Arbor, Michigan 48109-1382, USA

FRANCOIS TADDEI
Laboratoire de Mutagenese, Institut Jacques Monod, 2 place Jussieu, 75251 Paris cedex 05, France

ALAN TEMPLETON
Department of Biology, Washington University, St. Louis, MO 63130, USA

DAVID THALER
The Rockefeller University, 1230 York Avenue, New York, NY 10021-6399, USA

CLAUS WEDEKIND
Universität Bern, Zoologisches Institut, Abteilung Verhaltensökologie, CH-3032 Hinterkappelen, Switzerland

GEORGE C. WILLIAMS
Department of Ecology and Evolution, State University of New York, Stony Brook, NY 11794-5245, USA

WILLIAM L. WISHART
Neurogenomics Group, Building 360 Office 605, NOVARTIS Pharma AG, Basle, CH-4002, Switzerland

Part I: Introduction

1

INTRODUCING EVOLUTIONARY THINKING

Stephen C. Stearns

Human pathogens have evolved and are evolving drug resistance; emerging diseases have evolved and are evolving to infect us; we have evolved and are evolving in many ways, including symptoms, physiological responses, and behaviour patterns. However, medical research and medical education have paid relatively little attention to evolutionary biology. This book explains why medical doctors might want to consider evolutionary thinking as a standard part of their tool kit (also see Nesse and Williams 1994, one motivation for this book).

Evolutionary biology is a rich collection of well-developed alternative approaches to the interpretation of biological diversity and organismal design. It is not just selectionist thinking about adaptations, although it is certainly that. It is also the study of what genealogies and phylogenies can tell us about relationship and history. (The methods used to reconstruct history make use of the observation that much of the variation in DNA and proteins is neutral or close to neutral—that part of evolutionary biology is certainly not selectionist.) It is the study of conflicts: conflicts between hosts and parasites, between parents and offspring, between genes with different transmission patterns. Participants in conflicts often come off worse than they would have without the conflict. In such cases, evolutionary biology is as much the study of maladaptation as it is of adaptation. Evolutionary biology is also the study of genetic and phenotypic dynamics, regardless of whether they lead to adaptation or not. Sometimes they do not, and sometimes they cannot. No ideological monolith, it is rich in alternatives that can be played off against each other to provide a self-critical, well-tested, and increasingly re-

liable interpretation of the natural world, of which humans are a part.

This book describes how evolutionary thinking gives insight into human health and disease; this chapter summarizes evolutionary principles. We begin by describing how evolutionary biologists think, then give reasons why medical doctors might be interested in what evolutionary biologists have to say.

HOW EVOLUTIONARY BIOLOGISTS THINK

Evolutionary biologists want to understand how the variation in reproductive success that causes selection arises, how the genetic variation that enables a response to selection originates and is maintained, and how that response is constrained by geography, time, inheritance, conflicts, development, and history. Their insights apply to all organisms, including humans. Evolution contains many special areas of research in which different questions are asked and different approaches are used. There are many ways of thinking about evolution. Here are some of the more important ones. The views of each of these specialties are often useful alternatives to dogmas that develop in another.

Population geneticists think about microevolution, which occurs within populations over relatively short periods of time, about genes, about the different forms that one gene can take—its alleles—changing frequency or being held by various mechanisms in stable intermediate frequencies. Their central problem is what maintains genetic variation. Among the candidate explanations are natural selection, gene flow, and the drift of neutral genes. Population geneticists tend not to worry about the design of phenotypes. This area is well represented in medical research and education. For example, population genetic models of HIV make it clear that treating patients with several drugs at once is better than treating them with the same drugs sequentially, for such treatment prevents or delays the emergence of resistant mutants.

Evolutionary ecologists and anthropologists think about the design of phenotypes for survival and reproduction, particularly about traits like age and size at maturity, number and size of offspring, lifespan and ageing (these are all life history traits and this specialty is referred to as life history theory), various strategies of favouring offspring of one sex over the other (this is part of sex allocation theory), and the consequences of competition for mates and of choosing mates based on particular criteria (this is the field of sexual selection). They tend not to concern themselves with genetic details. This area has been largely missing from medical thought; it is represented here in Chapters 7, 8, 9, 10, and 22.

Molecular evolutionists think about history recorded in DNA sequences. Some of them examine parts of the genome that are not translated into proteins and that have no influence on the phenotype and processes occurring in non-coding DNA that have little to do with adaptive, selection-driven change. They tend not to worry about adaptive change, either in gene frequencies or in genotypes, except as distractions that need to be noted and controlled. This area is well represented in medical genetics and in Chapters 3, 15, 16, and 17.

Systematists think in terms of evolutionary trees, give great weight to history, and concentrate on variation among species. For them, the major evolutionary problem is to infer relationships among species so that they can reconstruct the structure of life on the planet, not to understand why gene frequencies change or how phenotypes are designed for reproductive success. This area may not appear directly relevant to medical research, but it has greatly aided the interpretation of sequence data from

viruses and bacteria (Chapters 15 and 16). The methods of systematics were used, for example, to trace the persons responsible for the transmission of HIV in Florida and Sweden. It has become a part of forensic medicine.

Paleontologists think in deep time and concentrate on large-scale trends and major events, such as adaptive radiations, mass extinctions, and increases in body size within large groups over long periods of time. They often do not notice processes occurring within periods of less than a hundred thousand years, but they do see the big picture with particular clarity. This area, not directly relevant to medical treatment, provides useful cultural background. For example, it tells us how old, in evolutionary terms, the different parts of our bodies are: our hands are very old (hundreds of millions of years), our chin is quite young (less than a million years).

That was a brief description of evolutionary biology in terms of research specialties. The field can also be divided by analytical approach. Evolutionary change in a population of organisms of both sexes, with each individual containing thousands of genes affecting many traits with impact on survival and reproduction, is too complex to analyse in detail. There are four major ways of simplifying it. Each leads to a different way of thinking about evolution with its own advocates, its own school, and its own focus.

One approach concentrates on changes in gene frequencies, focuses on genetic mechanisms, and simplifies the interactions of phenotypes with their environment. This is the genetic dynamics approach that finds wide application in population and quantitative genetics and in sexual selection theory. It seeks to answer the question: how will gene frequencies change? A second approach seeks to explain the design of whole organisms at evolutionary equilibrium by analysing the interactions of the phenotypic traits contributing to reproductive success and by simplifying their genetics. This is the optimization approach used in life history evolution and behavioural ecology. It seeks to answer the question: what is the state of the phenotype at evolutionary equilibrium? The third approach is used whenever selection is frequency dependent, when the fitness of one thing depends on its relative frequency in the population, when success depends on how an opponent responds. This is the game theory approach, which has had great success in explaining the evolution of behaviour, sex ratios, and evolutionary conflict. It also seeks to answer the question: what is the state of the phenotype at evolutionary equilibrium? However, it allows for more complex interactions than does optimization theory. A fourth approach is used when we want to understand the history of a population by studying the patterns in its genes and comparing them with those found in related populations. This phylogenetic approach uses the information in DNA sequences and the logic of modern methods implemented on computers to reconstruct with increasing reliability the history of populations and species, human and non-human. It makes the assumption that the DNA variation used is neutral, close to neutral, or under equivalent selection pressures in all lineages. It seeks to answer the questions: what is the history of this set of populations or species, and how can we reconstruct that history by inferring their evolutionary relationships?

All four approaches are legitimate simplifications of a complex process. None of them retain all important features of that process. Therefore one should remember, when adopting one of them, to check the consistency of assumptions, interpretations, and predictions with those that would have been made had one adopted another approach. (This sounds easy in theory but is surprisingly hard in practice, for both technical and psychological reasons.)

EVOLUTIONARY EVIDENCE: SPECIAL PROBLEMS IN HUMANS

Evolutionary biologists are accustomed to dealing with kinds of inferential evidence that are not commonly used in medical research, including phylogenetic reconstructions of trees of relationship, comparisons across species and higher taxa, and descriptions of unmanipulated field populations. They use these kinds of evidence because they study processes that, in many cases, cannot be subjected to experimental analysis. Their work is often made more rigorous by quantitative mathematical models that make testable predictions and strengthen the interaction of ideas with evidence. Procedures of inference are now in principle reliable; their strengths and weaknesses are well understood. You will encounter the results in many of the chapters of this book.

Experiments are also done in many branches of evolutionary biology. In a particularly strong form of experiment, experimental evolution, the assumptions and predictions of evolutionary models are tested by creating conditions under which certain traits are expected to evolve, then seeing whether they actually do evolve. The expectations are shaped by a theoretical analysis that can be quite sophisticated. This approach requires organisms with short generation times that can be cultured in large numbers, and until now it has only been applied to bacteria, single-celled algae and protozoa, and fruit flies. The results (e.g. Elena *et al.* 1996) promise to strengthen the empirical basis of evolutionary thought.

Humans have long generation times and are not ideal material for evolutionary studies. Selection pressures strong enough to change gene frequencies measurably in a single generation are quite unusual, and with human generation times of 25–30 years, direct observations of gene frequency changes driven by selection are particularly difficult to obtain in our species. Nevertheless, there is enough inferential evidence to lend credibility to many evolutionary principles in humans.

Some of the best evolutionary evidence pertaining to medical research and practice has been obtained on infectious diseases, particularly the evolution of antibiotic resistance and the evolution of virulence. In both cases, evidence from human medical research is also widely discussed among evolutionary biologists because it is some of the best data available on problems they find of central interest, including the potential rate of evolution and the possibility of quite local adaptation.

REASONS FOR MEDICAL DOCTORS TO THINK ABOUT EVOLUTION

1. Each human individual has had a slightly different evolutionary history, and each has a different genetic make-up. This leads to differences in the way that different human individuals react to drugs and to diseases, differences that can result in life or death (Chapters 3, 4, 5, and 6).

2. Micro-organisms and cancer cells rapidly evolve resistance to drugs. This has important implications for drug design and treatment (Chapters 11 and 13).

3. The vaccination of a human population against a disease exerts a strong selection pressure on that disease; it will show an evolutionary response. Evolutionary analysis of vaccine design and application helps to reduce the chances of unpleasant surprise (Chapter 12).

4. Evolutionary theory tells us why virulence evolves to a certain level and no further and what measures could be taken to reduce it. Changes in our lifestyle, in treatment, and in public health measures could all cause virulence to evolve, for better or for worse (Chapters 14 to 18).

5. Why are so many sperm needed for fertilization? Why are so many eggs ovulated but discarded? Why do both the placenta and the ovary make apparently excessive quantities of reproductive hormones during pregnancy? And why are some fetal proteins derived only from the father's genes while others are derived only from mother's genes (genomic imprinting)? The answers may come from the evolutionary analysis of genetic conflicts (Chapter 7).

6. Human sexual behaviour, reproduction, and the assurance of parenthood are affected by evolutionary forces, often with consequences for the welfare of sons versus daughters. Some of the reasons for the neglect and abuse of children are evolutionary. Understanding why such things occur should help us to prevent them (Chapters 8, 9, and 10).

7. Symptoms may be adaptations. They may also simply be by-products, the reactions of organisms to novel environments. In either case, the best treatment requires understanding of why symptoms evolved (Chapter 2).

8. The problems of ageing result from selection operating on the whole human life cycle, from conception to birth to maturity to death. Because evolution operates on reproductive success, selection pressures drop with age and disappear in postreproductive individuals. Because up to a point more fitness can be gained by investing in reproduction than in maintenance that would improve survival, most organisms must evolve senescence. By understanding why we age, we can understand the consequences of treating the symptoms of ageing and of attempting to prolong life (Chapters 19, 20, 21, and 24).

NATURAL SELECTION

We distinguish between selection, which depends on variation in reproductive success, and the reaction to it, a genetic change in the population, which depends on inheritance. Organisms vary in how many offspring they have, whether they have them earlier or later in life, and in how many of their offspring survive to reproduce. Natural selection is simply this variation in reproductive success; it is necessary for adaptive evolution to occur. A second condition is also necessary. Only if some of the variation in reproductive success is inherited will a response to natural selection take place, for only then will the offspring from the better-performing parents inherit some of the capacity for better reproductive performance.

Natural selection is deceptively simple. Although it is hard to believe that such a mechanism could design an eye or a brain, it is incredibly powerful and rapidly produces highly improbable states. Consider the 31 letters in the sentence, THERE IS GRANDEUR IN THIS VIEW OF LIFE, as a sequence of 31 genes each with 26 alleles. If we assembled such sequences at random, we would have to sort through 26^{31} combinations of letters to find this one—as though enough monkeys typing long enough did eventually produce

Hamlet, except in this case we are asking them to produce the first sentence of the last paragraph in *The origin of species.*

Note, however, that combinations of genes that are good in one context are often good in others, and inheritance of naturally selected variants will preserve them. Once part of the message is found, that part need not be lost. In our example, strong selection retains the correct letter whenever it occurs. If we start with any random sequence of 31 letters and retain all the letters that happen to be correct, then repeat the process by generating new letters at random for the ones that are not yet correct, we get to the right sequence in about 100 trials. This is 30 orders of magnitude faster than a random search. Natural selection is so efficient that in this case its performance was 15 orders of magnitude faster than the difference between the blink of an eye and the origin of the dinosaurs.

This apparently artificial example is readily translated into an experiment on the evolution of RNA in test tubes (Schuster 1993; Schuster *et al.* 1994). One can extract, from a virus that infects a bacterium, an enzyme that replicates RNA. Given some RNA and a supply of the four nucleotides from which RNA is made, this enzyme rapidly produces a large population of RNA molecules in solution in a test tube. At the start of the experiment, the lengths of the molecules, and the sequences of the nucleotides, are close to random. By transferring a drop of the solution into a new test tube every 30 minutes, one selects those RNA molecules that are present at highest frequency, for they are most likely to be transferred in the drop. Replication is quite good but not exact, and in about 1 in 10000 cases the wrong nucleotide gets substituted. Thus the growing population of RNA molecules becomes variable, and some variants are replicated faster than others. Two types of molecules have an advantage in this situation: small ones, and those with characteristic structures enhancing affinity to the replicating enzyme.

After many transfers—more than 100—a rather large, complex molecule dominates the population—which molecule in particular depends on the details of the experiment. One such dominant molecule, with a structure that greatly enhances its affinity to the replicating enzyme, is 218 nucleotides long, and it, among others, emerges repeatedly from such experiments. The chances of such a molecule appearing even once, randomly, are 1 in 4^{218} or 1 in 10^{131}. Since the experiment is set up in such a way that there are only about 10^{16} molecules in a test tube just before transfer, the procedure screens about 10^{16} molecules every half hour, and if the search were random, the chances of finding this molecule would be 1 in $10^{131}/10^{16}$ or 10^{115} per half hour of screening. This amounts to about 10^{112} years to find such a large, complex molecule using a random search. In contrast, selection produces the molecule that is best at getting itself replicated in about 2 days. It works because each step leads to a molecule that is better than the previous one. The improvements are preserved, and they accumulate.

Recall this example when people claim evolution cannot work because mutations occur at random. Although mutations do occur at random with respect to fitness, natural selection filters and preserves them remarkably efficiently. *Natural selection extracts order from randomness.*

The last example was pitched at the molecular level and helps to explain the origin of life; the origin could have been extremely rapid. Now consider an example involving whole organisms. If you were given the task of breeding race horses to run faster, you would select the fastest parents and breed them with each other. This method is effective and has been used for thousands of years to produce desired characteristics in domestic plants and animals. It makes an essential point: when some parents with certain characteristics have more offspring than others, those characteristics will spread if inherited. Artificial selection differs from natural selection in that the trait selected is determined by some human preference. *In natural selection the trait selected is always reproductive success.*

The design of organisms for reproductive success can only be changed by changing the stored genetic instructions; genetic instructions are changed by natural selection within a population when organisms with different genetic instructions vary in their reproductive success. Thus evolution occurs both in information and in matter. Organisms function both as *replicators*, replicating the information they have stored in their genes, and as *interactors*, interacting with their environments and with each other to survive, reproduce, and get their genes into the next generation (Williams 1992).

ADAPTATION

A response to selection occurs whenever heritable variation in reproductive success improves reproductive performance. If this improvement continues for enough generations, a *process* called adaptation, it results in a *state* also called an adaptation. The state of adaptation is normally recognized in a particular trait, for example: the opposable thumb is an adaptation for grasping objects with the hand. Claims of adaptation can be controversial, for we usually have not seen the process that produced it, and there are alternative explanations. For starters, we define adaptation as a state that suggests to us that it evolved because it improved survival or reproduction or both.

Adaptations can be incredibly precise and complex. If our ears were any more sensitive, they would detect the random noise of Brownian motion in the atmosphere. Our eyes can detect a match struck at a distance of 15 kilometres on a dark night. Our intermediate metabolism would make the engineers who design oil refineries shrink with envy.

Natural selection operates whenever there is variation in reproductive success, and there is virtually always some variation in reproductive success. Therefore natural selection has always acted and is currently acting in virtually all populations, including our own. *A trait only experiences selection pressure if variation in that trait is correlated with variation in reproductive success, and it only responds to selection if some of that variation is heritable.* When

both conditions are fulfilled, the *process* of adaptation begins. Whether it will ever result in the *state* we call adaptation depends on whether or not other factors are present that can constrain the response to selection. Since natural selection has acted and is acting on all traits that contribute to survival and reproduction, if such a trait is not well adapted, then something must be constraining its evolution. Several constraints are particularly important: gene flow, sufficient time, trade-offs, and historical accidents.

CONSTRAINTS ON ADAPTATION

Gene flow

Genes 'flow' from one place to another when organisms born in one place reproduce in another place, introducing their genes into the local gene pool. When natural selection favours different things in different places, movement of organisms transports genes that have been selected in one place to other places where they are not appropriate. Gene flow, like mutation, can introduce new genetic variants into local populations, and it can also produce local maladaptations. For example, the gene for sickle cell anaemia is adaptive where malaria is prevalent but maladaptive in other environments.

Despite gene flow, local adaptations do evolve when selection is strong, and selection is often strong. A classic example (Antonovics 1971) is heavy metal tolerance in plants. Plants on mine tailings grow on toxic soil and rapidly evolve adaptations to deal with it. Along a transect across a zinc mine tailing and into an uncontaminated pasture, the index of tolerance changed from 75 to 5 per cent in less than 10 m; the plants that were zinc tolerant flowered later, were smaller, and were more tolerant of inbreeding than those that were not zinc tolerant. Plants growing within 10 cm of a galvanized fence had significantly higher zinc tolerance than those just 20 cm from the fence. Zinc tolerance does not spread throughout the population because it is costly: when zinc is not present, it does not pay to be zinc tolerant.

Thus strong selection can produce very local adaptation despite gene flow; species often consist of genetically quite different populations each displaying different adaptations. For gene flow to prevent local differentiation, selection must be weak and the mean distance that genes move in each generation must be large. In modern humans this is often the case for many traits.

Sufficient time

With or without gene flow, it takes time for a population to adapt to an environmental change. Consider the absorption of milk sugar, lactose, by human adults (Simoons 1978; Durham 1991). Like all other mammals, human children come equipped with the biochemical machinery needed to digest milk. Most children lose that ability at the age at which they were normally weaned prehistorically: 4 years. A minority of humans retain the ability to digest fresh milk into adulthood, including the populations of northern Europe and some in western India and sub-Saharan Africa. The ancestral condition was the inability to digest fresh milk after the age of 4 years, and the new, recently evolved condition is the ability to do just that.

How long would it take that ability to evolve? The origin of dairying can be traced to between 6000 and 9000 years ago, and the ability to digest fresh milk after the age of 4 years has a simple genetic basis: it behaves as a single dominant autosomal gene. This is important, for dominant genes increase in frequency under selection much more rapidly than do recessive genes. Imagine a human population of about 10 000 people in which dairy milk production has begun. A mutation occurred that allowed people to utilize fresh milk after they were 4 years old. It had an advantage, for people who drink milk but cannot absorb lactose suffer from flatulence, intestinal cramps, diarrhoea, nausea, and vomiting, which reduces their reproductive performance. Lactose absorbers benefited from an additional high-quality food source, especially when other food was scarce and especially for

nursing mothers and growing children. Selection for such a mutation could have been very strong during serious famines.

Suppose that the ability to absorb lactose conferred a selective advantage of 5 per cent, so that for every 95 surviving and reproducing children of non-absorber parents, the same number of absorber parents produced 100. Initially the gene was rare, and simply because it was rare, it could only increase slowly, for very few people carried it, enjoyed its advantages, and produced a few more surviving children than did those who did not carry it. As the gene increased in frequency, and more people carried it, it began to spread more rapidly through the dairying culture. However, when it became common, its rate of spread decreased, for then most people carried it, and there were very few who suffered from the disadvantage of not having it. How long did it take to increase from a single new mutation to a frequency of 90 per cent? The answer is about 350 to 400 generations or 7000 to 8000 years. This answer would change if we assumed a weaker selective advantage. If the estimated age of the milk-drinking habit is accurate, then it must have conferred substantial benefits if we are to explain its current frequency in northern Europe. Even for a gene under strong selection—and a 5 per cent advantage is strong selection—time is a constraint.

Trade-offs

A trade-off exists when a change in one trait that increases reproductive success causes changes in other traits that decrease reproductive success. For example, some organisms have to pay for improved reproduction with decreased chances for survival. This trade-off exists in the fruit fly *Drosophila melanogaster*, where flies that mate and reproduce have shorter lifespans than virgins. We do not know whether this trade-off exists in humans—it might—for we cannot do the kinds of experiments on humans that we can on fruit flies. However, the evolutionary design of human biology is certainly subject to other trade-offs. For example, the immune system protects us against pathogens, but it does so at the cost of autoimmune diseases. Evidence from birds also suggests that one of the major costs of reproduction is a reduction in immunocompetence and increased susceptibility to pathogens and parasites.

Whenever one analyses some feature of organismal design in terms of the costs and benefits of changes in traits, trade-offs are involved. They place limits on how much fitness can be improved by changing traits, for most traits are connected through genetics, development, and physiology and cannot be changed by selection independently of one another.

Historical constraints

Organisms are not soft clay out of which natural selection can sculpt arbitrary forms. Natural selection can only modify the variation currently present in the population, variation that is often strongly constrained by history, development, physiology, and the laws of physics and chemistry. Natural selection cannot anticipate future problems, nor can it redesign existing mechanisms and structures from the ground up. We illustrate this principle with two examples.

The first concerns the vertebrate eye, often cited for its astonishing precision and complexity. It contains, however, a basic flaw (Goldsmith 1990). The nerves and blood vessels of vertebrate eyes lie between the photosensitive cells and the light source, a design that no engineer would recommend, for it obscures the passage of photons into the photosensitive cells. Long ago, vertebrate ancestors had simple, cup-shaped eyes that were used only to detect shades of light and dark, not to resolve fine images. These simple eyes developed as an out-pocketing of the brain, and the position of their tissue layers determined where the nerves and blood vessels lay in relation to the photosensitive cells. If the layers did not have the correct position, relative to one another, then the mechanisms that induce differentiation would not function correctly, for they rely on an inducing substance produced in one layer that diffuses into the neighbouring layer. Once such a developmental sequence evolved, it could not be changed without seriously damag-

ing optical performance in the intermediate forms that would have to be passed through on the way to a more 'rationally designed' eye.

The second example concerns the length and location of the tubes connecting the testicles to the penis in mammals (Williams 1992). In the adult cold-blooded ancestors of mammals, and in present day mammalian embryos, the testicles are in the body cavity, near the kidneys, like the ovaries in the adult female. Because mammalian sperm (for some unknown reason) develop better at temperatures lower than those found in the body core, there was selection to move the testicles out of the high-temperature body core into the lower-temperature periphery and eventually into the scrotum (in some species they only drop into the scrotum during the breeding season). This evolutionary progression in adults is replayed in the developmental progression of the testes from the embryo to adult, and as they move from the body cavity towards the scrotum, they wrap the vas deferens around the ureters like a person watering the lawn who gets the hose caught on a tree. If it were not for the constraints of history and development, a much shorter vas deferens would have evolved that did the job just as well or even better.

THE NEUTRALIST–SELECTIONIST PROBLEM

Natural selection has been responsible for some of the variation in human DNA and proteins, but some of that variation has been caused by random drift, where selective advantages were small or lacking. Two mechanisms introduce randomness into evolution: mutations and meiosis. The randomness of mutations with respect to fitness was discussed above.

The randomness of meiosis consists of the 50 per cent chance that each copy of a chromosome has of getting into a particular gamete. Since only some gametes succeed in forming a zygote, developing, and reproducing, the random effects of meiosis are particularly important in small populations. This can be seen by the limiting case of a population of two individuals, one male and one female, that produces just two offspring. Consider a gene sitting on a chromosome in the female. It has just two chances to get into the next generation—one chance represented by each offspring—and each chance is determined by the flip of a fair coin. Thus even if a gene is strongly selected, there is a $0.5 \times 0.5 = 0.25$ probability that it will be lost from such a small population. As population size increases, so do the number of chances that each gene has of making its way into the next generation (many organisms will carry the same genes), and the effects of drift diminish. However, even in large populations, drift is the only force acting on portions of the genome that do not experience any selection pressure; thus drift is not an exclusive characteristic of small populations. However, it is only in small populations that drift is important enough to overcome the effects of strong selection.

In contrast to the randomness introduced by meiosis, the randomness introduced by mutations increases with population size, but not nearly fast enough to compensate for the declining effects of meiotic drift.

Current evidence is not sufficient to decide what proportion of genetic variation is caused by selection and by random processes. In humans, the amount of genetic variation to be explained is large: the human genome contains between 50 000 and 100 000 structural genes coding for proteins, of which about 30 per cent may be polymorphic. In many proteins only a few amino acids are critical to their function; substitutions at many other positions may be selectively neutral or close to it. On the other hand, the fact that no selective function is known for most human polymorphisms does not mean that selection has been absent, for modern civilization has changed our nutrition, our diet, and has eliminated or reduced many pathogens that were selective agents in the past. Not all functional differences among enzymes must have influenced fitness in the past, but without such an assumption it is hard to explain, for example, why the rare variants of polymorphic enzymes generally show lower activities. 'The neutral hypothesis, when applied to the study of human polymorphisms, might even have a counterproductive effect if it discourages the search for sources of natural selection' (Vogel and Motulsky 1996).

EVOLUTIONARY CONFLICTS

An evolutionary conflict occurs when genes have different patterns of transmission but interact, directly or indirectly, in the organism or organisms that carry them. A simple conflict occurs between the genes of hosts and the genes of parasites, especially parasites that can be transmitted horizontally, from host to host, as opposed to vertically, from host parent to host offspring. The host evolves mechanisms to reduce the damage inflicted by the parasite, and the parasite evolves adaptations to extract resources from the host and to improve the chances that its descendants will be transmitted to infect new hosts. The result is a co-evolutionary arms race in which both species introduce measures and countermeasures in an open-ended escalation. Both host resistance and parasite virulence evolve, both are often costly, the costs of resistance and virulence ensure that both evolve to some intermediate level, rather than escalating open-endedly, and the result is a state of reduced adaptation in both host and parasite—reduced relative to the state we would have observed if the conflict had not been present.

Conflicts can occur between species, between relatives, and between genes within individuals. For example, mitochondria, which are inherited through the female line, and Y chromosomes, which are inherited through the male line in mammals, have different transmission patterns than autosomal nuclear genes and are therefore potentially in conflict with them.

Natural selection does not always produce adaptations. When evolutionary conflicts are not resolved, all parties suffer.

However, sometimes the conflicts are resolved. One case is suggested by the answer to the question: why are there no parthenogenetic mammals? In mammals, development requires one egg-derived and one sperm-derived nucleus. Sperm- and egg-derived nuclei are marked differently by DNA methylation (genomic imprinting). Early development requires the expression of some genes derived from the father and some genes derived from the mother, which can be recognized by their sex-specific imprinting. If all the genes came from the mother, some genes would not be turned on at the right time, and development would fail—a strong constraint on partheno-genesis. Why did it evolve? It appears to be the result of a conflict between nuclear genes and mitochondrial genes that the nuclear genes have won. Mitochondrial genes have zero fit-ness if they occur in males, whose sperm trans-mit no mitochondria, and mutant mitochondrial genes that induce parthenogenesis increase in frequency. The strong constraint on partheno-genesis in mammals suggests that genomic imprinting may protect nuclear genes from sub-versive feminization by rogue mitochondria.

PRINCIPLES OF INFORMATION TRANSMISSION

All evolutionary change is based on genetic change; here is a short summary of the basic principles of genetic information transmission.

Sex versus asex

Information is transmitted in fundamentally different ways in asexual and sexual organisms. In asexual organisms (to simplify a bit, for there are many types of asexual reproduction), the entire genome is transmitted as a unit, and the only genetic differences between parents and offspring are the result of mutations. In sexual organisms (to simplify a bit, for there are many types of sexual reproduction), off-spring are a 50:50 mixture of genes derived from each parent, and the primary con-sequence of sexual reproduction is to produce genetically diverse offspring. A sexually repro-ducing population generates much more genetic variation in each generation than does an asexually reproducing population, where the influx of new variation is limited by the mutation rate. Because the response to selec-tion depends on the amount of genetic varia-

tion available, sexual populations may evolve more rapidly than asexual populations. They do so by combining from different parents favourable mutations that can be selected and disadvantageous mutations that can be elimi-nated by natural selection. Especially in small asexual populations, the favourable mutations must accumulate gradually, one after the other, and the opportunities for eliminating dis-advantageous mutations are limited.

This may make it sound like human hosts, which are sexual, would have the upper hand over their principle pathogens, viruses and bacteria, which are asexual. For three reasons, this conclusion is too optimistic. First, some pathogens are actually sexual, for they fre-quently recombine genetically (e.g. *Neisseria gonorrhoeae*, which appears to be panmictic—see Chapter 16). Second, they have enormous populations, which greatly increases the proba-bility that an advantageous mutation will arise in a short period of time. Third, they have short generation times, especially at 37°C inside the human body. Recombination, enormous numbers, and short generation times could combine to give pathogens an evolutionary advantage over their hosts.

There are also reasons why the pathogens do not always win. First, most hosts are sexual, and sexual reproduction plays an important role as a defensive measure that creates a mov-ing genetic target (Jaenicke 1978). Second, all hosts have some kind of defence system, and vertebrate hosts like humans have very sophis-ticated defences in the immune system, which functions according to evolutionary principles on a generation time as short as that of pathogens. Third, finding a new host is an extremely risky business for pathogens; most die in the attempt. On balance, both hosts and pathogens have a fighting chance; neither can dominate for very long in evolutionary terms; but sometimes a pathogen can dominate long enough to drive a host population to extinction.

Meiosis conserves gene frequencies

The information stored in the genes is copied precisely both as a DNA sequence, where repli-cation and repair are very accurate, and in the population, where gene frequencies do not change from generation to generation if there is no selection, mutation, gene flow, or variation due to random sampling in small populations ('genetic drift'). The fairness of meiosis, the large size of most populations, and the accurate replication of genes as material units are what make information transmission in sexual populations a fundamentally conservative affair.

Why gene frequencies do not change from generation to generation is a basic principle, the Hardy–Weinberg law, that can be found in any genetics text. Here the important point is not why they do not change, but *that* they do not change. With Mendelian inheritance, the genetic variation necessary for a response to selection is preserved in populations, not destroyed. If gene frequencies changed very much, very often for reasons that had nothing to do with natural selection, then systematic change in response to selection would be impossible, beneficial changes could not accumulate, and adaptive evolution could not occur. The background noise would be too great for the signal to emerge (as happens in genetic drift). Thus the conservatism of Mendelian inheritance is the stable foundation of all adaptive evolutionary change in sexual organisms.

That is not only an important principle, it is also rather peculiar. Why did a complicated mechanism like meiosis that can be so fair to all the genes, distributing them with precisely equal chances to the gametes, ever evolve? Current evolutionary thinking suggests that the fairness of meiosis resulted as a defensive strategy of the nuclear genes to counteract the distorting effects of rogue genes that have the effect of overrepresenting themselves in the gametes at the cost of other genes— meiotic drivers (Hurst 1992).

SELECTION DESIGNS PHENOTYPES FOR REPRODUCTIVE SUCCESS

Fitness is relative reproductive success

The basic insight of population genetics, that the evolutionary process can be reduced to the analysis of the factors that increase or decrease the number of copies of a gene in a population from one generation to the next, has great simplicity and power. It is a good starting point. However, focus on gene frequency change is not sufficient to explain phenotypic evolution. If we want to understand why organisms are designed in some particular way for reproduction and survival, then we must analyse the organism as an interactor representing the genes in all activities contributing to survival and reproduction—as an organism with an ecology, with food to find, predators and parasites to avoid or combat, and mates to convince: as a phenotype with a certain lifetime reproductive success, with a certain fitness.

Natural selection has several components

The analysis of reproductive success begins with the factors determining the number of offspring produced by a single individual over its lifetime. This is the most general component of reproductive success, *individual fitness*. Selection on offspring number per lifetime is called *individual selection*. In growing populations, offspring produced earlier in life contribute more to fitness than offspring produced later in life, for they produce grandchildren earlier, and the effects of shorter generation time accumulate multiplicatively. Thus individual fitness is not just lifetime reproductive success; it often depends on the timing as well as on the amount of reproduction. This type of selection is often sufficient to account for the states of many traits which appear to be shaped to increase the number of offspring per individual per lifetime, and it is sufficient to account for ageing and senescence.

In sexually reproducing organisms, individual selection contains an important component associated with mating success, with interacting with a partner of the opposite sex to produce offspring. This component of natural selection is called *sexual selection*. Traits under sexual selection are subject to evolutionary changes that improve mating success but may reduce survival. For example, the male peacock's tail improves his reproductive success by making him attractive to females but reduces his chances for survival by making it harder for him to fly. Sexual selection involves the two sexes in a complex interaction with fascinating properties. Females have evolved preferences for certain kinds of males and by mating with them transfer the preferred traits to their sons and their preferences to their daughters. The results can be surprising. Mothers may pass on their preference for mates that take risks to their daughters, while their sons inherit the risk-taking trait itself. This would explain why car insurance rates are higher for 20–25-year-old males than they are for females (Daly and Wilson 1985).

Organisms living in groups of related individuals experience a third kind of selection, one that has resulted in deep insights. *If what matters to evolution is the relative number of copies of genes that exist in the population in the next generation, then it does not matter through whose reproductive activities those genes were replicated—directly, by an individual, or indirectly, by its relatives.* Thus if an individual can influence the reproductive success of its kin, it should do so if the benefits—the increase in number of genes in the next generation through the reproductive activities of relatives—exceed the costs—the reduction in the number of genes in the next generation it gets through its own reproductive activities (Hamilton 1964). This is called *kin selection*, and it has helped us understand the evolution of apparently self-sacrificial, co-operative, altruistic, and other kin-related behaviour. Its empirical success has convinced most evolutionary biologists that their focus on genes is probably correct (Williams 1966; Dawkins 1976).

The gene-centred point of view also explains why senescence is a property of the soma, not of the germ line. If the gene-centred view is correct, then evolution 'cares' about the germ line—the genes—whereas medical doctors treat the soma, which is, from the point of view of evolution, disposable.

Traits do not evolve for the good of the species

Formerly one often heard that some adaptation had evolved for the good of the species,

helping it to avoid extinction. As a general explanation this is fundamentally wrong, and that statement represents the broad consensus of the evolutionary community. Traits evolve because they improve the reproductive success of individuals and their kin, and if the species to which those individuals belong happen to survive longer because of those changes, their longer survival is a by-product of the essential process and not the reason for it.

This insight was achieved through a fascinating episode whose main result can be summarized in a single phrase: *selfish mutants invade.* If a trait did evolve that benefited the species at the cost of the individual, some mutant that selfishly exploited the more altruistic individuals would invade and take over the population. It could do so because selection on individuals is much stronger than selection on species. Individuals have much shorter generation times than species, and in the time that it takes for new species to form and go extinct, a process spanning many thousands of individual generations, hundreds of millions of the individuals that form those species have lived and died. For that reason, selection has much greater opportunity to sort among individuals than it does to sort among species, and species selection simply cannot shape adaptations (Maynard Smith 1964; Williams 1966).

BIOLOGICAL CAUSATION: PROXIMATE AND ULTIMATE

Biologists want to understand all the features of living organisms. One natural approach is to study the immediate causes. How does respiration work? What determines the sex of an organism? What causes senescence? These are a few examples of questions about immediate causes answered by physiology, genetics, biochemistry, development, and related fields. Here the aim is to identify the factors that cause the trait or process during the lifetime of a single organism through the study of *mechanism* or *proximate causation*. Much of biology is devoted to it.

Evolutionary biologists ask different questions and investigate different kinds of causes. Why does respiration occur in the mitochondria and not in the cell nucleus? Why do most species have approximately equal numbers of males and females? Why do many animals senesce, but many plants and fungi hardly at all? Why are so many sperm needed to traverse the female reproductive tract, and why do millions of ova develop when less than a thousand are ovulated in the course of a female's lifetime? These are some questions asked by evolutionary biologists. They are also questions about causation, but on a time scale of many generations and at the level of populations and species rather than individuals. This is the study of *evolutionary* or *ultimate causation*. Whereas in mechanistic analysis the causes can be described as biochemical and physical processes, in evolutionary analysis one often describes the causes as how natural selection, evolutionary conflicts, historical contingencies, or chance events shaped the trait under study.

All traits have both types of causes; therefore a complete biological explanation demands the analysis of both. It would be a strategic error to isolate the two kinds of analysis from each other. We should be able to see the world both ways—from the bottom up, from molecules to populations, and from the top down, from selection to molecules.

SUMMARY

Some key points:

1. In natural selection the trait selected is always reproductive success.
2. Natural selection has great power to shape precise adaptations; it can rapidly produce highly improbable states.
3. Natural selection does not always lead to adaptation. In situations of evolutionary conflict, it is often the case that all parties suffer.
4. Traits do not evolve for the good of the species.
5. Natural selection cannot anticipate future problems, for evolution proceeds by tinkering with what is currently available.
6. It does not matter through whose reproductive activities genes are replicated—directly, by an individual, or indirectly, by its relatives.
7. Because selfish mutants invade, arguments that traits evolved for the good of the species are usually invalid.

Evolution combines with physics and chemistry to explain all biological phenomena, and it is the only part of biology containing basic principles not implicit in physics and chemistry. It has three major principles—natural selection, inheritance, and history—and one fundamental property: selection acts on organisms, but the response to selection occurs in stored information. The first principle, natural selection, is a great law of science, the only mechanism known that can maintain and sometimes increase the complexity of organisms, extracting order from randomness, producing systems organized to overcome, locally, the dissipative effects described by the second law of thermodynamics. Variation among organisms in reproductive success produces natural selection; this happens in physical and chemical material. Populations of organisms respond to selection when some of that variation is genetically based; this happens in stored information. The result is a genetically based change in the phenotypic design of offspring from the more reproductively successful parents. Genetic changes also occur at random as mutations and as the consequence of meiotic sampling and they persist in parts of organisms that are not under selection and in small populations.

Those processes have produced the complexity of all living creatures.

ACKNOWLEDGEMENTS

Comments by Dieter Ebert, Ed LeGrand, Randy Nesse, Beverly Strassmann, Jacob Koella, Arno Motulsky, Andrew Read, and an anonymous reviewer improved a draft.

2

RESEARCH DESIGNS THAT ADDRESS EVOLUTIONARY QUESTIONS ABOUT MEDICAL DISORDERS

Randolph M. Nesse and George C. Williams

Diseases result usually from webs of interacting causes of enormous complexity, while human minds seek explanatory principles of extraordinary simplicity. This conflict gives rise to a central problem for medicine. Explanations of disease, and most programmes of medical research, tend to emphasize a single cause, while most diseases result from multiple environmental factors interacting with several sources of vulnerability. An evolutionary approach fosters clear thinking about the complex origins of disease. It is, however, easy to underestimate the change in perspective that an evolutionary view of disease offers and requires. Many evolutionary research questions differ qualitatively from those usually pursued in medicine, and the routine application of traditional modes of medical inference may prove inadequate to the task of testing them. Addressing the formulation and testing of evolutionary hypotheses may seem like abstract philosophy, but it is essential if this enterprise is to succeed.

New evolutionary perspectives on basic mechanisms are leading to substantial discoveries in established areas of medical research, such as virulence, senescence, and genetic variation, as documented by the chapters in this book. Equally important is the potential for integrating multiple causes of specific diseases provided by an evolutionary approach. The benefits of this approach will be delayed, however, if its initial applications are poorly done. It is easy to make up evolutionary stories about adaptation but harder to test them. This chapter outlines how evolutionary hypotheses about specific diseases can be formulated and tested. It is premature to promote any such system as definitive, and we hope that readers will sympathize with the difficulties of trying to impose some order, however preliminary and arbitrary, on an unruly tangle of questions.

This chapter proceeds in two steps. First, we list several distinctions that are essential to defining the objects of explanation and the kinds of explanations proposed, and second, an outline of how hypotheses about vulnerability to diseases can be formulated and the kinds of evidence that can be used to test them. Examples are provided in other chapters; Chapter 23 applies this framework systematically to the major psychiatric disorders.

DISTINCTIONS FOR DARWINIAN DOCTORS

The most common and serious difficulties in pursuing an evolutionary perspective on disease arise from lack of clarity about fundamental distinctions (Table 2.1). While some of them will be familiar, this outline will be worthwhile if it spares some unnecessary misunderstanding and argument.

The distinction between proximate and evolutionary explanations for an adaptive trait is critical. A proximate explanation describes a trait's ontogeny and the resulting anatomy and physiology and how they work to accomplish the trait's function. An evolutionary explanation, by contrast, describes a trait's phylogeny and how the trait gives a fitness advantage

Table 2.1 Some distinctions for Darwinian doctors

---

1. Proximate versus evolutionary explanations
2. Explanations of individual differences (why some individuals get a disease when others do not) versus traits individuals share (why all members of a species are vulnerable to a disease)
3. Manifestations of disease that arise from defects versus defences (or other adaptations)
4. Explanations of a disease itself versus explanations of why the body is designed so it is vulnerable to a disease
5. Group versus individual versus gene levels of selection
6. Most traits are involved in tradeoffs that force compromises
7. The different meanings of adaptation, fitness, defence, and evolution in different fields
8. The several distinct kinds of evolutionary explanations for vulnerability to disease

---

(Mayr 1982). Tinbergen (1963) clarified the distinction when he listed four questions that must be answered to explain any adaptive trait completely:

1. What is the proximate mechanism of the trait—its structure and operation—at all levels from chemistry, to anatomy, to physiology and interactions with other traits?
2. What is the ontogeny of the trait, from the zygote to the mature individual?
3. How does the trait give a fitness advantage that can account for selection for the genes that give the trait its current form? What adaptive functions does it serve and how do those functions increase Darwinian fitness?
4. What is the phylogeny of the trait? What are its precursors and what forces and intermediate stages gave rise to its present form?

The answers to the first two questions are both parts of the proximate explanation; the answers to the last two questions are both parts of an evolutionary explanation. Proximate and evolutionary explanations are distinct and complementary; both are necesary for a full understanding of any trait shaped by natural selection.

Proximate explanations seem more straightforward than evolutionary explanations because they refer to observable physical structures and because often only one hypothesis can be correct. DNA either is or is not a double helix. Cortisol either does or does not come from the adrenal cortex. Evolutionary explanations, by contrast, refer to past events whose traces are found in fossils, in gene frequencies, and in the resulting body structures, and they refer to mechanisms of natural selection involving competition of phenotypes and traits within populations.

The obvious difficulties of investigating past events and population processes are compounded by the possibility of multiple explanations. Eyebrows may both keep sweat out of the eyes and signal conspecifics. The tongue may both process food and form words. Beyond these difficulties, still other problems arise. What qualifies as a trait? How can we tell if we are carving the body at nature's joints? How can we tell if a trait is an adaptation shaped by natural selection or something else? The temptation is to give up, but the problems are important. Solving them is not impossible, just difficult. Physiology has a well-developed tool kit for determining the functions of an organ: observing the effects of extirpation, determining if form matches function, natural experiments, and more (Vander *et al.* 1985; Schmidt-Nielsen 1990). Behavioural ecology also has well-developed methods (Krebs and Davies 1997). A comparable tool kit for evolutionary medicine is still being assembled.

A second crucial distinction is between explanations of why one individual gets sick and another does not, versus an explanation of why all members of a species share some vulnerability to disease. The reasons why some individuals on early ship voyages developed scurvy long before others are different from the reasons why all humans require vitamin C. Explaining why some members of modern populations will develop disease from arterial occlusion and others will not is a problem separate from that of explaining why atherosclerosis is so common. Medical research has focused, understandably, on trying to explain why one person beomes sick and another does not. Such questions differ fundamentally from evolutionary questions about why all humans are vulnerable to certain diseases.

A third distinction is between manifestations of disease that reflect defects and those that are aspects of defences. Many medical problems, like seizures, jaundice, and hallucinations arise from defects in the body's machinery. Others, like fever, cough, pain, and anxiety, are defensive responses shaped by natural selection that remain latent until aroused by a threat. Because they are usually unpleasant and associated with an unfavourable situation, it has been easy to confuse these defences with the diseases that arouse them. But as we develop new drugs that block defences more effectively, we need to recognize the functions of these defences and understand when it is, and is not, safe to block them.

A fourth distinction is between explanations of a disease and aspects of the body that make us vulnerable to it. Diseases are, in general, not shaped by natural selection and have no adaptive explanation. But diseases arise from some vulnerability in the body's design. These design features are shaped by natural selection, directly or indirectly, and they do have evolutionary explanations. This small change in focus, from a disease itself, to the aspects of the body that make it vulnerable to disease, makes the evolutionary approach to medicine viable.

A fifth distinction is between levels at which selection may operate, especially between the gene, the individual, and the group. Despite vivid expositions on selfish genes (Dawkins 1976), many physicians remain unaware of the difficulty with explanations based on naïve group selection (Williams 1966). The main difficulty is that selfish mutants, or selfish phenotypes, continually arise and invade populations of organisms that are sacrificing their reproductive interests for the good of some larger unit. It takes rather special and implausible circumstances for a refined group-selection theory (e.g. Wilson and Sober 1994) to work at all. Among those who grasp the possibility that genes can persist at the expense of individuals,

many imagine that mutations in somatic cells can somehow help themselves by inducing cancerous cell division. In fact, of course, an individual's somatic cells are, for the most part, genetically identical and so they do not have conflicting interests (with a few possible exceptions), while different individuals have different genes, and therefore different interests that give rise to conflicts.

Related individuals do, of course, have some proportion of their genes in common, so if a gene decreases an individual's reproductive success but increases that of relatives (say, for instance, a gene that tends to make parents protect their children) then the frequency of the gene can increase by kin selection—so long, that is, as the cost to the actor is less than the benefit to the relatives multiplied by the percentage of genes identical by descent (Hamilton's rule, of Hamilton 1964). Many people, on first grasping kin selection, are initially distracted by Washburn's fallacy, thinking that common interests are based on the percentage of genetic code that is identical, say the 97 per cent of the nucleotide sequences that humans and chimpanzees have in common. In fact, the relevant factor is the percentage of genes in common by immediate genealogy, 50 per cent for first-degree relatives, 25 per cent for second-degree relatives, and so forth (Williams 1996: 43–51).

Sixth, selection can influence gene frequency in several ways. The traditional dichotomy between natural selection and sexual selection conceals their underlying common mechanism, and the different ways that effects of genes can increase their representation in future generations (Andersson 1994). The trade-offs between these different 'strategies' are at the core of understanding how natural selection shapes the organism (Stearns 1992; Krebs and Davies 1997). Survival is essential, but it is selected only in the service of direct and indirect reproduction. When a gene increases net reproductive success at the expense of individual health or survival, whether by effects on sexual competition, display, fertility, or parenting, that gene will tend to increase in frequency. This may underlie the shorter lifespan of men than of women. Also important are trade-offs between mating effort and parenting effort, between investing in self and in kin, and between investing much in a few offspring versus less in more offspring.

Seventh, one must carefully specify the meanings of several words that are used differently in different fields. Darwinian fitness, or reproductive success through oneself and one's relatives, is quite different from everyday meanings of fitness and health. Adaptation in biology is very different from the several meanings of adaptation in psychology or neurology. An inducible defence, in biology, is a special protective trait or a state that is evoked by certain threats, a very different matter from the notion of defence in psychoanalysis, and only somewhat similar to the idea of defence in war, sports, or law. Even the word evolution itself most often refers in medical databases not to the process of biological descent with modification, but to all manner of gradual changes in organisms, organizations, and beliefs.

Finally, it is essential to distinguish the main possible kinds of evolutionary explanations for vulnerability to disease (Williams and Nesse 1991; Nesse and Williams 1994: 8–11). These categories (Table 2.2 and Fig. 2.1) give us a foundation for analysing vulnerability to disease.

POSSIBLE EXPLANATIONS FOR VULNERABILITY TO DISEASE

Each of the categories in Table 2.2 generates hypotheses that are somewhat different. These hypotheses, in turn, require different tests that

Table 2.2 Categories of evolutionary explanation for vulnerability to disease

1. Defence—what we think is a disease or defect is actually an adaptation
2. Conflict with other coevolving agents, such as pathogens
3. Novel aspects of the current environment
4. Genetic quirks that are harmful only in a novel environment
5. Design trade-offs at the level of the gene
6. Design trade-offs at the level of the trait
7. Historical legacy and path dependence
8. Random factors

pose different special problems. Much of the confusion in evolutionary approaches to disease arises, we believe, from lack of clarity about the exact kind of evolutionary hypothesis being considered and the different ways in which different kinds of hypotheses can be tested. Each category is addressed in turn, with a brief description, followed by the most characteristic predictions (formulated as 'If … then …' statements), and a note, if necessary, about special problems that arise when testing hypotheses in each category.

The first possibility is that an adaptation has been misinterpreted as a defect, usually because it is a defence against some threat. Inducible defences are readily confused with defects and with causes of diseases, for they are often unpleasant and are manifested mainly in association with disease. Examples include fever, cough, pain, vomiting, anxiety, and jealousy. Dysregulation of defences often results in disease; examples include autoimmune diseases, pulmonary embolism, seizures from high fever, anxiety disorders, and depression.

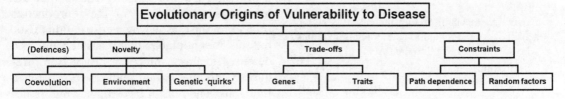

Fig. 2.1 Evolutionary origins of vulnerability to disease.

Hypotheses about defences assert that a trait is an evolved adaptation of a special kind; testing them involves all the complexities and judgement required to assess the status of a trait as an adaptation (Rose and Lauder 1996: Part 1). No uniform standards of evidence exist for making such assessments, but when hypotheses are formulated as 'if trait T is a defence, then ...', several kinds of predictions can be made (Table 2.3).

Testing hypotheses about *defences* involves still other inherent difficulties. First, looking for the utility of defences is counterintuitive because they usually are unpleasant, seem harmful, and are observed only in conjunction with danger. Second, they usually *are* somewhat harmful, otherwise they would be expressed all the time. Third, because many defences are inexpensive compared with the harms they protect against, selection shapes regulation mechanisms, according to the principles of signal detection theory, that tolerate many false alarms to ensure a response when it is essential (the 'smoke-detector principle', see Nesse and Williams 1994: 159). Because of this, we often see no ill effects when drugs block defences, and this fosters the illusion that the defences are not useful. Finally, in the relative safety of the modern environment, many defences are excessive. For some, like the fight–flight response, most instances of expression are now pathological (panic attacks, destructive violence). All of this has very practical implications as new drugs that block aversive emotions are discovered and widely used.

A recent example of a test of the idea that an evolved feature is a defence is the proposal that menstruation serves to rid the reproductive tract of infections introduced by sperm (Profet 1993). Profet's arguments were based largely on category 4 tests in Table 2.3, but she also made or implied other predictions, for instance in category 5. She noted that danger from pathogens introduced by semen should be greatest in those species with the most frequent mating and greatest diversity of mates. That the most profuse menstruation is found in women and female chimpanzees accords nicely with this expectation. Unfortunately for Profet's theory, a more inclusive phylogenetic analysis of menstrual phenomena in relation to mammalian mating systems, recently carried out by Strassmann (1996), showed that Profet's expectations are not borne out.

Strassmann's work clearly shows that defence against infections of the reproductive tract is not the reason why menstruation exists, but this does not mean that it has no value as a defence against infection. Evolution frequently modifies extant features to serve new functions. The reason birds have wings is not the same as the reason vertebrates have fore-

Table 2.3 If trait T is a defence, then:

1. Individual differences in the degree of expression of the trait should influence the degree of protection against the threat
   (a) Inherited defects
   (b) Inherited variation
   (c) Surgical excision
   (d) Drug manipulation
2. The form of the trait should match its function—details of the state should match those that defend against a threat
3. If there are subtypes of the defence, their characteristics should match the demands posed by specific threats, and the regulation system should express the correct subtype in the face of different threats
4. Details of the trait's proximate mechanisms should match expectations based on its function
5. Phylogenetic relatives should have homologous adaptations that vary according to the kinds and amounts of the threats they experience
6. Unrelated species that face similar threats may have analogous adaptations
7. The mechanisms that regulate the expression of the defence should express the kind, degree, and duration of response that is appropriate to the threat
8. Facultative mechanisms may increase the sensitivity of the regulation system in individuals who are more exposed to the threat
9. The proximate mechanisms of the regulation system should match the challenges posed by the threat

limbs. Some of the features that Profet noted, for instance differences between menstrual and circulating blood, might well relate to defence against pathogens, and if so, they are of medical importance. Strassmann's alternative to Profet's theory is that menstruation is a special mammalian aspect of a general vertebrate adaptation for energy conservation. Keeping the reproductive tract constantly ready for use, even when no use is possible, is more expensive than intermittently shutting it down, even when the shutting down requires considerable loss of blood. We have recently learned of impending challenges to both Profet's and Strassmann's ideas. The adaptive significance of menstruation is a topic for which such attention is long overdue.

A second possible explanation is *conflict* with other organisms. Pathogens evolve faster than their hosts, ourselves, and have evolutionary interests that are often in conflict with ours. Thus pathogens pose a constantly novel environment in which our adaptations are always out of date (Ewald 1994). Parent–offspring conflict (Trivers 1974) is a related example of importance to psychiatry (Slavin and Kriegman 1992), and a special type of parent–offspring conflict, maternal–fetal conflict, is of potential major medical importance (Haig 1993 and Chapter 7). Explanations invoking conflicts are covered in detail in other chapters. Here we note only that such arms races can cause disease in apparently unrelated systems; for instance, heart damage caused by rheumatic fever because of immune responses to a streptococcus that mimics human antigens. Also, the social environment may involve arms races as challenging as those that result from competition with microbes, and may create designs of remarkable complexity and fragility, perhaps shaping human tendencies for mental conflict and psychopathology (Alexander 1987; McGuire *et al.* 1992).

A third explanation is the *mismatch* between our bodies and novel aspects of the environment; novel, that is, since the development of agriculture, and especially since the industrial revolution. Natural selection cannot change the body's design fast enough to cope with rapid environmental changes. The resulting diseases include atherosclerosis, breast cancer, substance abuse, eating disorders and probably depression. If a disease is caused by a mismatch between our bodies and novel factors in the environment, then:

(1) groups more exposed to the novel factor should have higher rates of the disorder;

(2) the disease should be much more prevalent in modern than ancestral environments;

(3) within groups, individuals more exposed to the factor should have increased rates of the disorder; and

(4) it should be possible to demonstrate the proximate mechanism by which the factor causes disease.

Fourth, diseases that result from novel environmental factors may affect only some individuals because of differences in exposure and because of genetic variation. 'Genetic quirks' that vary between individuals and cause no harm in the ancestral environment may dramatically increase the risk of disease in modern environments. It makes no sense to call these variations 'defects', for they were harmless, or even helpful, until the past few thousand years. Examples include genetic factors in diabetes, coronary artery disease, depression, and substance abuse. If a disease is postulated to result from a genetic quirk, then:

(1) individuals with the responsible allele will have normal reproductive success in the ancestral environment, but decreased health in the modern environment;

(2) the degree of decreased fitness among those with the allele will be related to the degree of exposure to the novel factor; and

(3) individuals without the allele will have normal phenotypes in both ancestral and modern environments.

A fifth explanation is trade-offs at the level of the gene. Some individuals have alleles that give them fitness advantages (at least in some environments or genotypes) but at the cost of increased vulnerability to disease.

Manic depressive illness will be addressed in this perspective. Often the frequency of such genes varies between groups because different environments pose different threats. The haemoglobinopathies that protect against malaria are the best known examples. Most mutations that confer an advantage will also give disadvantages. If selection pushes such a gene to fixation, all members of a species may be vulnerable to a disease. If the disease is prevalent and severe enough, there will be further selection for modifier genes that decrease the adverse effects. The complexity of the resulting interaction is probably responsible for maintaining considerable genetic diversity and susceptibility to disease (Sing and Hanis 1993). In the simplest case, if a disease is caused by a gene that was selected for despite causing disease (such as the gene that causes cystic fibrosis or, perhaps, manic depressive illness), then:

(1) it may be possible to identify the specific benefit that confers this advantage;

(2) individuals with the gene should have a net inclusive fitness advantage in some environments (unless the allele is maintained by segregation distortion); and

(3) the mechanism by which the gene confers the advantage should be explicable.

The sixth explanation is trade-offs at the level of complex traits. The design of every trait is constrained by the effects on other traits. If a trait gives rise to disease because of a trade-off with other costs and benefits, then alternative designs that seem superior will, in fact, result in an overall decrement in fitness. This applies to every trait in every organism and helps to explain why nearly everything that can go wrong in a body sometimes does go wrong. A simple example is the limited strength of our hands, the price of dexterity. For mental disorders, Chapter 23 will consider the advantages and disadvantages of anxiety versus risk taking.

Seventh, historical constraints result from path dependence. Modern technological examples of such problems are vividly described in the phrase 'why things bite back' (Tenner 1996). The design for some fundamentally flawed traits cannot be corrected, because forms intermediate between the current design and a superior one have decreased fitness. The inside-out design of the vertebrate eye is an example. Human patterns of attachment and cognition may be others. If vulnerability to a disease is hypothesized to result from historical constraints, then forms intermediate between the current design and a superior alternative should give lower fitness.

Finally, random factors affect the gene pool and developmental processes. Gene frequencies are determined both by selection and by random factors including mutation, drift, and bottlenecks. These random factors are sufficient to explain most rare genetic disease, but the simple model of mutation balanced by selection is often applied to common genetic diseases where it is may well not apply. At what point does a gene's frequency, penetrance, and effects on fitness suggest that we should begin to look for selective forces that might maintain its frequency? The answers are very much up in the air (Chadwick and Cardew 1996); the issue is addressed by several chapters in this volume. If a disease results from genes that are present for reasons other than natural selection, then the balance between rates of mutation and the force of selection in the ancestral environment should be consistent with the persistence of the genes, except in the presence of special factors like segregation distortion.

These eight categories clarify hypotheses about the different ways natural selection can explain vulnerability to disease and how they can be tested. Addressing each kind of evolutionary explanation separately clarifies how they can combine to explain a specific disease.

SUMMARY

Evolutionary biology is providing new insights into the specific mechanisms of disease, including host–pathogen contests, genetic variation, and senescence. It also offers a framework for understanding why the body is vulnerable to specific diseases. Applying evolutionary approaches to medical problems involves questions about adaptation for which canons of evidence are not well established. Progress will be possible, however, if we can specify precisely what is to be explained, distinguish the possible evolutionary explanations, and formulate and test hypotheses appropriate to each kind of explanation.

Part II: Human history and human genes

3

THE HISTORY AND GEOGRAPHY OF HUMAN GENETIC DIVERSITY

Guido Barbujani and Laurent Excoffier

Anatomically modern humans (*Homo sapiens sapiens*) have been present on Earth for at least the past 150 000 years (Klein 1992). Related human species like *Homo erectus* and *Homo sapiens neandertalensis* were present before and during that time, the last having disappeared only relatively recently (Groves and Lahr 1994). Over this period humans have left traces in the form of bones, teeth, and artefacts through whose analysis palaeontology and archaeology have taught us much about our evolutionary and demographic history. The genes of past people have also left a trace, albeit not an obvious one. The genomes of living humans, including the alleles responsible for hereditary diseases, are derived from the genomes of earlier humans, modified by mutation and recombination. Therefore, many questions about the origin and distribution of current pathologies and disease risk can be addressed by understanding how current genetic diversity came to be distributed the way it is. Often, such historical reconstructions involve analyses of geographical patterns of genetic diversity interpreted with evolutionary models that identify the history of human settlement. This, in turn, may allow inferences of medical relevance. For example, the distribution among human populations of genetic variation in drug response appears to be strongly associated with the history of human dispersal and divergence (Chapter 4).

NON-GENETIC INFORMATION ON EARLY HUMAN HISTORY: A BRIEF OVERVIEW

There is broad consensus that the genus *Homo* originated in Africa. A plausible date is around 2.5 million years ago for the first traces of *Homo habilis*, whose fossils are all restricted to Africa. While the oldest fossils of *Homo erectus* have been found in Kenya (dated to 1.7–1.8 million years), the oldest *Homo erectus* specimens out of Africa seem to be 1.5–1.8 million years old and have been found in Java (Swisher *et al.* 1994) and in the Caucasus (Gabunia and Vekua 1995). More recent fossils have been found in Europe and Asia. Thus *Homo erectus* colonized a large part of the Old World.

Similarly, fossils that can be attributed with certainty to *Homo sapiens sapiens* appear to be oldest in Africa (Klein 1989; Bräuer *et al.* 1997). This observation, together with comparisons of skull shape in human remains of various ages and provenance (Waddle 1994), support a modern human origin in Eastern Africa followed by expansions in other continents (Stringer and Andrews 1988). Although alternative scenarios have been suggested, notably the possibility that the populations of *Homo erectus* of Africa, Asia, and Europe evolved in parallel into anatomically modern humans (the multiregional hypothesis: see Wolpoff 1989), they have received little recent support. Neither palaeontological (Lahr and Foley 1994) nor, as we shall see, genetic data seem to agree with them. There is a strong

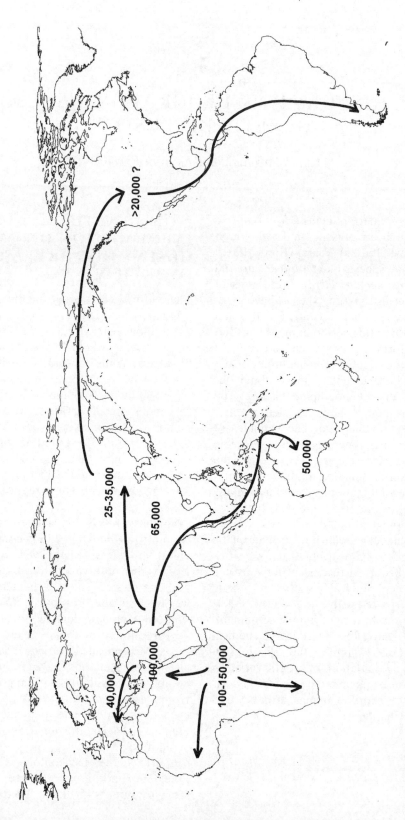

Fig. 3.1 Early migrations and arrival times (years ago) of *Homo sapiens sapiens* in different parts of the world.

theoretical objection to multiregional evolution: for a new species to evolve over a range as broad as a continent, unreasonably high levels of gene flow would be necessary (Livingstone 1992). It is simpler to imagine that we developed from a limited group of initial founders whose exact number remains open to debate (Takahata 1993).

The first traces of modern human presence outside Africa, in the Near East, have been dated at between 90 000 and 115 000 years ago (Schwarcz *et al.* 1988; Valladas *et al.* 1988). Estimated dates of first arrival on other continents suggest successive expansions in all directions, perhaps along two different routes: one from Ethiopia to Arabia and then to South-East Asia and Australia, and the other from North Africa into the Near East, Europe, northern Asia, and the Americas (Lahr and Foley 1994). Modern humans seem to have reached China perhaps 67 000 years ago, Australia 50 000 years ago, Europe around 40 000 years ago, northern Asia some 35 000 to 25 000 years ago, and the Americas a little later, while Oceania was settled only within the last 4000 or 5000 years (Howells 1992; Cavalli-Sforza *et al.* 1994: 60–7). Figure 1 sketches these early migrations. These dates are minimum estimates, for older specimens may not yet have been discovered; some dates, especially for the early settlement of the Americas (Meltzer 1993), are still controversial. Also, the first inhabitants of a certain area are not necessarily the ancestors of the present populations, for more recent immigrants may have partially or completely overridden past populations. None the less, the pattern emerging from the study of human fossils is one in which groups derived from African ancestors spread over the entire world, replacing pre-existing populations of African and Asian *Homo erectus* and European *Homo sapiens neandertalensis*. How rapid such a replacement was, whether or not it involved some admixture with other human species, and how large the populations were that underwent these processes, including the initial modern human group(s), continue to be debated (Takahata *et al.* 1995).

The development of farming and animal breeding was probably the most important episode in human population history after the colonization of the five continents. Palaeobotanical and archaeological data suggest that agriculture developed about 10 000 years ago in various regions of the world (Harlan 1971). Evidence for farming, characterized by neolithic implements, comes from a wide area of western Asia (Renfrew 1991), South-East Asia, and Central America. In Africa, the neolithic seems to have arrived later, some 4000 years ago in the Sahara and in East Africa, and most Australian populations do not seem to have developed agriculture in prehistoric times.

The neolithic revolution led to a substantial population increase. Biraben (1979) suggests that the entire human population did not exceed 0.5 to 1 million in the late palaeolithic, reached 5 million 10 000 years ago, and has increased steadily ever since. The last two millennia have seen periods of reduced growth, even demographic crises, but the exponential growth that started a few hundred years ago continues. The major migrational episodes of the last five centuries have also certainly affected the genetic make-up of many populations.

PHYLOGENETIC METHODS

Many of the conclusions reached in this chapter depend on modern methods for constructing phylogenies, or evolutionary trees. A major method is cladistics, also called phylogenetic systematics. It provides clear logic for the analysis of qualitative characters. The ready availability of DNA sequences now contributes an almost unlimited number of independent characters to the analysis of relationships. A DNA sequence can be analysed either as a large set of qualitative traits or as a single quantitative trait. To handle the latter case, statistical methods have been developed. In contrast to cladistic methods, where one assumes that a rigorous logical approach will produce the best tree, statistical methods assess phylogenies by using probability models to estimate the reliability of any given tree, or part of a tree. The methods of systematics yield a phylogeny, which is a hypothesis about relationships expressed as a tree-like diagram.

PATTERNS OF ALLELE FREQUENCIES

What can we infer about our species' early history from genetic evidence, and what can we learn about the origin and distribution of heritable disease? Next we review some genetic studies based on different types of markers and ask: how well do these studies agree among themselves and with palaeontological, archaeological, and linguistic results? There are some important contradictions, both between different fields and between inferences based on different biological traits in the same populations. Population structure inferred from different kinds of data contains information about evolutionary processes occurring at different time depths. Allele frequency change can be comparatively rapid, but DNA base mutations take a long time to accumulate.

Population classifications

Global analyses of allele frequency distributions indicate that the human population is not simply divided into a few clearly distinct groups. Lewontin's (1972) analysis of 15 markers in seven major groups showed that 85 per cent of genetic diversity is represented by

differences among individuals within populations, whereas differences among groups accounted for less than 7 per cent. Additional studies (Table 3.1) confirmed these figures (Barbujani et al. 1997). The differences among major human groups, no matter how they are defined, are only a small fraction of the global genetic variation of our species, even when the frequencies of DNA markers, rather than protein variants, are considered. Thus the traditional concept of human races does not reflect actual patterns of human genetic variation (for another view see Chapter 6).

Studies of allele frequencies show that the differences between populations, albeit limited, are not random (for another view see Gill and Evett 1995). In this section we focus on four findings:

1. Africans differ from non-Africans at many loci.

2. Population trees based on genetic and on linguistic distances are similar at a global level, suggesting that the same demographic processes helped to shape both of them.

3. Many zones of sharp genetic change overlap with barriers to population dispersal, geographical or otherwise.

4. Wide clinal patterns of allele frequencies suggest past expansion processes in several regions of the world.

Table 3.1 A summary of studies partitioning human genetic diversity

Polymorphism	Average variance components (%)				
	No. of loci	No. of groups	Within samples	Among samples within groups	Among groups
Protein (Lewontin 1972)	17	7	84.5	8.3	6.3
Protein[a] (Latter 1980)	18	6	85.0	5.9	9.1
Protein (Ryman et al. 1983)	25	3	86.0	2.8	11.2
DNA[b] (Barbujani et al. 1997)	109	4–5	84.4	4.7	10.8
Y chromosome[c] (Poloni et al. 1997)	1	4	70.1	6.9	23.0
mtDNA RFLPs[d]	1	4	65.5	7.1	27.5
mtDNA RFLPs[e]	1	4	77.4	7.3	15.3

[a] Average based on three different statistical techniques.
[b] Nuclear loci; average of 30 microsatellite and 79 RFLPs.
[c] 58 populations analysed for Taq I/p49a,f RFLPs.
[d] Weighted average over 100 polymorphic sites. Excoffier (unpublished).
[e] Variance components were computed from RFLP haplotypes frequencies only. Excoffier (unpublished).

Nei and Roychoudhury (1993) constructed population trees based on 26 populations and 29 polymorphic loci. The four African groups appear clearly distinct from their non-African counterparts. Among the latter, European and western Asians seemed to form a cluster, whereas all other Asian populations appear to belong to a group including Australians, Oceanians, and Amerindians. Their trees suggest an initial differentiation in Africa, followed by the subdivision of the ancestral human population into groups corresponding to the ancestors of current African and non-African populations.

A tree analysis of genetic diversity assumes that human populations evolved mainly by a branching process, which may not be the case in regions with extensive admixture and migration. However, the main results of this analysis have been confirmed in several other studies. Moreover, evolutionary trees allow comparison of population relationships inferred from biological and linguistic data. The use of evidence from different sources to test hypotheses on human evolutionary history has been explored for thirty years, often corroborating the finding of one discipline with consistent, if not fully overlapping, support from independent sources.

Relation between genetics and linguistics

One of the broadest comparisons of genetic and non-genetic data was made by Cavalli-Sforza et al. (1988), who compared the human linguistic tree with a tree constructed using allele frequencies from 42 loci, which differed somewhat from the tree obtained later by Nei and Roychoudhury (1993). Africans were still separated from non-Africans by the deepest branch, but the same cluster included all European, northern Asian, and American groups, and was distinct from a South-East Asian cluster that included Australians and Oceanians. The details of the tree and the technical reasons that may account for the discrepancies exceed the limits of this chapter.

Notable, however, is the similarity between the population relationships inferred from that tree and from a comparative analysis of the major linguistic families of the world (Ruhlen 1991). Although the correspondence is not perfect, the fact that linguistically homogeneous populations also tend to cluster genetically suggests that linguistic and allele-frequency divergence proceeded in parallel. Both genes and languages tend to diverge in isolation and to converge when populations have contacts. The population splits and expansions that most deeply affected the distribution of allelic variants also seem to have had a major impact on linguistic change. Molecular data (see below) support this conclusion.

A correlation between linguistic and genetic diversity has been found in several regions (Sokal 1988; Forster et al. 1996), suggesting that a common cause—probably the history of human settlement—has shaped both. When a population splits, the two resulting groups evolve independently and tend to diverge genetically and culturally. But does a causal relationship also exist between language and genetic differentiation? The most significant zones of sharp genetic change tend to occur where gene flow is restricted by factors such as mountain chains, deserts, and seas. However, in some areas of Europe (Barbujani and Sokal 1991), Africa (Excoffier et al. 1991), the Caucasus (Barbujani et al. 1994a), Japan (Sokal and Thomson 1997), and elsewhere, sharp local genetic differences are evident with no obvious obstacles to gene flow. In many such cases, one can identify a linguistic, rather than a physical, barrier to dispersal. Therefore, cultural differences including language barriers appear to contribute to population subdivision and to enhance the role of genetic drift.

Recent expansion of linguistic groups

Genetic variation in Asia and Europe is largely clinal, which means that gene frequencies change gradually along geographical transects (for a review see Cavalli-Sforza et al. 1994). Analysis of allele frequency data within and among major linguistic families suggests mechanisms that probably caused this geographical structure. Menozzi et al. (1978) first argued that the gradients of allele frequencies common in Europe may be the result of the expansion of a Near Eastern population accompanied by little or no gene exchange (Barbujani *et al.* 1995) with the local populations encountered. Their model implied different growth rates for farming groups expanding out of the Near East and contacting stable groups of early European hunter–gatherers. The process has been termed neolithic demic diffusion, and its timing has been estimated by comparison with archaeological evidence on the spread of neolithic industries (Ammerman and Cavalli-Sforza 1984; Sokal *et al.* 1991)

The European regions into which the early agriculturists expanded roughly correspond to the current range of Indo-European languages in Europe. Renfrew (1987) proposed that a single episode of population expansion could account for the gene-frequency gradients, for the archaeologically documented diffusion of farming techniques, and for the linguistic structure of Europe. Among his predictions was the existence of similar allele-frequency gradients in three linguistic regions of the world, corresponding to the areas where similar, simultaneous processes of farming expansions were postulated (Renfrew 1991). An alternative view held that the diffusion of Indo-European languages in Europe was much later than, and thus uncorrelated with, the neolithic demographic processes (Gimbutas 1979).

A study of the distribution of allele frequencies at 15 loci in eight linguistic families of Europe and Asia found clines within three language families where Renfrew predicted them: Indo-European of Europe; Elamo-Dravidian and Indo-European of Asia; and Altaic. The results for a fourth language family (Afro-Asiatic) are still unclear. Conversely, language families which, according to Renfrew, do not owe their present distribution to neolithic expansions from the Near East (Sino-Tibetan, Uralic, Caucasian, and Austric) had less regular distributions of allele frequencies in space (Barbujani and Pilastro 1993).

Linguistic information has also been fruitfully applied to data from sub-Saharan Africa, where the allele-frequency similarity does not reflect geographical proximity (Excoffier *et al.* 1987, 1991). This has been attributed to exten-

sive, recent population movements, including the Bantu expansion within the last 4000 years (Ehret 1973; Vansina 1985), which disrupted a prior correlation between genetic and geographical distances. In contrast to geography, the classification of the sub-Saharan populations into four major language families (Greenberg 1963) proved to be a good predictor of their genetic relationships. The evolutionary relationships among distant populations are still recognizable in their common genetic and linguistic characteristics.

MOLECULAR DIVERSITY

Gene genealogies versus population trees

Allele frequencies suffer some drawbacks, among them the fact that different combinations of evolutionary pressures can produce similar allele-frequency distributions, making it difficult to infer the underlying process from the pattern of genetic diversity. An advantage of molecular data is that one can quantify the amount of divergence between any two alleles by simply counting the number of mutations that occurred since their divergence. Given an estimated mutation rate and the assumption that this rate is constant through time (the molecular clock hypothesis), one can estimate how much time elapsed since the most recent common ancestor of any two genes. When more than two genes are compared, one must reconstruct their complete genealogy to identify the common ancestor of them all. This phylogenetic reconstruction process is seldom trivial; much effort has been put into the development of efficient methods. Once a reliable genealogy is obtained, however, the number of mutations separating the root of the genealogy (the molecular ancestor) from its tips can be used to date when this most common recent ancestor lived. Reliable genealogies allow us to discuss the relative age of genes, to speak of old genes and young genes.

In populations that have remained constant in time, however, there is generally no relationship between the age of the most recent common ancestor of a sample of genes and the age of the population itself (Tajima 1983; Pamilo and Nei 1988; Saitou 1996). Usually the genes are much older than the populations in which they are found: one often observes old genes in recently founded populations. A closer correspondence between gene trees and population trees is only expected when populations have been founded by so few individuals that the founders could be considered monomorphic.

Natural human populations are generally highly dynamic, experiencing demographic fluctuations, migrations, and gene flow, so that gene genealogies may differ considerably from population trees in both branch lengths and topology. One should not expect strong geographical consistency in gene trees until populations have been separated for very long times, about six times their effective population size in number of generations (Tajima 1983). For human populations with an effective size of 10 000 individuals, this means separation times of about 1.2 million years, which clearly exceeds the estimated age of our species. Thus the presence of related genes is expected in very divergent populations without necessarily implying recent episodes of gene flow.

Mitochondrial DNA diversity

Recently many studies have reported on the molecular diversity of human mitochondrial DNA (mtDNA). This extranuclear DNA, which represents only 1/200 000 of our genome, has attracted attention for several reasons. It has a simple haploid mode of transmission, is transmitted through females, and escapes recombination. Some minute paternal contribution, and therefore recombination, is possible (Ankels-Simons 1996; Howell et al. 1996; Lunt and Hyman 1997) but is too rare to make much difference. Because mtDNA also mutates at least 10 times as fast as nuclear DNA, populations that diverged recently are expected to harbour different mitochondrial mutations.

An analysis of 144 mtDNAs sampled in different human groups (Cann et al. 1987) has been widely publicized. On the basis of a minimum-length tree inferred from those sequences, and from the amount of sequence diversity, and assuming that the separation between the human and the chimpanzee lineage occurred 4–5 million years ago, the authors argued that the molecular ancestor of all human mtDNA molecules was approximately 200 000 years old. This result has been confirmed by other studies (e.g. Vigilant et al. 1989, 1991; Hazegawa et al. 1993; Tamura and Nei 1993), and most human population geneticists, including ourselves, agree with this estimate. Note that this divergence time depends on our estimate of the mutation rate, which itself depends on the age of the human–chimpanzee split (Templeton 1993). The claim that the ancestral mtDNA molecule was present in an African woman (the African Eve theory: Cann et al. 1987; Vigilant et al. 1991) has been more strongly disputed (e.g. Excoffier et al. 1989; Templeton 1992). It has been argued that the geographical location of the root could not be inferred with any statistical confidence from intraspecific mtDNA phylogenies, for equally plausible mtDNA phylogenies point to a European or an Asian female (Maddison 1991; Maddison et al. 1992; Templeton 1992). However, a recent comparison of complete hominoid mtDNA sequences suggests that human mtDNA variation is rooted in Africa (Horai et al. 1995). Although an African origin for our species appears the most likely way to reconcile palaeontological, archaeological, and genetic data, the African origin of all our mtDNAs is neither implied by that view nor statistically proved by the available evidence.

While the location of our common mitochondrial ancestor remains under debate, more mtDNA diversity is found within sub-Saharan African populations (Vigilant et al. 1991; Chen et al. 1995; Graven et al. 1995; Watson et al. 1996) than in European, Asian, or Amerindian populations. Two non-exclusive hypotheses have been advanced to explain why sub-Saharan Africa contains so much mitochondrial polymorphism. First, all else being equal, the amount of diversity maintained in a population depends on its effective size (Kimura and Crow 1964). The greater mtDNA differentiation in Africa could result from

African populations having been larger than non-African populations for much of our evolution (see Harpending *et al.* 1993). Second, longer local differentiation in sub-Saharan Africa would also produce greater diversity, implying that human evolution started earlier in Africa than anywhere else.

The effect of sudden demographic expansions

Human mitochondrial data also allow us to recognize the effects of recent demographic processes on the pattern of molecular diversity within some populations (Slatkin and Hudson 1991; Rogers and Harpending 1992; Rogers 1995). Recent, sudden demographic expansions starting from a small number of individuals leave recognizable signatures in the pattern of molecular diversity, particularly in what is now called the mismatch distribution, the distribution of pairwise differences between all sequences of a sample. Demographic expansions generate star-like gene genealogies because many present individuals share a common ancestor who lived just before the expansion (Slatkin and Hudson 1991).

Molecular signatures suggesting expansions have been found mainly in populations from Europe, the Middle East, and Asia, are less clear in American and African populations, and seem absent from populations who have been hunter–gatherers until recently, like the Bushmen of southern Africa and the Pygmies of the equatorial forest (Di Rienzo and Wilson 1991), the Saami of Finland (Sajantila *et al.* 1995), and the Mukri of India (Mountain *et al.* 1995). Thus population expansions, probably but not necessarily in the neolithic, have deeply affected the pattern of mitochondrial diversity in many human populations.

The effect of unequal population sizes

Different demographic histories may have another important consequence for the genetic similarity of human populations. Genetic drift, the random change of gene frequencies from generation to generation, affects small populations much more than large ones. For the same time of divergence, two small populations should differ more from each other than two

large populations. Hence, genetic distances between pairs of populations are proportional to divergence times and inversely proportional to population sizes. Allele frequencies in small populations are thus expected to differ considerably from those in large populations and even more from other small populations. This pattern was recently observed in a comparative study of mitochondrial diversity in many sub-Saharan African populations (Watson *et al.* 1996). Hunter–gatherer populations such as Pygmies or Khoisan were also genetically distant from all other populations. In general, one should not conclude that the most differentiated populations are the ones that separated first, unless it is reasonable to assume that all of them have had comparable sizes through most of their history.

Keeping the effect of demography in mind, global analyses of mtDNA, restriction fragment length polymorphisms (RFLPs), and control region sequences agree in suggesting:

(1) much divergence between Eurasian populations and two very distinct groups of African populations belonging to two distinct language families (the Khoisan and the Niger-Congo families);

(2) little divergence between populations speaking Indo-European languages;

(3) remarkable sequence affinity between two Amerindian linguistic groups (Amerind and Na-Dene) and populations of northeastern Asia (Forster *et al.* 1996) or Mongolia (Kolman *et al.* 1996; Merriwether *et al.* 1996), suggesting recent shared ancestry; and

(4) frequency gradients of one 9-base-pair deletion in the Pacific, supporting a South-East Asian origin of Oceanian populations (Lum *et al.* 1994; Redd *et al.* 1995; Melton *et al.* 1995) without any recognizable genetic input from Amerindians (Hagelberg *et al.* 1994).

Y-chromosome polymorphisms

Transmitted only from father to son, the Y chromosome is the male counterpart of mtDNA. It allows us to follow the transmission and reconstruct the genealogies of male lineages and can provide information on the age of the most recent common ancestor of all Y chromosomes, revealing the genetic affinities of the males of different populations. However, its main interest for the study of the history of human settlement resides in direct comparisons with mitochondrial DNA, which can reveal asymmetrical patterns of gene flow between the sexes and differential contributions of males and females to the gene pool of a given population.

The mode of transmission of the Y chromosome is essentially symmetrical to that of mtDNA, for its non-telomeric parts do not recombine and are transmitted as a single unit from father to son. However, the Y chromosome is not very polymorphic: sequencing studies on large segments of DNA in several individuals have revealed either the complete absence (Dorit *et al.* 1995) or very low levels of polymorphisms (Hammer *et al.* 1995; Whitfield *et al.* 1995). Despite this scant diversity, the age of the most recent molecular ancestor, which one would be tempted to call the molecular Adam, has been estimated to be around 180 000 years (Hammer 1995; Donnelly *et al.* 1996; Fu and Li 1996; Weiss and von Haeseler 1996), similar to that of the molecular Eve.

Can we take this close agreement as a proof of a recent origin of our species? Unfortunately not. Speciation could have occurred either before or after the origin of the genetic polymorphisms, although the similar genealogical age of many nuclear genes suggests that the age of our species cannot be too different. For the present, it is safest to say that similar ages for the trees inferred from mitochondrial and Y-chromosome variation indicate that male and female populations have had similar sizes over the course of our history. The average age (in generations) of the most recent common ancestor is about twice the effective number of gene copies in a population (Hudson 1990). As the effective size of human populations has been estimated to have fluctuated around 10 000 (Nei and Graur 1984; Li and Sadler 1991), or about 5000 males (Goldstein

et al. 1996) and 5000 females (Vigilant *et al.* 1989; Graven *et al.* 1995) for much of our evolutionary history, mitochondrial and Y-chromosome ancestors would have been present some 10 000 generations or about 200 000 years ago with a generation time of 20 years.

Our ability to infer the past is thus limited by the long-term effective size of the populations. We cannot identify phenomena that affected the early history of the human species more than 200 000 years ago by studying mtDNA or Y-chromosome polymorphism alone. Only nuclear DNA polymorphisms allow us to go further back in time, for there are four times as many copies of nuclear genes as there are of mtDNA or Y chromosomes in a given population.

Global analyses of male-mediated genetic affinities have only been possible so far on the basis of the RFLPs of Y-chromosome segments hybridizing with probes p49a and p49f, a polymorphism that has now been analysed in more than 70 populations worldwide (Excoffier *et al.* 1996; Poloni *et al.* 1997). The pattern revealed by this polymorphism agrees well with those inferred from mtDNA and conventional nuclear markers: a clear distinction between a group of Eurasian populations and two linguistically defined sub-Saharan groups, the Khoisan and the Niger-Congo language families. Within the Eurasian group, two subclusters also corresponding to language families emerge: a cluster of Afro-Asiatic speaking populations including North African, East African, and Middle Eastern populations; and an Indo-European cluster.

In some cases, the comparison of Y-chromosome genetic affinities with those from mtDNA or molecular markers also reveals populations where males and females seem to have a different origin. The Bantu speaking Lemba from South Africa appear to have a Y chromosome of Middle Eastern origin, the Dama from Namibia speak a Khoisan language but have Y chromosomes indistinguishable from Niger-Congo speakers, and the Basques have Indo-European mtDNA but Y chromosomes very different from their neighbours. This is consistent with a scenario in which men from one group always married women from another group, but other scenarios are possible.

Recently, Hammer *et al.* (1997) surveyed the polymorphism of an Alu element insertion (YAP), point mutations, and microsatellite loci on the Y chromosome in 1500 males from 60 world-wide populations. All major haplotypes were found in sub-Saharan Africa, whereas only a subset were found in Europe, Asia, or Australia. The greater African diversity suggested either an older origin or a larger past effective population size of Africans. Northern and southern Asian populations also clearly differed, and the lowest levels of diversity were found among northern Europeans. They suggested that an important part of YAP variation originated in Asia and was introduced by a subsequent migration back to Africa; human migration patterns may have been more complex than hypothesized under the simple out-of-Africa scenario.

Nuclear DNA diversity

There have been many fewer nuclear DNA studies than mtDNA or Y-chromosome studies, mainly because it is hard to infer which loci or DNA sequences will be found together in a single gamete when the linkage relationships of several sites polymorphic within the same individuals are not known. This difficulty has been evaded either by devising complex analytical tools to amplify single chromosomes or by multiple single-locus analyses that sacrifice the information carried by the gametic phase.

Note that in the following discussion, different ages are discussed for the common ancestors of different genes, and that some of these will appear to conflict with the ages inferred for the common ancestors of Y chromosomes and mitochondria because they are much greater than 200 000 years. To see that this conflict is only apparent, recall that the age of the molecular ancestor is not the age of our species. Genes differ in mutation rates and experience different evolutionary processes. Moreover, population sizes are smaller for genes transmitted only through females (mtDNA) or only through males (Y-chromosome genes), and the

impact of genetic drift is greater. Thus our genome is a mosaic whose parts may well reflect different evolutionary histories.

In a pioneer study of the β-globin polymorphism, Wainscoat *et al.* (1986) proposed the hypothesis of an original expansion out of Africa accompanied by a substantial reduction in population size (a bottleneck). Although the hypothesis was later disputed, in part because of different definitions of 'bottleneck' (Jones and Rouhani 1986; Takahata 1993), that study opened the door to increasingly complex analyses of nuclear, autosomal polymorphism most conveniently summarized by the type of marker they used.

Random nuclear RFLPs

A set of 80 random nuclear markers has been analysed in about 10 populations from all continents but America (e.g. Bowcock *et al.* 1991; Poloni *et al.* 1995). Most of these markers are a combination of anonymous probes and RFLPs scattered over autosomes. African populations (two Pygmy samples and one Senegalese Mandenka sample) show the most divergence from all other populations, and Europeans appear intermediate between the African cluster and an Oriental cluster itself divided into a north Asian branch and an Austronesian branch. Globally, a large and significant correlation is found between geography and genetics, suggesting that early migrations and isolation by distance have played a major role in shaping the pattern of genetic affinities between these populations. Unfortunately, no comparison between genetic and linguistic affinities could be carried out due to the small number of populations involved in the analysis.

β-*Globin gene cluster RFLPs*

The region of chromosome 11 harbouring the β-globin gene cluster has been analysed through RFLPs and a few direct sequencing studies. RFLP haplotype frequencies in 23 populations from all continents shows a pattern of population clustering in close agreement with geography (Fig. 3.2). African populations are found at one extreme of the distribution and Ocea-

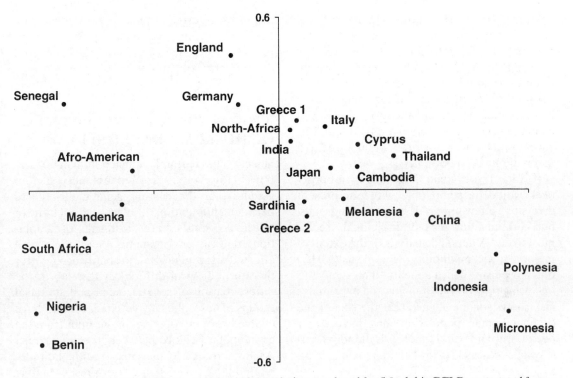

Fig. 3.2 Multidimensional scaling analysis of 23 populations analysed for 5 β-globin RFLPs, computed from a matrix of genetic distances between populations (pairwise linearized FST values).

nian populations at the other, with European populations being clearly intermediate between them. It thus seems that there is a gradient of genetic differentiation between Africa and South-East Asia.

The β-globin gene encounters balancing selection where malaria is endemic. A single nucleotide substitution producing an amino acid change from Val to Glu in the sixth amino acid of the β-globin peptide leads to sickle-cell anaemia in the homozygous state. This mutation can reach frequencies of 20 per cent in some regions with a high prevalence of malaria. A detailed analysis of the surrounding haplotypic background suggests that the sickle-cell mutation occurred at least five times independently, four times in Africa and once in India (Trabuchet *et al.* 1991). Despite the selection that has acted on it, the pattern of genetic affinities inferred from this region is compatible with a pure geographical differentiation process, a result consistent with the hypothesis that *falciparum*-mediated malaria

selection began within the last 2000–5000 years (Flint *et al.* 1993*b*; Fullerton 1996). Selection appears not to have had enough time to disturb pre-established differentiation patterns.

β-Globin sequences

A Melanesian sample has been analysed for the sequence polymorphism of a 3-kb region spanning the β-globin gene (Fullerton *et al.* 1994). The age of the common ancestor of those sequences is estimated at more than 400 000 years, much older than the ancestors of mitochondrial or Y-chromosome lineages. The same team has recently extended this data set to more than 300 sequences from nine populations from Africa, Asia, and Europe (Harding *et al.* 1997). The time to the most recent common ancestor was estimated as about 800 000 years, implying no severe bottleneck at the origin of present modern humans. The occurrence of Asian lineages more than 200 000 years old was taken to support world-wide late Pleistocene migrations and a mixed Asian and

African ancestry for all modern humans. These nuclear polymorphisms are revealing complex patterns of differentiation at the world-wide scale that may be difficult to reconcile with unidirectional migration out of Africa (Harding *et al.* 1997). Because nuclear lineages have a much older common ancestor, the recent common origin of our haploid mtDNA and Y-chromosome lineages is probably not due to a severe population bottleneck.

It is hard to evaluate how narrow a bottle-neck would be compatible with our present diversity—a few hundreds or a few thousands, but certainly not tens of individuals seems most likely. A detailed analysis of the extent of major histocompatibility complex (MHC) polymorphism (which appears to be subject to balancing selection) in humans and in primates also shows some evidence that the size of the original human population may have never been less than several thousands of individuals (Ayala *et al.* 1994, 1995; Takahata *et al.* 1995).

Microsatellites

Because nuclear microsatellite polymorphism is a relatively new technique, only a few studies of a few populations are available. The first study comparing patterns of diversities among human populations showed that microsatellite polymorphism patterns are geographically consistent; most of the genetic diversity is due to differences among individuals within popu-lations (Bowcock *et al.* 1994). A more recent study comparing the diversity pattern of 20 microsatellite loci among 16 populations from all over the planet showed that the genetic affinities inferred from microsatellite data depend on the choice of the evolutionary model for the emergence of microsatellite mutations (Pérez-Lezaun *et al.* 1997). Popula-tion trees revealed close affinity between African and European populations, the African populations being either intermixed among European populations, or clustered together but branching very close to European popula-tions, depending on which genetic distance was used to measure population relatedness. These data did not suggest a strong bottleneck out of Africa. Owing to the high variance of genetic

distances inferred from microsatellite data (Zhivotovsky and Feldman 1995), it seems prudent to wait for more studies before drawing firm conclusions.

THE ACTION OF SELECTION ON HUMAN DIVERSITY

Selection has certainly affected human popula-tions. However, it is often not necessary to assume that the current genetic differences between populations reflect the action of selective agents or are the results of adapta-tion to local environments. Selection which affected a few genes can probably account for the morphological differences that led to the differentiation of several racial and subracial groups. These phenotypic changes may have evolved very rapidly (e.g. in the Inuit from the circumpolar region, between the South-East Asians and the Melanesians, or in the latitude-related gradient of skin colour present among Amerindians). Therefore, the perceived pheno-typic diversity among human groups, besides being somewhat arbitrary, does not always correlate with genetic diversity measured by polymorphic markers and can mislead when inferring population affinities.

Adaptive selection has certainly affected many traits in human populations and may have led to convergent evolution for traits like skin colour, stature, and some disease-related genes (e.g. sickle- cell anaemia), even in very distant, and evolutionarily loosely related, populations. Because of convergence, these traits are unlikely to lead to accurate recon-structions of the global affinities among popu-lations. The simplest example is the dark skin of Africans and Australians, who are the most differentiated continental groups at the allele-frequency level (Cavalli-Sforza *et al.* 1994).

The demographic expansions of human pop-ulations have allowed the emergence of many infectious, microparasite-mediated diseases that have caused genetic reactions in the in-fected populations. However, these population expansions are recent on the time-scale of human evolution, and human population affinities inferred from selected loci such as β-globin (Fig. 3.2) or HLA (Tiercy *et al.* 1992) still show traces of differentiation processes that occurred during paleolithic times. Therefore, recessive pathological alleles may often have spread together with the rest of the genome of the individuals carrying them, and hence their distribution may reflect the same demographic phenomena that affected most non-pathological alleles. The distribution of the δ-F508 mutation for cystic fibrosis in Europe is an example of the similar impact demographic changes had upon the distributions of both pathological and non-pathological alleles.

In conclusion, to understand the origin and distribution of hereditary pathologies and lia-bilities to disease, we have to interpret correctly the broad features of human history, as reconstructed from non-pathological markers. In particular, the relationships among current populations may date back to extremely remote prehistorical times.

SUMMARY

The evidence suggests that the present genetic affinities among human populations are the result both of a complex settlement history and of a series of historical events which punctuated that history. Table 3.2 summarizes some of them, with special attention to their genetic consequences. Human populations seem to

Table 3.2 An outline of historical events that have affected the present genetic affinities among human populations

Historical event	Long-term genetic consequences	Example
Early dispersal of small human groups with founder effects	Deep genetic differences between geographically separate groups	Large differences between Africans and Australians
Long-range mass migration	Genetic homogeneity over large regions	Bantu and Indo-European expansions
	Discordance between genetic and geographical affinities	
	Gradient of gene frequencies	
Gene flow between neighbouring populations	Establishment of isolation-by-distance pattern	Short-range differentiation in limited areas
	Genetic continuity	
Long-range sporadic migration	Discordance between genetic and geographical affinities	Hungaric speakers
Secondary contact between distinct genetic pool	Zone of sharp genetic gradient	Genetic differences at linguistic boundaries, e.g. between Khoisan and Niger-Congo speakers
Admixture	Enhanced genetic diversity within populations	African-Americans, East Africa, Japan
	Gene frequencies intermediate between those of source populations	
	Linkage disequilibrium, even between unlinked loci	
Sudden demographic expansions	Apparent lack of genetic diversity	Neolithic expansion of Indo-European, Altaic, and Elamo-Dravidian speakers
	Star-like gene genealogies	
	Linkage equilibrium between linked loci	

have separated when much of the current genetic diversity was already present. Many alleles are not restricted to one or few populations: they occur in wide geographical areas. Recent processes of gene flow have also helped to spread alleles far from their origins. With the possible exception of societies in which different immigrating groups have not substantially mixed, markers such as stature and skin colour are not useful for the kind of classification that would be needed to identify individuals at increased risk for certain diseases.

Although genetic diversity among populations is limited, its pattern reflects both geographical proximity and genealogical relationships between populations; the latter may sometimes be inferred from their current linguistic affinities. Departures from the pattern of similarity between genetic, geographical, and linguistic distances have allowed us to identify, and sometimes to reconstruct, some major colonization and expansion processes. The details may be hard to infer from genetic data, but some agreement between gene diversity measures and archaeological and palaeontological findings is undeniable.

Molecular data have allowed us to refine the results obtained from allele-frequency studies. They have confirmed the known relationships and extent of differences between groups of populations defined on linguistic grounds, and in some cases they have allowed us to trace specific mutations along migration pathways, as in the Pacific Ocean or in America. A major achievement has been the demonstration of a recent common origin of all human populations, even though the geographical origin of the ancestral population and its exact genetic composition cannot yet be reconstructed very precisely: East Africa and the gene pool of East Africans are likely candidates. The hypothesis of multiregional evolution is clearly refuted by mitochondrial and Y-chromosome data, as much older molecular ancestors would have been found if more than one branch of *Homo erectus* gene pools had been transmitted to us.

The genetic study of human populations is done through a discrete sampling process, but genetic differentiation is a continuous process in both time and space. Even if we recognize distinct groups and clusters of populations in our analyses, smoother transitions are expected in the field. Our present classification and clustering systems will certainly need to be refined when additional populations are included in the analyses.

ACKNOWLEDGEMENTS

We thank André Langaney, Alicia Sanchez-Mazas, and Giorgio Bertorelle for their helpful comments and suggestions on this manuscript, and David Roessli for his help in setting up the figures. L.E. was supported by Swiss NSF grants No. 31–039847.93 and 32–047053.96, and G.B. by the Italian CNR grant No. 96.01182.PF36 within the project Beni culturali.

4

MEDICALLY RELEVANT GENETIC VARIATION OF DRUG EFFECTS

Urs A. Meyer

The study of genetically determined variations in drug response is commonly referred to as pharmacogenetics. Many examples of strikingly exaggerated heritable responses to drugs, novel drug effects, or lack of effectiveness of drugs have been described and comprehensively reviewed (Kalow 1992; Meyer 1992; Evans 1993; Gonzalez and Idle 1994; Vogel and Motulsky 1996). The objective of this chapter is to discuss genetic variation in drug responses that occur at high frequencies and have clinical consequences.

Interethnic differences in reactions to drugs and chemicals, sometimes called pharmacoanthropology or ethnopharmacology, represents another area of clinical interest. Population differences may be produced by different environmental effects on drug action or have a genetic origin (Kalow and Bertilsson 1994). A question of particular interest is how genetic diversity between individuals and populations will affect future drug therapy and drug development. Although studies of overall patterns of genetic variation suggest that there is no genetic basis to the traditional concepts of human races (Chapter 3), those studies are based on many genes. Individual genes do vary in frequency among human populations, and some of them affect drug response.

The field of ecogenetics is concerned with the dynamic interactions between genotype and environmental agents such as industrial chemicals, pollutants, plant and food components, insecticides, and pesticides. Examples include interindividual differences in ethanol sensitivity, sensitivity to milk because of lactase deficiency, development of pulmonary emphysema in patients with α_1-antitrypsin deficiency, and many other interactions between the environment and the genetic make-up of the individual. Similarly, toxicogenetics describes individual predisposition to carcinogenic, teratogenic, and other toxic effects of chemicals. These aspects of genetic variation are beyond the scope of this review.

Genetic variation of drug effects is caused predominantly by changes in drug metabolism. In some subjects, however, unexpected, novel, or unusual reactions to drugs may be inherited without the hereditary defect being directly associated with the pharmacokinetic behaviour or with the usual response of organs, tissues, or cells to a particular drug. In fact, pharmacogenetics had its initial impact on medicine predominantly through the discovery of such unexpected alterations in drug response. The prototype was the syndrome of drug-induced haemolytic anaemia in subjects with deficient glucose-6-phosphate dehydrogenase. Some inherited metabolic diseases, such as acute hepatic porphyria or malignant hyperthermia, are uncovered or dramatically precipitated by the administration of drugs—the unusual drug response serves as a phenotypic marker of the genetic disease (for reviews see Meyer 1992; Evans 1993).

GENOTYPE AND PHENOTYPE

A key issue in individual variation in drug response is the differentiation of genetic and environmental factors. Obviously, the interaction of genetic and environmental influences is difficult to disentangle in a single patient. Variations in drug response, whether controlled by genes, environmental factors (diet, smoking, disease), or both, may occur at sites of absorption, distribution, protein binding, drug–cell interaction, metabolism, and excretion. Moreover, several independent variables may modulate more than one of these discrete processes. Only when one of the numerous interactions of a drug with the product of one aberrant gene assumes decisive importance for drug action can we expect its easy recognition and its transmission by classic Mendelian inheritance. Under these circumstances, the frequency distribution of a measure of drug response will be multimodal, each subpopulation corresponding to a different genotype. This is the case for most classical examples of pharmacogenetic conditions, for example the slow acetylation of isoniazid or the prolonged apnoea after succinylcholine. Family studies easily reveal the mode of inheritance of the inherited variant in these disorders.

More often, since drug action depends on numerous events controlled by different gene products, several genes interact with environmental factors to result in a particular variation in drug effect. This polygenic inheritance is more difficult to detect, and the distribution curve does not segregate pharmacogenetic subpopulations. Rather, the interaction of the multiple environmental and genetic factors results in an apparently statistically normal, unimodal, and continuous distribution of drug response. The analysis resembles that for complex diseases, and careful studies of families may still disclose genetic control of the suspected pharmacogenetic trait. However, even pedigree studies involving several generations frequently do not uncover the polygenic contribution to observed variation, for the necessary comparison of subjects of different age, sex, and environment results in large 'nongenetic' contributions to variation that hide the genetic components.

An important and strikingly simple method that distinguishes between hereditary and environmental components of variation is the comparison of small series of mono- and dizygotic twins in whom the variation within pairs can be analysed by established statistical methods (Galton 1875; Vesell 1990). However, methods using twins do not permit a distinction between Mendelian or polygenic modes of inheritance. Application of the 'twin method' has demonstrated important genetic factors in the pharmacokinetic behaviour of a large number of commonly used drugs.

PHARMACOGENETIC POLYMORPHISMS

The concept of genetic polymorphism arose from the finding that many phenotypic traits such as blood groups, histocompatibility antigens (HLA system), and enzyme variants exist in the population in frequencies that could not be explained by spontaneous mutations. A genetic polymorphism is defined as a Mendelian or monogenic trait that exists in the population in at least two phenotypes (and presumably at least two genotypes), neither of which is rare, that is, neither of which occurs with a frequency of less than 1–2 per cent (Vogel and Motulsky 1996). The definition of a phenotype frequency of more than 1 per cent as 'common' or 'polymorphic' is arbitrary and has practical reasons, for a frequency of 0.1 or 0.01 per cent would be hard to detect in the usually small groups of volunteers tested. Moreover, the original definition did not specify if the rare phenotype is of heterozygous or homozygous genotype for the variant allele. An allele-based definition of pharmacogenetic polymorphism therefore has been proposed (Meyer 1991).

Genetic polymorphism has evolutionary impact, for the variation in the population responds to a change in the environment (e.g. exposure to toxic chemical or nutritional components) and changes the structure of the population, increasing its chance of survival. In this regard, the interaction between a poison and an organism has many parallels with the interaction between pathogenic micro-organisms and their host. The result of both interactions is population survival of catastrophic exposures.

Table 4.1 Genetic polymorphism of enzymes of drug metabolism of clinical or toxicological relevance

Butyryl cholinesterase
Paraoxonase/arylesterase
Debrisoquine/sparteine hydroxylase (CYP2D6)
Mephenytoin hydroxylase (CYP2C19)
Cytochrome P450 (CYP1A1)
Cytochrome P450 (CYP2E1)
Alcohol dehydrogenase (ADH)
Aldehyde dehydrogenase (ALDH)
Dihydropyrimidine dehydrogenase (DPD)
N-Acetyltransferases NAT1 and NAT2
Glucuronosyltransferase
Thiol methyltransferase
Catechol-*O*-methyltransferase
Thiopurine methyltransferase
Phenol sulphotransferase
Glutathione-*S*-transferases (GSTM1, GSTT1)

With the exception of glucose-6-phosphate dehydrogenase deficiency, clinically relevant pharmacogenetic polymorphisms concern drug-metabolizing enzymes. Table 4.1 lists polymorphic drug-metabolizing enzymes. Only the three most common polymorphisms that involve clinically used drugs will be described here.

EVOLUTION OF DRUG-METABOLIZING ENZYMES

It is likely that 'drug'-metabolizing enzymes evolved during plant–animal coevolution. Animals have been eating plants and plants have been defending themselves by developing new toxins (alkaloids, terpenes) for more than 1000 million years. In response, animals developed new or modified enzymes able to detoxify these 'phytoalexins' (Gonzalez and Nebert 1990). This scenario is best illustrated by the analysis of the evolutionary divergence of cytochrome P450 isozymes, the best studied and largest enzyme system for drug metabolism, also called the P450 gene superfamily (Nelson *et al.* 1996). Approximately 600 P450 genes are known in 85 eukaryote (including vertebrates, invertebrates, fungi, and plants) and 20 prokaryote species, and comparison of their amino acid sequences allows classification into families and subfamilies. Phylogenetic trees calculated by various algorithms are consistent with a very ancient ancestral gene that originated long before the prokaryote/eukaryote division and thus predated animal–plant interactions, combustion of organic matter, or pharmaceuticals. Many gene duplications with a major expansion of the number of genes in the families preceding drug-metabolizing enzymes (CYP2, 3) occurred approximately 800–1200 million years ago, consistent with plant–animal warfare. Most of today's drugs still have structural similarities to plant molecules such as opioid alkaloids, glycosides, and ergot alkaloids, and this may explain the emergence of multiple enzymes able to detoxify plant and possibly other biological toxins. P450s have many other functions, including the metabolism of endogenous compounds such as vitamins, steroid hormones, cholesterol, and fatty acids. Estimates of the number of individual P450 genes in any mammalian species range from 60 to 200. Other drug-metabolizing enzyme families (UDP-glucuronosyl transferases (UGT), methyl transferases, *N*-acetyl transferases) also show considerable diversity.

Drug-metabolizing enzymes carry out myriad transformations on small lipid-soluble molecules. Their substrate specificities are limited; it is not unusual for a drug to be metabolized by several different enzymes to varying degrees and with different affinities. A single P450 or conjugating enzyme thus can metabolize numerous, structurally diverse chemicals. The multitude of enzymes and their overlapping substrate specificities provide the organism with the capacity to metabolize and detoxify countless substances in the diet and the environment. One major clinical consequence of the evolution of these enzymes is polymorphism in metabolism.

CLINICALLY RELEVANT POLYMORPHISMS

The acetylation polymorphism concerns the metabolism of a variety of arylamine and hydrazine drugs and chemical carcinogens by a cytosolic *N*-acetyltransferase, NAT2. Seven mutations of the NAT2 gene occurring singly or in combination define numerous alleles associated with decreased (never totally absent) function (for reviews see Vatsis *et al.* 1995; Meyer and Zanger 1997). Slow acetylation is recessively inherited and the slow acetylation phenotype varies remarkably in different populations (see below). A second related enzyme, NAT1, also is polymorphically expressed, but its clinical importance is not clear.

Slow NAT2 acetylators are at risk of developing adverse reactions to sulphonamides, isoniazid, and several other drugs; rapid acetylators are at risk of developing side-effects to the cytostatic drug amonafide because a toxic metabolite is formed at a higher rate (Evans 1993). The association of the slow NAT2 acetylation genotype/phenotype with an increased risk for urinary bladder cancer, particularly with occupational exposure to aromatic amines, is well documented (Vineis *et al.* 1994; Risch *et al.* 1995). A recent study also observed a markedly increased incidence of breast cancer in cigarette-smoking postmenopausal women who were slow NAT2 acetylators (Ambrosone *et al.* 1996). All these studies suggest that the capacity to detoxify carcinogenic aromatic amines in cigarette smoke is associated with cancer risk.

The debrisoquine/sparteine polymorphism of drug oxidation affects the metabolism of numerous drugs. The poor metabolizer phenotype is caused by homozygosity for 'loss-of-function' alleles of the cytochrome P450 CYP2D6 gene (for review see Meyer and Zanger 1997). On the other hand, 'ultrarapid metabolizers' have a duplication or amplification (up to 13 copies) of a CYP2D6 gene on chromosome 22 (Johansson *et al.* 1993). Intermediate metabolizers are often heterozygotes or carry alleles with mutations that only partially decrease enzyme activity. The unravelling of the complex and highly variable CYP2D cluster of genes and pseudogenes on chromosome 22 (Heim and Meyer 1992) has characterized 48 mutations and 53 alleles of the CYP2D6 gene (Daly *et al.* 1996; Marez *et al.* 1997).

The clinical importance of this polymorphism was initially questioned because the drugs involved in its discovery were soon either obsolete or not widely distributed. Then many other drugs were discovered that are inefficiently metabolized in poor metabolizer subjects. Over 40 clinically used drugs have now been identified whose metabolism *in vivo* cosegregates with that of debrisoquine/sparteine. They include: antiarrhythmics such as *N*-propylajmaline, flecainide, propafenone, and mexiletine; antidepressants such as amitriptyline, clomipramine, desipramine, fluoxetine, imipramine, maprotiline, mianserin, paroxetine, and nortriptyline; neuroleptics such as haloperidol, perphenazine, and thioridazine; antianginals (perhexiline); and opioids (dextrometorphan, codeine), as well as amphetamines such as methamphetamine and methylenedioxymethamphetamine (ecstasy) and many other drugs. Moreover, a large number of β-adrenergic receptor antagonists are influenced in their elimination by this polymorphism (Meyer 1992; Evans 1993; Kroemer and Eichelbaum 1995).

Numerous case reports and clinical studies have demonstrated that for some of these drugs the polymorphic oxidation has therapeutic consequences, leading either to a higher propensity to develop adverse reactions at conventional doses (as demonstrated by its discovery) or to decreased drug effects, such as absence of the analgesic effect of codeine in poor metabolizers (Sindrup and Brøsen 1995) or therapeutic failure at normal antidepressant doses in ultrarapid metabolizers (Bertilsson *et al.* 1993). The most significant effects are seen with drugs acting on the cardiovascular and central nervous systems. Moreover, epidemiological studies have indicated that poor metabolizers and extensive metabolizers may be at higher risk of developing certain diseases including some forms of cancer, though many of these associations are weak and disputed. Eichelbaum and Gross (1990), Meyer (1992), and Evans (1993) provide detailed reviews of the clinical and toxicological implications of the debrisoquine polymorphism.

The mephenytoin polymorphism is caused

by mutations in another cytochrome P450 isozyme. Poor metabolizers are homozygous for loss-of-function alleles of CYP2C19 (for reviews see Wilkinson *et al.* 1992; Goldstein and de Morais 1994; Meyer and Zanger 1997). The polymorphism was discovered because of side-effects of poor metabolizers to the anticonvulsant drug mephenytoin, which is no longer in clinical use. Additional substrates for the enzyme are mephobarbital, hexobarbital, diazepam, the antidepressants imipramine and citalopram, the proton-pump inhibitors omeprazole, lansoprazole, and pantoprazole, and the antimalarial prodrugs proguanil and chloroguanil. The clinical consequences of impaired mephenytoin hydroxylation have not been studied in detail.

INTERETHNIC VARIABILITY AND EVOLUTION OF POLYMORPHISMS OF DRUG METABOLISM

NAT2 acetylation polymorphism

The frequency of the slow acetylator phenotype varies considerably among ethnic groups (for review see Evans 1993). Genotyping data of the common NAT2 alleles trace the molecular history of the acetylation polymorphism (Table 4.2). Two groups can be distinguished by the incidence of the mutations T341C and C282T/G857A of the NAT2 gene. Caucasians and Africans are very similar and have high frequencies of T341C (more than 28 per cent) and low frequencies of C282T/G857A (more than 5 per cent). Japanese, Chinese, Korean, and other Far Eastern populations have a low incidence of T341C mutations (more than 7 per cent) and a higher incidence of C282T/G857A mutations (10–18 per cent). The difference in the incidence of the slow acetylator phenotype between Caucasians/Africans (40–70 per cent) and Asian populations (10–30 per cent) is thus mostly due to the low incidence of the alleles carrying the T341C mutation (alleles NAT2*5A, B, C) in Asians. Amerindians appear to be similar to Asian populations in their high proportion of rapid acetylators and the T341C or C282T/G857A relationship. Hispanics appear to be intermediate between the Caucasian/African and the Asian groups. The data are consistent with an ancient African origin of the acetylation polymorphism, before divergence of human populations (see Chapter 3). The practical absence of the NAT2*12 and *14 alleles (A803G and G191A) in Caucasians compared with the more than 8 per cent incidence of the G191A mutation in Africans suggests either that these mutations occurred after the African/non-African split approximately 100 000 years ago or, if they occurred earlier, that they were maintained by selection in Africa and eliminated by selection or drift outside Africa.

There are no clues to the mechanism that maintains the high frequency of the slow acetylator phenotype, and no natural or endogenous substrates of these enzymes are known.

CYP2D6 polymorphism

Significant ethnic differences exist in the frequency of the poor metabolizer phenotype, which appears to be rare (1–2 per cent) in most populations except Caucasians (5–10 per cent). Interethnic variability of CYP2D6 alleles is less well studied than the NAT2 polymorphism, but the existing data allow some conclusions on the origin of the most common alleles (Table 4.3; for review see Meyer and Zanger 1997). The most extreme variabilities are observed for the two allele families CYP2D6*4 and *10. The practical absence of the common Caucasian B-allele (CYP2D6*4) from east Asian populations explains the very low prevalence of the poor metabolizer trait in the latter. A similarly low frequency of this allele was described for two African populations. In most African studies, poor metabolizer frequencies were also found to be low. One explanation is that the CYP2D6*4 allele appeared in Causasoids after their separation from Asian/Amerindian groups about 35 000 years go. The intermediate frequency of *4 alleles in the American black population is probably the result of penetrance of Caucasian

Table 4.2 Interethnic comparison of NAT2 alleles[a]

Ethnic group	Number of individuals analysed	Rapid acetylator alleles (%)[b]	Slow acetylator alleles (%)[b]			
		NAT2*4 (wt)	NAT2*5 A/B/C (T34IC)	NAT2* 6A/B (G590A)	NAT2* 7A/B (C282T/ G857A)	NAT2* 14A/B (G191A)
Caucasians, USA	421	24	43	31	2	0
Caucasians, Europe	434	26	46	26	2	0
Caucasians, Spain	504	22	44	26	1	1
African Americans	214	35	30	23	5	8
Native Africans (Gabonese)	102	27	40	22	2	9
Hispanics	148	40	28	18	14	1
Japanese	224	67	1	22	10	0
Chinese	254	53	5	30	12	0
Koreans	85	68	2	18	11	1
Filipinos	100	40	7	36	18	0
Aborigines (Australia)	49	41	2	17	40	–
Amerindians						
Ngawbe	71	74	2	0	24	0
Embera	101	65	10	4	21	0

[a] For details and references see Meyer and Zanger (1997); allele nomenclature as in Vatsis *et al.* (1995).
[b] The sum of all alleles is not necessarily 100 per cent because rare alleles were not considered in all studies.

genes. However, the high frequency (17 per cent) observed in an Amerindian tribe by Jorge *et al.* (1993) challenges the above hypothesis.

Another major interethnic difference is a shift in the frequency distribution of the urinary metabolic ratio of debrisoquine to 4-hydroxy-debrisoquine to higher values in Chinese populations than Caucasians. This means that the metabolism is slightly slower in Chinese than in Caucasian extensive metabolizers. The molecular basis for this phenotypic difference is a very high prevalence in Orientals of *10 alleles with slightly lower enzyme activity due to the C188T (P34S) mutation.

The two major functional *1 and *2 alleles appear to be frequent in all populations, but one should note that the frequencies of these alleles are determined by the absence of other mutations, and the discovery of new mutations may change the picture considerably. Compared with Caucasians, Ethiopians have a con-

siderably higher incidence of gene duplication and amplification alleles (Aklillu *et al.* 1996). The authors discuss the interesting possibility that duplicated and amplified L-alleles (CYP2D6* 2xN) may have originated in Africa and were transmitted to Caucasians recently during the immigration of Arabs into Spain, where a higher frequency is observed than in the rest of Europe.

It is obvious that humans can exist without functional expression of the CYP2D6 gene. However, in some studies the poor metabolizer genotype and phenotype had a higher vitality, alertness, and efficiency than did extensive metabolizers (Bertilsson *et al.* 1989; LLerena *et al.* 1993). This could suggest a role of CYP2D6 in the bioactivation or inactivation of endogenous compounds, such as neurotransmitters. Another study suggests that the intrinsic perception of pain may differ in the two phenotypes (Sindrup *et al.* 1993).

Table 4.3 Interethnic comparison of CYP2D6 allele frequencies

Ethnic group	No. of individuals analysed	Functional alleles (%)				Non-functional alleles (%)		
		*1	*2	*2 × 2	*10	*3	*4	*5
Caucasians	Compiled	34	33	1.5–3.5	3	1	20	5
Chinese	Compiled	23	20	1	50	0	<1	5
Japanese	Compiled	42	12	0	33	0	0	13
Koreans	21	33	24	2	36	0	0	5
Ngawbe Amerindians	30						17	0
Sinhalese	77					0	9	
African Americans	127					2.4	8.5	5.5
Ethiopians	115			13	10	0	1.3	
Zimbabweans	114					0	1.8	3.9

Data compiled and modified from different sources reviewed in Meyer and Zanger (1997); allele nomenclature as in Daly *et al.* (1996).

CYP2C19 polymorphism

The poor metabolizer phenotype associated with homozygosity for two loss-of-function alleles of CYP2C19 occurs with a frequency of 2.5–6 per cent in Caucasian populations but with much higher frequency in Japanese (18–23 per cent) and Chinese (15–17 per cent) subjects. Eighteen phenotyping studies have recently been analysed by Evans *et al.* (1995) and include Saudi Arabian, Filipino, and Indian populations. An interesting relationship between longitude and the frequency of poor metabolizers was observed, suggesting a cline containing a step between Riyadh and Bombay. The Saudi population resembled Europeans in the frequency of poor metabolizers for mephenytoin, but was similar to Orientals in the frequency of poor metabolizers for debrisoquine. Recent genotyping data have extended the database. The m1 mutation (de Morais *et al.* 1994b) is the predominant defective allele in African-Americans (Edeki *et al.* 1996), in black Zimbabwean Shonas (Masimirembwa *et al.* 1995), and in Caucasians, but in all these populations and in Orientals accounts for only 75–85 per cent of poor metabolizer alleles. The second described mutation of CYP2C19, m2 (de Morais *et al.* 1994a), explains the remaining 20 per cent of defective alleles in Orientals but is very rare in

Caucasians (Brøsen *et al.* 1995). It appears therefore that the CYP2C19 m1 mutation is very ancient in the evolution of human populations and that the m2 mutation arose subsequent to the emergence of the South-East Asian lineage. Moreover, the CYP2C19 locus appears to be much more stable genetically then the CYP2D gene cluster responsible for the debrisoquine polymorphism. Whether this indicates different initial selection mechanisms remains an interesting question.

GENETIC VARIABILITY OF RECEPTOR SYSTEMS

At first sight it seems surprising that genetic polymorphism of drug-metabolizing enzymes is the rule and that genetic variation of receptor function appears to be rare in healthy populations. In other words, if receptors vary, the result is disease and not just pharmacogenetic variability. One major difference between receptor systems and drug metabolism is that drug-metabolizing enzymes are 'low affinity' systems whereas receptor interactions are high affinity chemical interactions with stringent structural requirements. Mutations in such systems are therefore frequently either fatal or cause severe disease. This concept is supported by the observation that there are considerable species differences in drug-metabolizing capacity but little interspecies variation in receptor systems. In addition, until the recent development of recombinant DNA techniques, it was difficult to clone receptor genes and study their sequences. More recent studies now suggest that variation in receptor sequence may cause subtle variations in response to drugs. As an example, a combination of mutations of the dopamine D4 (DRD 4) receptor gene apparently predicts the variable response to the antipsychotic drug clozapine in patients with schizophrenia (Kennedy *et al.* 1994). Each of the five dopamine receptors subtypes has a distinct pharmacological profile and a unique tissue distribution. Present attempts therefore try to associate polymorphic variants of these receptor subtypes with features of human personality (Benjamin *et al.* 1996; Ebstein *et al.* 1996). Similar studies in other receptor systems, such as the serotonin 5-HT2A receptors involved in the effect of hallucinogens (Arranz *et al.* 1996), or adrenergic receptors involved in cardiovascular and respiratory physiology, are presently underway in several laboratories.

CLINICAL IMPACT OF PHARMACOGENETICS

The discovery and understanding of genetic variability of drug-metabolizing enzymes has contributed substantially to the understanding of the factors that determine individual and population differences in the pharmacokinetics and pharmacodynamics of many existing drugs. Furthermore, these concepts are already generally applied in drug development in the pharmaceutical industry (e.g. Cohen 1997). However, in contrast to the detailed knowledge of molecular mechanisms, the clinical relevance of these polymorphisms is poorly documented and is deduced largely from anecdotal reports and studies of small numbers of patients. There is an obvious discrepancy between the detailed molecular information and the rudimentary clinical evaluation of the

consequences of these polymorphisms. The availability of genotyping tests should enable prospective clinical studies of adverse drug reactions and drug toxicity that will help adjust drug doses to individual metabolic capacity.

In a recent retrospective study in which the CYP2D6 genotype was determined in patients with adverse responses to tricyclic antidepressants, the frequency of mutant alleles of CYP2D6 was more than twice that of psychiatric patients without adverse effects or in the general population (Chen *et al.* 1996). Similarly, phenotyping for CYP2D6 variants by the dextrometorphan metabolic ratio predicted the steady-state plasma levels of the antidepressant desipramine, identifying subjects at risk for concentration-dependent adverse effects (Spina *et al.* 1997). Other therapeutic areas with evidence for clinical significance of decreased metabolism capacity for drug toxicity involve cardiovascular drugs such as perhexiline, encainide, propafenone, procainamide, and hydralazine (Bucher and Woosley 1992; Arcavi and Benowitz 1993; Kroemer and Eichelbaum 1995).

Why then do most physicians not consider genetic variation in treating patients? There are many reasons. Among them are imprecise dose–response relationships, vague therapeutic endpoints, and the variable contribution of environmental factors in an individual patient. But the most important reason is the lack of convincing prospective studies in large populations. We need well-planned prospective studies. In psychiatric patients, for example, does early individualization of drug doses with genotyping tests lead to shorter hospital stays, less side-effects, and lower treatment costs?

Preventing adverse effects of drugs by early individualization of drug doses may be the most important aspect of pharmacogenetic testing. The future use of modern chip technologies to determine metabolic genotypes rapidly (Cohen 1997) should convince physicians of applying this knowledge to individualize therapy. As with all genetic testing, ethical issues must be considered before large-scale testing can be recommended.

Population differences in drug effects are

another area with clinical impact. They obviously are important when new drugs are introduced world-wide with dosage recommendations that may not be ideal for all populations. The dosage recommendations for omeprazole may differ for Orientals, for their increase in gastrin levels is more pronounced at the same dose. This is because the incidence of poor metabolizers of omeprazole (CYP2C19 polymorphism) and of heterozygotes with intermediate metabolism is very high in Oriental populations (18–23 per cent homozygous, 49–50 per cent heterozygous): the majority of the population has markedly or moderately impaired metabolism.

Finally, the associations of drug metabolism polymorphisms with individual risk of developing certain diseases suggest that the polymorphic enzymes activate or inactivate environmental carcinogens, other xenobiotics, or endobiotics. But with the exception of amine carcinogens as substrates for acetyltransferases, the search for chemical carcinogens that may be activated or inactivated by CYP2D6 or CYP2C19 has not been successful. The role of drug-metabolizing enzymes in the metabolism of endogenous substances, for instance of 'endogenous' cocaine and morphine by CYP2D6 (Mikus et al. 1994) or of hormones and neurotransmitters, as inferred from indirect observations, has not been evaluated. Perhaps the most intriguing aspect of the evolution of drug-metabolizing enzymes and their genetic and functional diversity is that this field is still in its infancy. The 'drug'-metabolizing function of these enzymes is obviously only one of their most recent functions and earlier and present roles of these enzymes in critical subcellular processes, such as during embryonic and fetal development, have not yet been discovered.

SUMMARY

Genetic variation of drug effects is predominantly caused by polymorphism in genes for drug-metabolizing enzymes (probably over 100 per organism). These enzymes for foreign compound metabolism probably evolved during plant–animal warfare, for most drugs are related to plant molecules. Clinically relevant genetic variation of drug metabolism is caused by the acetylation polymorphism involving the enzyme N-acetyltransferase 2 (NAT2), the debrisoquine/sparteine or cytochrome P450 CYP2D6 polymorphism, and the mephenytoin or CYP2C19 polymorphism. Polymorphisms of many other drug-metabolizing enzymes are known but have not been studied to the same extent. These polymorphisms result in subpopulations of so-called poor metabolizers who are at higher risk of developing adverse drug reactions, are more susceptible to certain toxic chemicals, and are at higher risk of developing certain diseases, including cancers. The frequency of the mutant alleles of the polymorphic genes and consequently the frequency of poor metabolizers varies considerably between populations of different ethnic origin. Simultaneous analysis of multiple genes and their mutations can predict these individual risks.

5

GENETIC VARIATION AND HUMAN DISEASE: THE ROLE OF NATURAL SELECTION

Adrian V.S. Hill and Arno G. Motulsky

The current investment in analysis of the human genome is providing a vast database of potential value in understanding the evolution of human genetic diversity. Already the human genetic map is complete and a physical map of the genome is beginning to emerge. Before too long a sequence of one entire human genome will be available. Initially, this torrent of new information will provide more questions than answers for those interested in the evolution, causes, and consequences of human genetic diversity. However, the potential utility of the current mapping tools for identifying and quantifying the underlying evolutionary forces and mechanisms contributing to human genetic diversity is enormous. Mapping studies of the human genome are localizing and beginning to identify a multiplicity of genes affecting susceptibility to polygenic disease and responsible for non-disease-associated phenotypic variation.

These studies will provide new opportunities to search for possible selection pressures that have contributed to the prevalence of disease-associated genes in human populations. Such work will ultimately help to resolve the ongoing debate about the relative importance of selective and non-selective (neutral) mechanisms in the maintenance of human genetic diversity. Our objective in this brief overview is to catalogue some genes for which there is already evidence of the action of natural selection in maintaining the frequency of allelic variants in human populations, particularly those where the relevant selective agent has been identified. We pay particular attention to the likely role of selection by infectious patho-

gens on the maintenance of polymorphism because infectious diseases have probably played a major role in maintaining human polymorphic diversity and have provided some of the best examples of ongoing selection. A question of particular interest is the extent to which selection for resistance to infectious diseases in the past may have contributed to the prevalence of alleles that predispose to common chronic diseases of modern societies. Finally, we suggest some areas and diseases where future genetic analysis is likely to uncover new polymorphisms that have been maintained by specific selective agents.

We emphasize the role of infectious disease as a possible selective factor and provide a variety of specific examples below. In a more general sense, the role of host genes in infectious disease mortality was shown in an adoption study in Denmark. Sorensen *et al.* (1988) calculated the risk of death from infectious disease among adoptees (between the ages of 16 and 58 years) when the cause of death (before the age of 50) for both their biological and adoptive parents was known. When the biological parents had died of infections, the relative risk of infectious disease death for the adoptees was 5.81 (95% confidence limits: 2.5–13.7). In contrast, the relative risk of infectious disease mortality among the adoptees did not differ from unity when their adoptive parents had died from infectious causes, thereby demonstrating the role of host genes in influencing infectious disease mortality. Genetic influences on cancer mortality were also studied by this design and were much less impressive.

SELECTION OR FOUNDER EFFECT?

Genetic diseases and their frequencies differ among populations. Selection by malaria clearly has shaped the distribution of the genes for sickle-cell anaemia, the different α- and β-thalassaemias, and the glucose-6-phosphate dehydrogenase (G6pd) deficiencies (see below). A characteristic and unique pattern of some relatively common autosomal recessive diseases has been observed in Finland. The history of Finland and the founding of its population and expansion over the past 1000 years are well described (de la Chapelle 1993). Genetic disequilibrium is due to persistent cosegregation of linked, non-homologous genes over many generations. Study of marker genes (haplotypes) located around different Finnish disease genes allows inferences as to the number of generations since the original mutation. Study of several Finnish autosomal recessive diseases has shown that these conditions originated from single founders with subsequent expansion of the population (Hastbacka *et al.* 1992; de la Chapelle 1993). No selection needs to be postulated to account for the relatively high frequency of heterozygotes. Thompson and Neel (1997) have pointed out on statistical grounds that a high degree of genetic disequilibrium between two non-homologous alleles less than 0.5 cm apart is an expected result in any expanding population in the absence of selection. In view of their findings, they caution against postulating selection to explain the high heterozygote frequencies of common autosomal recessive diseases, such as phenylketonuria.

The high frequency of certain diseases in Ashkenazi Jews has often been discussed. Twelve different diseases (largely autosomal recessive) are much more common in Ashkenazi populations than in other groups. They include the lysosomal lipid storage disorders, Tay–Sachs disease, Gaucher's disease, and Nieman–Pick disease, which are somewhat related metabolically. Also, some variants of the BRCA1 and 2 breast cancer genes are also common in Ashkenazi populations (Greene 1997). It often has been suggested that these conditions might be caused by a common selective advantage unique to Ashkenazi populations that gave abnormal gene carriers a survival advantage in the past (see Motulsky 1995). Various endemic or recurrent acute infectious diseases, such as tuberculosis or influenza, have been suggested as the selective agents and were related to the crowded living conditions in European ghettos where the ancestors of Ashkenazi Jews lived.

Recent molecular genetic analysis of 'Jewish' diseases suggests that a single mutation caused the large majority of cases of a given disease, while rarer mutations accounted for the rest. A recent analysis of an autosomal dominant Jewish disease, torsion dystonia, based on genetic disequilibrium analysis makes a strong case for a relatively small number of founders of the current Ashkenazi Jewish populations some 300–400 years ago (Risch *et al.* 1995b). A single mutation then expanded to lead to the current frequency of that disease. The authors suggest that probably all Ashkenazi Jewish diseases owe their current frequencies to a similar mechanism and do not require selection. While not absolutely conclusive, the recent population and molecular studies, as well as theoretical considerations, make a strong case for expansion of single mutations without selection.

Definitive laboratory proof of selection in humans will be difficult to obtain. Experimental models may give clues. Thus, heterozygote knockout mice without the normal gene that is mutated in cystic fibrosis may be less likely to die under conditions mimicking the electrolytic disturbance of cholera (Gabriel *et al.* 1994; Cuthbert *et al.* 1995). This finding would be compatible with a selective advantage of the cystic fibrosis gene *vis-à-vis* cholera.

RED BLOOD CELL VARIANTS

The ABO blood group polymorphism has been known longer than all other variations in blood group. Differences in ABO blood groups in a variety of world populations have been extensively described (see Vogel and Motulsky l996). Blood group O, for example, is usually frequent in populations that have been isolated for many generations, such as in Central and South American Indians. Since these and other populations with high blood group O frequencies differ in other genetic markers, such as the Rh blood groups, natural selection has been suggested to explain the high O frequencies, but a founder effect cannot be excluded. Infectious diseases can have a major impact on gene frequencies if associated with significant mortality in children and young adults over many generations. Candidate diseases in temperate climates are infantile diarrhoeas, recurrent epidemics such as plague, cholera, and smallpox, and chronic infections such as tuberculosis and syphilis. In fact, a variety of data suggest that the immunological response to syphilitic infections, severity and/or mortality of cholera, plague, and infantile diarrhoeas, and possibly of smallpox, depend on ABO blood types (Vogel and Motulsky 1996). However, the data are not definitive and the mechanisms for these associations remain unknown.

Another ABO blood group association with a non-infectious disease has led to a mechanistic explanation. Blood group O is definitely associated with peptic ulcers. In recent years it has been shown that the micro-organism *Helicobacter pylori* is involved in the pathogenesis of peptic ulcers. Eradication of the organism with appropriate antibiotics cures the ulcer and recurrences. Adherence of *H. pylori* to gastric epithelial cells is mediated by fucosylated blood group antigens associated with blood group O (Boren *et al.* 1993) and may explain the higher frequency of that blood type among patients with peptic ulcers. Conversely, associations with *H. pylori* infections and blood group O have not been observed in all series (Hook-Nikkane *et al.* 1990; Mentis *et al.* 1991). However, peptic ulcers are only weak selective agents since the disease rarely affects children and its mortality and fertility effects are likely to be small.

The blood group polymorphism for which we have strongest evidence of selective agents is that affecting the secretion of ABO blood group antigens into saliva. The relevant gene,

fucosyltransferase-2 (FUT2), has recently been cloned. The molecular basis of the polymorphism has now been found to be a stop codon mutation in non-secretors that appears to be the predominant mutation in both Africans and Caucasians (Kelly *et al.* 1995). Non-secretors are homozygous for this point mutation and are present at frequencies of 15–30 per cent in most populations. Early studies provided evidence that non-secretors are at increased risk of infectious diseases caused by bacteria and fungi, such as meningococci and recurrent bacterial urinary tract infections (Blackwell 1989; Sheinfeld *et al.* 1989). In 1991 Raza *et al.* studied upper respiratory tract infections in Scotland and found a lower frequency of non-secretors amongst patients with several diverse viral infections. This suggested an attractive mechanism for maintenance of the non-secretor polymorphism through protection against viral pathogens but with increased susceptibility to some bacterial and perhaps fungal pathogens.

The study of globin gene variants provided the earliest and best evidence of natural selection of human genetic variants by an infectious disease, *Plasmodium falciparum* malaria. Protection of heterozygotes for the haemoglobin S variant against *P. falciparum* malaria remains the textbook example of a balanced polymorphism in human populations. The data on haemoglobin S have facilitated further studies of malaria resistance genes in several ways. Notably, the accumulated data on haemoglobin S demonstrate that protection is greatest against severe malaria and death and less strong against uncomplicated clinical malaria (e.g. Allison 1964; Hill *et al.* 1991). The protective effect on parasite density is less marked and weakest of all for parasite rates. Furthermore, the protective effect is seen most readily in young children and is less marked in adults, when acquired immunity appears to become more important. These observations have facilitated the design of studies to detect less powerfully protective resistance genes (see below).

The molecular diversity of the thalassaemias, uncovered by extensive studies in the 1980s, provides in itself a compelling argument for a selective force having elevated the frequencies of a great variety of thalassaemia mutations. There is substantial population specificity of the mutations. The background globin gene haplotypes on which these variants are found show (at least in Europe, Asia, and Melanesia) generally similar levels of diversity, compatible with a relatively synchronous time of onset of selection in these continents (Flint *et al.* 1993*b*).

Surprisingly, in view of the extensive research effort in analysing the molecular basis of the thalassaemias, there is still only one published case–control study that has demonstrated a protective effect against malaria. This was undertaken by Willcox *et al.* (1983), in Liberia, where a protective effect of about 50 per cent was found against clinical malaria episodes. More recent studies of α-thalassaemia in Kenya and the Gambia have found a smaller protective effect of about 25 per cent for α-thalassaemia homozygotes against severe malaria, but no detectable protection for heterozygotes (Yates 1995). This implies that high frequencies of the common single gene deletion forms of α-thalassaemia (-α/) may mainly reflect selection of homozygotes (-α/-α) by malaria and supports the suggestion that α-thalassaemia is a transient polymorphism undergoing weak directional selection, rather than a balanced polymorphism.

So, despite the considerable interest paid to globin genes and malaria by evolutionary biologists, several important question remain outstanding. To what extent do haemoglobins E and C protect against malaria? Do any of the haemoglobinopathies protect against *Plasmodium vivax* malaria? What is the extent of protection provided by the thalassaemias and how long have they been under selection by *P. falciparum* malaria? The lack of diversity of the background haplotypes of β-thalassaemia mutations and the lack of fixation of mild α-thalassaemia variants appear to place an upper limit on the time depth of malarial selection in human populations, but whether this is 5000 or 50 000 years remains unknown. Recently, uncertainty has been generated by a very surprising result from a cohort study of children in Vanuatu. Melanesian children with

homozygous α-thalassaemia were more, not less, likely to develop both *P. vivax* and *P. falciparum* malaria than those of other genotypes (Williams *et al.* 1996). If this can be confirmed as a general finding, it raises new questions as to the mechanism by which the thalassaemias may protect against malaria and suggests that the magnitude of protection afforded by the thalassaemias might vary temporally and geographically according to local ecological conditions.

Glucose-6-phosphate dehydrogenase (G6pd) deficiency of the red cell is a common X-linked trait in tropical and subtropical populations (Beutler 1994, 1996). The condition is usually benign, with a normal red cell lifespan, but is associated with red cell destruction under certain conditions (Motulsky and Stamatoyannopoulos 1966). A variety of drugs, some infections, and eating of fava beans can produce acute haemolysis. Neonatal jaundice is commonly seen among male G6pd-deficient infants.

Over 100 different mutations of G6pd have been described using DNA techniques, but many more (over 400) have been distinguished by biochemical characteristics such as electrophoretic mobility and various enzymological properties (Beutler 1994). A large number of supposedly genetically unique variants distinguished on biochemical grounds alone turned out to be identical on DNA analysis, thereby establishing the superiority of DNA analysis over the more complex enzymological techniques, where slight differences in enzyme characteristics suggested spurious genetic heterogeneity. A group of G6pd deficiencies is associated with chronic haemolytic anaemia in the absence of any environmental agents. None of these variants occur in polymorphic frequencies since the deleterious clinical effects of blood destruction outweigh any selective advantage from malaria (see below). The majority of G6pd variants are associated with lesser degrees of enzyme deficiency, and often occur in polymorphic frequency. Some polymorphic variants affect introns with no phenotypic effects and are only detected at the DNA level. Practically all expressed G6pd deficiencies are caused by missense mutations. Large deletions and major rearrangements have not been detected. Since G6pd is a housekeeping enzyme necessary for function of all cell types, completely absent G6pd activity presumably is lethal to the embryo.

The distribution of various G6pd deficiencies in tropical and subtropical countries suggested a selective advantage of the trait *vis-à-vis* a common environmental agent. The ubiquity of malaria and the demonstration of the role of malarial selection for the sickle-cell trait pointed to a similar role for malaria (Motulsky 1960). The geographical correlation of G6pd deficiency and malaria all over the globe was also demonstrated by micromapping in narrower geographical areas, such as in Sardinia, Greece, East Africa, and Papua New Guinea. In these locations, populations with malarial exposure in previous generations had a significantly higher frequency of G6pd deficiency when compared with genetically similar populations who had not been exposed to malarial selection. As an example, the results from Sardinia showed a high G6pd frequency in the plains where endemic malaria had been frequent and a low frequency of the trait in the non-malarial hills (Siniscalco *et al.* 1961). The populations from the hills and plains had similar frequencies of other genetic markers.

The finding of *different* common G6pd variants of independent mutational origin in populations from Africa (G6pd A⁻), the Mediterranean basin (G6pd-Med), and Asia (several G6pd variants) (Vulliamy *et al.* 1993) was further evidence for the role of malaria as a selective agent. Other evidence was the excellent correlation between the frequency of the G6pd A⁻ deficiency and the sickling trait in Africa and between G6pd deficiency of the Mediterranean type and β-thalassaemia in Sardinia. Since G6pd deficiency is X linked, and haemoglobin S and β-thalassaemia are autosomal traits, selection for all these traits by malaria is the simplest explanation. Direct measurement of parasite density suggested lower parasite counts among G6pd-deficient males, but results were often ambiguous. However, *in vitro* studies of erythrocytic cultures of

Plasmodium falciparum showed impaired growth in G6pd-deficient red cells (Roth *et al.* 1983).

The dynamics of selection of an X-linked trait are more complex than those of autosomal traits as there are five different genetic classes (normal and mutant males; normal, heterozygote, and homozygote females). Female heterozygotes, because of X inactivation, on average have 50 per cent normal red cells and 50 per cent G6pd-deficient red cells. Lowered malaria growth in the deficient cells was postulated to provide a survival advantage due to lower mortality from malaria. Recent work in Kenya and the Gambia (Ruwende *et al.* 1995) demonstrated a 50 per cent reduced risk of developing severe *falciparum* malaria for both G6pd-deficient female heterozygotes and G6pd-deficient males. However, the selective disadvantage of haemolysis is likely to be more severe in G6pd- deficient males, where all red cells lack enzyme activity. In contrast, female heterozygotes with 50 per cent normal red cells would develop less severe red cell destruction. Consequently, a lesser mortality compared with G6pd-deficient males would be expected. The net 'biological fitness' of G6pd-deficient female heterozygotes, therefore, must have been higher than that of G6pd-deficient males. Under such conditions, the G6pd-deficiency allele would increase in frequency and could reach equilibrium conditions at gene frequencies of 0.5. Using different values for biological fitness, various scenarios of the increase of G6pd frequencies over many generations in non-expanding populations have been modelled (Lisker and Motulsky 1967; Ruwende *et al.* 1995).

The G6pd A⁺ mutation is common in populations of African origin. The G6pd A⁻ variants presumably arose on chromosomes carrying the A⁺ variant. At least four other G6pd variants exist that also carry the A⁺ mutation, suggesting that the A⁺ variant is ancient (Beutler *et al.* 1989). It has been concluded that more ancient variants, such as G6pd A⁺, spread by drift and migration, while more recent variants, such as A⁻ and G6pd-Med, were selected for by malaria. In most populations with a high frequency of G6pd deficiency (except for Africa), multiple polymorphic alleles of G6pd deficiency are found, each presumably selected for by malaria.

Work with G6pd deficiency illustrates some of the difficulties in proving the malaria hypothesis. Unlike sickle-cell anaemia and thalassaemia major, where there is strong selection against affected homozygotes, the selective disadvantages in G6pd deficiency are less severe. Therefore, a smaller selective advantage than that necessary to balance autosomal traits (such as sickle cell and thalassaemia) is needed to produce significant increases in X-linked gene frequency. Actual proof of a gene protecting against dying will be difficult to document in population studies and protection against severe malaria is the best surrogate. Mortality data will be difficult to obtain as deaths from malaria are less likely to occur when good medical care is available.

HAEMOCHROMATOSIS

A common autosomal recessive genetic disease among populations of European origin is haemochromatosis (Bothwell *et al.* 1995). This condition is an iron storage disease and is caused by increased intestinal iron absorption over many years. The resultant iron excess can cause liver fibrosis, cirrhosis, liver cancer, cardiomyopathy, diabetes, arthropathy, hypogonadism, and generalized ill health. Men develop clinical findings in middle life. However, not all homozygotes develop clinical symptoms. Women who lose iron by periodic menstruation have a markedly lower frequency of the disease. About 8–12 per cent of the European population are heterozygotes with resultant homozygote frequencies ranging between 1/300 and 1/600. Heterozygotes have slightly increased iron absorption and may have somewhat increased iron stores, which are never sufficiently severe to produce organ damage and clinical findings. The condition can easily be treated with frequent venesections that remove excess iron from the body. Treatment must be initiated before liver damage occurs.

The condition is caused by a specific mutation (Cys282Tyr) of a novel HLA-like gene (HLA-H) located on chromosome 6 (Feder *et al.* 1996). The mechanism by which this defect increases iron absorption is related to β_2-microglobulin function. Knockout mice without β_2-microglobulin develop excessive liver and other organ iron storage due to failure of transfer of iron from mucosal cells to the plasma (Santos *et al.* 1996). About 85 per cent of all patients with haemochromatosis in the USA and Europe are homozygotes for the Cys282Tyr mutation (Feder *et al.* 1996), but this mutation appears to be very rare in other populations (Merryweather-Clarke *et al.* 1997). In Australia, where the population is largely of British and Irish origin, 100 per cent of patients with haemochromatosis have this mutation (Jazwinska *et al.* 1996). Another mutation at this locus (His63Asp) rarely causes haemochromatosis in the compound heterozygote state (Cys282Tyr/His63Asp). No other haemochromatosis mutations have yet been identified at this or any other locus.

Homogeneity for a specific mutation is not seen in most genetic diseases, where many different mutations at a disease locus can usually be detected. The finding of a single mutation in haemochromatosis is similar to that observed in sickle-cell anaemia, where one nucleotide alteration of the haemoglobin gene has increased by natural selection so that all affected patients carry this mutation.

The high frequency of a single haemochromatosis mutation in large population groups also suggests a common mutational origin followed by selection. The selective disadvantage of homozygous haemochromatosis presumably is small since clinical manifestations occur only after reproduction has ceased. Testicular damage and reduced libido in homozygotes conceivably might reduce fertility but no data are available. Increased iron absorption in heterozygotes may protect female carriers against iron deficiency under conditions of nutritional deprivation, when women, particularly with multiple pregnancies, can develop severe iron-deficiency anaemia. The haemochromatosis allele by increasing iron absorption would reduce the prevalence of common iron-deficiency anaemia, thereby providing a fertility advantage that would lead to an increased gene frequency. Considering the haemochromatosis gene as deleterious has been considered a male-chauvinist point of view. 'Increased iron absorption is good for most women. From an unbiased evolutionary view, the death of some post-reproductive males is a small price to pay for protecting many women in the same population against anemia' (Rotter and Diamond 1987).

The haemochromatosis allele has additional deleterious effects. Porphyria cutanea tarda is a blistering skin disease with photosensitivity associated with liver cell damage and precipitated by various environmental agents such as alcohol, oestrogens, iron compounds, and other chemicals. About 15–20 per cent of patients with this disease were found to be homozygotes and 25–30 per cent were heterozygotes for the characteristic Cys282Tyr mutation of haemochromatosis (Roberts *et al.* 1997). The haemochromatosis allele, therefore, appears to be a genetic risk factor for porphyria cutanea tarda. It has been known for some time that multiple venesections may produce remissions of porphyria cutanea tarda, presumably related to removal of excess iron. Iron causes liver damage and inhibits the activity of uroporphyrinogen decarboxylase, an enzyme with characteristically reduced hepatic activity in this disease.

PATHOGEN RECEPTORS

Mutation of the receptor that a parasite uses to invade cells is a major mechanism for host defence against infectious micro-organisms. The best known example is the Duffy blood group that is the erythrocyte receptor for the malaria parasite *Plasmodium vivax*. Most sub-Saharan Africans lack the Duffy blood group on their red cells and Miller *et al.* (1976) discovered that these individuals are resistant to *P. vivax* malaria, explaining the virtual absence of *P. vivax* in Africa. Recently, the relevant gene has been cloned and the inactivating

mutation identified. Interestingly, in Duffy-negative Africans, the Duffy antigen is expressed on endothelial cells (Iwamoto *et al.* 1996). A point mutation in the promoter of the gene is found among Africans. This is at the binding site for an erythroid-specific enhancer protein, compatible with the tissue specificity of the antigen loss (Tournamille *et al.* 1995).

Another polymorphic receptor that has attracted recent interest is the complement receptor-1, CR1 (also CD35), that like the Duffy antigen is present on both red and white blood cells. Furthermore, it likewise displays a polymorphism only of red cell expression, so that the copy number of CR1 molecules varies on red cells but not monocytes and macrophages. A polymorphism in the CR1 gene has been associated with the erythrocyte phenotype with a low receptor density (Wilson *et al.* 1986). This genotype has been associated with the autoimmune disease systemic lupus erythematosis in some (Wilson *et al.* 1987) but not all (Kumar *et al.* 1995) studies. CR1 is one of the receptors for entry of mycobacteria into macrophages (Schlesinger 1996), suggesting that variants of this gene might affect susceptibility to tuberculosis or leprosy.

There has recently been enormous interest in the many newly discovered chemokine receptor genes. One of these, the CC chemokine receptor gene-5 (CCR-5), has been found to be a coreceptor for the entry of macrophage-trophic strains of HIV-1 into cells. An inactivating 32-bp deletion of this gene is found at high allele frequencies (0.1) in populations of European origin. Homozygotes, who comprise 1 per cent of the population, are very resistant to HIV-1 infection. Heterozygotes show slower rates of disease progression to AIDS but are not protected from infection (Dean *et al.* 1996; Huang *et al.* 1996; Liu *et al.* 1996; Samson *et al.* 1996). This mutation has not been found in populations of African or Asian origin. These findings raise fascinating questions about the origin of this deletion and why it has reached high frequencies in Europeans. It is clear that there has not been enough time or mortality for HIV-1 to have been the selective agent. There are no obvious selective agents amongst known present-day pathogens but candidates can probably be assessed using *in vitro* assays as well as in case–control studies. The search for additional genes conferring resistance to HIV infections is likely to be rewarding in explaining the occurrence of resistance to HIV-1 infections in non-European populations.

The likelihood that HIV-1 has not been responsible for the prevalence of its major identified resistance gene suggests a fresh perspective on malarial selection of the Duffy polymorphism in Africa. Interestingly, like CCR-5, the Duffy blood group is a chemokine receptor, but for a great variety of CC chemokines (Horuk *et al.* 1993). It is unclear why, unlike all other malaria resistance alleles, the Duffy negative variant should have almost reached fixation in much of Africa. This is especially puzzling because the *P. vivax* parasite is far less virulent than *P. falciparum* against which all the other malaria resistance genes offer protection. One possibility is that in the past *P. vivax* was more virulent and that it reached Africa before *P. falciparum* leading to the near fixation of Duffy negativity and, in turn, the elimination of *P. vivax* from most of the continent. However, the selection of the CCR-5 variant by a pathogen that is not HIV-1 suggests another possibility. The Duffy negative allele might have been selected in Africa by a virulent infectious pathogen other than *P. vivax* that may have since disappeared. The consequence of this may simply have been that *P. vivax* never became established in Africa because most of the population were already completely resistant to it (Livingstone 1984).

IMMUNE RESPONSE GENES

Probably more disease association studies have been undertaken involving HLA genes than any other loci. Most of these are with diseases that are unlikely to have a major selective impact. However, in the last few years, more studies have been undertaken of infectious diseases so that a clearer picture of the impact of MHC polymorphism on infectious disease is emerging. Conversely, the magni-

tude and nature of the possible impact of infectious disease on the diversity of MHC genes can now be estimated, although many central evolutionary questions remain unanswered.

Since HLA associations with infectious diseases have been reviewed by one of us recently elsewhere (Hill 1997), we shall simply outline some general conclusions. First, there is evidence from family studies of several infectious diseases that susceptibility is genetically linked to the MHC. This was demonstrated more than 20 years ago for leprosy (de Vries *et al.* 1976) and more recently for tuberculosis (Singh *et al.* 1983), malaria (Jepson *et al.* 1997*b*), and disease progression in HIV infection (Kroner *et al.* 1995). These studies are important because they provide some overall estimate of the importance of the MHC in susceptibility and they avoid the concern in case–control studies that certain associations may be spurious and result from population stratification.

HLA association studies of infectious diseases have been pursued since the 1970s but most studies have been too small to provide convincing evidence on whether associations exist with particular HLA class I or II alleles. The introduction of molecular typing methods has both improved the resolution of these studies and allowed large sample sizes to be studied. The most consistent association of an HLA type with an infectious disease is that of HLA-DR2 with susceptibility to leprosy (Todd *et al.* 1990) and the same class II antigen has been associated with susceptibility to tuberculosis in several Asian populations. However, the HLA-DR2 association with leprosy and tuberculosis is weaker or non-existent in non-Asian populations. Large-scale studies of severe malaria in Africa and of persistent hepatitis B virus infection have also provided strong evidence of HLA associations with these diseases. In the Gambia, both an HLA class I allele, HLA-B53, and a class II allele, HLA-DRB1*1302, were associated with resistance to forms of severe malaria (Hill *et al.* 1991). The largest study of persistent hepatitis B virus infection found that the same HLA-DR type (DRB1*1302) was associated with viral clearance in the Gambia (Thursz *et al.* 1995). However, as for mycobacterial diseases, there is evidence of geographical heterogeneity in associations with both malaria and hepatitis B virus infection. The numerous studies of HLA associations with disease progression to AIDS also reveal no consistent allelic association but much evidence that MHC polymorphism plays a complex role in susceptibility.

There are many possible reasons for the heterogeneity in these studies of infectious disease. At least some heterogeneity is real and cannot be explained by poor study design and inadequate sample size. Polymorphism in the infectious pathogen may be a factor. Both antigenic polymorphism and antigenic variation are well documented in HIV and malaria and may contribute to interpopulation variation in HLA associations. Diversity of HLA subtypes may also play a role, but with increasing use of high-resolution typing methods should be less of a factor in the future.

In the last few years there have been several studies of the MHC class III gene encoding tumour necrosis factor (TNF) that provide evidence that promoter polymorphism in this gene may be subject to selection by various infectious pathogens. An association between homozygosity for a promoter variant of this gene and susceptibility to cerebral malaria was found in a Gambian study (McGuire *et al.* 1994). Subsequently, associations have been identified for the same variant with mucocutaneous leishmaniasis (Cabrera *et al.* 1995), lepromatous leprosy (Roy *et al.* 1997), trachomatous scarring (Conway *et al.* 1997), persistent hepatitis B virus infection (Thursz, M. *et al.* unpublished data), and fatal meningococcal meningitis (Levine *et al.* 1996). Intriguingly, the associations identified thus far are all with disease susceptibility rather than resistance. Protective associations may exist with other infectious diseases, particularly those caused by certain viral pathogens. This TNF variant may be associated with higher levels of TNF production that are protective against some pathogens yet deleterious following infection by other species or strains so that a balanced

polymorphism is maintained. Interestingly, a recent study by Westendorp *et al.* (1997) reported that a familial tendency to low TNF production was associated with meningococcal disease whereas high levels of TNF have previously been associated with meningococcal septicaemia.

Although immunologists traditionally regard the MHC as almost synonymous with genes controlling the immune response, there is now good evidence that, collectively, non-MHC genes are at least as important genetic determinants of responsiveness. In a large study, Jepson *et al.* (1997) measured both humoral and cellular immune responses to a variety of *P. falciparum* antigens and to purified protein derivative in Gambian twins universally exposed to malaria and mycobacteria. Comparison of monozygotic with dizygotic twins allowed an estimate of the heritabilities of these responses. Heritability varied between antigens but was generally high. HLA typing of the dizygotic twins allowed an estimation of the proportion of the genetic variance attributable to MHC genes. This was in almost all cases less than 50 per cent, averaging about 20 per cent.

Interspecies comparisons of the rate of evolution of immunologically important molecules encoded outside of the MHC have indicated that such genes appear to be subject to relatively strong selection pressures (Murphy 1993), probably by infectious pathogens. Although there is further evidence from many animal studies of the importance of such non-MHC immune response genes, very few have been identified in either animals or humans. In the past few years the introduction of whole-genome approaches to the mapping of genes affecting complex phenotype, using informative microsatellite markers, has provided an opportunity to map and subsequently identify such major non-MHC immune response genes and these, once identified, should provide important examples of the action of natural selection (Hill 1996*a*).

Whilst awaiting the outcome of such genome scanning studies, many investigators have been assessing the relevance of polymorphism in a

Table 5.1 Some polymorphic candidate genes for infectious disease susceptibility

Mannose-binding protein	ICAM-1
Inducible nitric oxide synthase	HLA
Duffy chemokine receptor	Interleukin-12
Tumour necrosis factor	Interleukin-1 receptor
CCR5	MxA
Vitamin D receptor	CCR3
Fcg RII	IgE-RII
Km allotypes	Nramp1
T-cell receptor variants	IgH allotypes
Interferon-γ receptor	Fcγ RIII
Interleukin-4	TNF-receptor 55
Fucosyl transferase-2	Interleukin-10

plethora of genes that encode components of the innate and adaptive immune system. Table 5.1 lists some of the candidate genes that have been studied. Many of these are the genes for cytokines, chemokines, and their receptors. We shall discuss briefly just three loci that appear to have been subject to natural selection.

Mannose-binding ligand (MBL) is a component of the innate immune system and functions by activating complement and opsonizing carbohydrate-bearing infectious pathogens. It has been proposed that individuals heterozygous or homozygous for dysfunctional MBL alleles are at increased risk of recurrent infections or other manifestations of immunodeficiency (Summerfield et al. 1995). Remarkably, such alleles are found at high frequencies in most human populations. For example a codon 57 variant is present at an allele frequency of about 0.25 in the Gambia, and homozygotes for this variant have almost no MBL detectable in serum.

Recently, interest has converged from two quite distinct research fields on a gene (or genes) located on the short arm of chromosome 5. Linkage studies of families with atopy and asthma provided strong evidence of a linkage to this chromosomal segment in some but not all series (Marsh et al. 1994; Daniels et al. 1996). Interestingly, this map location contains a cluster of candidate genes such as interleukins-4, -5, and -9 that may be determinants

of a T-helper type 2 (T_{H2}) immune response. Such an immune response has been implicated in the pathogenesis of atopy and asthma, and also in protective immunity to helminthic infections such as schistosomiasis. It is therefore of particular interest that a recent genome scan for susceptibility genes in Brazilian families with high schistosome egg counts found a possible linkage to the same region of chromosome 5 (Marquet et al. 1996). These data are consistent with a frequently discussed hypothesis that genes which are associated with T_{H2}-like immune responses and predisposition to atopy may have been selected for resistance to helminthic infection. However, a full assessment of this possibility requires identification of the relevant gene or genes on chromosome 5.

Another gene that might influence the quality of immune responses to infectious pathogens is that encoding the vitamin D receptor (VDR). Variation in the chromosome 12 gene encoding this intracellular hormone receptor has been implicated in susceptibility to osteoporosis in several populations (Morrison et al. 1994). Homozygotes for the less common allele of a silent polymorphism in exon 9 that alters a *Taq I* restriction site generally have lower bone mineral density, but there is evidence of population heterogeneity in this association. Because the active metabolite of vitamin D ($1,25-D_3$—the ligand for the VDR) impairs mycobacterial growth in human macrophages (Rook 1988), Bellamy et al. (manuscript submitted) assessed the role of this variant in susceptibility to pulmonary tuberculosis and other diseases in a large case–control study in the Gambia. Homozygotes for the variant t allele, that may be overexpressed compared with the common T allele, were at reduced risk of tuberculosis and of persistent hepatitis B virus infection, but not at altered risk of severe malaria. Further studies have suggested that VDR genotype may affect risk of HIV-1 infection in Uganda (Ali, S. et al. unpublished) and the type of leprosy developed (Roy, S. et al. unpublished). The vitamin D receptor gene thus appears to affect susceptibility to several infectious diseases and the tt genotype, associated with osteoporosis, may have been selected to high frequencies by intracellular infectious pathogens such as *Mycobacterium tuberculosis*. Future identification of the major genes for common degenerative as well as major infectious disease will allow assessment of the extent to which susceptibility genes for the former—scourges of modern health-care budget managers—may have been selected by the plagues of past ages.

FUTURE DIRECTIONS

The potential of whole-genome analysis is just beginning to be exploited. Genome scans have been completed for a small number of non-infectious and just two infectious diseases of humans, schistosomiasis and tuberculosis. However, although genes have been localized using this approach, none have yet been identified. It is clear that the ease with which genes can be mapped and identified using a whole-genome approach depends on many factors. The magnitude of the genetic component to the complex disease and the numbers of families or affected sib pairs available for analysis are of course important, but so are the number of genes contributing to the genetic component, the degree of epistasis between these genes, and the extent of interpopulation heterogeneity.

An encouraging recent genome scan was reported by Satsangi *et al.* (1996) who found several strong linkages for inflammatory bowel disease in northern European families. It will be of interest to determine whether these genes will affect resistance to any gastrointestinal infectious pathogens. In contrast, three whole-genome studies of multiple sclerosis failed to identify non-MHC regions unequivocally linked to susceptibility to this disease (Bell and Lathrop 1996). The more limited power of family linkage studies compared with whole-genome association studies has recently been pointed out and quantified. Another quantum leap in technological power to allow such studies is required before minor genes will be detectable by such association studies for complex disorders (Risch and Merikangas 1996).

The potentially greater power of association studies implies that an important place remains for educated guesses as to which candidate genes may affect susceptibility or resistance to particular diseases. Clues from population frequency differences might be utilized. For example, the alcohol sensitivity due to an aldehyde dehydogenase variant that is prevalent in many oriental populations and the well-documented population differences in lactose absorption causing lactose 'intolerance' may contribute to more complex susceptibility phenotypes. Other phenotypes that are likely to have been selected are enzyme polymorphisms of hepatic detoxification. Some of these may well have evolved to adapt to local toxins and recent evidence suggests that two, epoxide hydrolase and glutathione-*S*-transferase M1, affect risk of hepatocellular carcinoma (McGlynn *et al.* 1995).

There has been much enthusiasm for ascribing gene frequency differences to selection by major infectious pathogens following the paradigm of malaria and the haemoglobin disorders. However, it is not unlikely that many alleles may have been selected by pathogens that have become extinct and we shall thus never find the selective agent. Smallpox, which is now eradicated, is likely to have been a major selective agent, and individuals differ markedly in susceptibility to poliomyelitis, which has now been eliminated by vaccination from the New World. But a far larger number of selective agents will have disappeared naturally leaving their trace only in the genome of their hosts. Perhaps cystic fibrosis was selected by a gastrointestinal pathogen that disappeared thousands of years ago.

The major initial investment in genome scans has been for prevalent diseases with a strong undefined familial component. However, it may be of at least as much evolutionary interest to define the genetic factors responsible for normal intra- and interpopulation phenotypic variation. Most progress has been made in elucidating the molecular basis of common variations in colour vision (Motulsky and Deeb 1995). More recently, the first indication of a molecular basis for skin pigmentation has appeared (Valvede *et al.* 1995). A recent genome scan has mapped a possible gene for human obesity to chromosome 2 (Comuzzi *et al.* 1997) and several other obesity genes exist. Genes controlling variation in normal stature would also be of great interest to the genetic anthropologist. We live in an exciting era of new gene discovery. The challenge of explaining the marked genetic variation between individuals and populations is becoming ever greater.

SUMMARY

A broad overview is presented of current knowledge of human disease susceptibility and resistance genes that may have been subject to natural selection. It is often hard to distinguish between a founder effect and genetic drift, but some cases provide strong evidence from population genetic data, *in vitro* analyses, or clinical studies that infectious pathogens have selected some alleles to high frequencies. The best examples of selection come from studies of red blood cell variants, immune response genes, and pathogen receptors.

Recently, two new molecular variants have been identified at high frequency in northern Europeans that are much less common in other populations. These are the main causative mutation for haemochromatosis in an HLA-like gene and a chemokine receptor (CCR5) variant that, in the homozygous state, provides almost complete resistance to HIV-1 infection.

Until now, genes studied in detail have been selected on the basis of their function and the geographical distribution of their alleles. With the introduction of methodologies for whole-genome family linkage studies and gene identification, there is considerable potential for the identification of new genes affecting risk of complex multifactorial disorders. It will be interesting to determine how many of these new loci will affect both infectious and non-infectious disease, and thus assess the extent to which selection for resistance to infectious pathogens may have contributed to the burden of non-infectious disease today. Similar methodologies may be applied to identifying the molecular basis of non-pathological variation in phenotype that appears to have been subject to strong selection.

6

HUMAN GENETIC VARIATION AND ITS IMPACT ON PUBLIC HEALTH AND MEDICINE

*Adrian V.S. Hill, Alicia Sanchez-Mazas (co-rapporteurs), Guido Barbujani,
Georgia Dunston, Laurent Excoffier, John Hancock, Jan Klein, Urs A. Meyer,
Arno G. Motulsky, Silvio Presciuttini, William L. Wishart, and André Langaney (chair)*

Human genetics seeks to identify and explain the nature and evolution of inherited variation in all human characteristics including susceptibility to disease. New techniques in molecular biology have revolutionized our ability to characterize human DNA sequence variation while population genetic studies of diversity have been fruitful. Consensus has emerged on the likely origins and major population movements of humans, and the molecular basis of many diseases has been elucidated. The new dense linkage maps allow us to perform genetic dissections of complex traits on a large scale and to detect genetic polymorphisms associated with diseases even when allelic effects are small at the phenotypic level. Since this kind of genetic variation may affect a substantial fraction of mankind, these studies will have a major impact on public health.

This chapter considers how these advances allow us to address several key points. First we ask: How should we classify human populations? Can we do better than using 19th century labels for races? What utility have such classifications for clinical practice? Then we consider how population differences may explain geographical variation in disease frequencies and review some of the major environmental agents that appear to have affected human gene frequencies, particularly those implicated in diseases. Genes involved in defence against infectious pathogens evolve relatively rapidly, suggesting that natural selection by infectious disease has had a significant impact on some regions of the human genome. The remarkable polymorphic genes of the major histocompatibility complex (MHC) are then considered. These genes illustrate the action of natural selection but nevertheless are useful population markers.

Throughout, we suggest areas for future research. In particular we endorse the objectives of the Human Genome Diversity Project which should provide a valuable and unique resource for genetic studies of human evolution.

POPULATIONS: ORIGINS, CLASSIFICATIONS, AND THE CLINIC

Palaeontology and genetics suggest a relatively recent common origin of modern humans between 100 000 and 200 000 years BP (before present), probably from a unique source in Africa (Lahr and Foley 1994). The ancestors of today's 6 billion people then spread to colonize the Old World, Oceania, and the Americas. They acquired different gene frequencies by founder effects, genetic drift, and differential natural selection, as influenced by migrations, effective population sizes, and local environments (Chapter 3). Pigmentation, body size, and body proportions seem to have evolved rapidly during late prehistory, producing convergence in physical traits among populations with different gene frequencies such as black Africans, south Indians, and Melanesians and divergence for skin colour and facial traits between genetically similar groups such as

Orientals and coastal Melanesians (Cavalli-Sforza *et al.* 1994).

The extent of migration during prehistory is the subject of an old and unresolved controversy in human population genetics. If subsequent migration was negligible between continental groups, the first dispersions and colonizations by remote ancestors were the main source of population differentiation. Isolated continental genetic pools then would have evolved towards significant racial differentiation.

The other view is that human movement since the neolithic period (about 10000 years ago) has been sufficient to induce a world-wide structure of populations fitting an 'isolation-by-distance' model in which the differences in gene frequencies between populations increase with geographical distance (Malécot 1948, 1966; Morton *et al.* 1972). If this view is correct, migration between settlements wiped out the effect of the first colonization and the world population can be perceived as a 'network' of subpopulations exchanging migrants (Langaney and Sanchez-Mazas 1987). Geographical or communication distances structured a continuous clinal variation of gene frequencies in which borders between populations and races can only be arbitrary.

Many studies (Morton *et al.* 1972; Sanchez-Mazas and Langaney 1988; Excoffier *et al.* 1991; Barbujani *et al.* 1994*b*; Cavalli-Sforza *et al.* 1994) suggest that clinal variation of gene frequencies is the rule for most genetic markers, including DNA polymorphisms (Bowcock *et al.* 1991). How patterns of molecular genetic variation between populations reflect the action of natural selection or simply population movement and random changes in gene frequency is unresolved. Most analyses of population affinities assume that selection is absent or negligible, but some of the most useful population markers, such as HLA (see below), are subject to selection.

Human populations differ in their genetic make-up. The variation among individuals within populations and the lack of correlation between physical traits and genetic characters leads to very different patterns of similarity/dissimilarity between populations for different genes and traits. This makes classification of populations into races difficult. Different criteria can lead to very different classifications of populations into races, making any racial classification of humans to some extent arbitrary.

There was disagreement within our group, as in the scientific community, about the scientific significance of any racial classification. A rough classification can be based on external phenotypes such as skin colour, hair colour and texture, and facial characteristics. It yields a rough correspondence between these population groups and underlying genotype frequencies for several genes. Although approximate, such a classification is useful for assessing risks of genetically influenced diseases and for predicting differential drug responses. Such empirical racial categories depend on the local history of immigration and admixture (Marks 1995) and are not transposable from the USA to European or other countries.

Some classification of individuals is important for disease risk assessment in clinical practice. Categories based on origin and ethnic and linguistic identity would be more informative than such poorly defined terms as 'Caucasian, Hispanic, African-American, or Asian' used in the USA. These labels do not reflect recent progress in the molecular genetic definition of populations, which promises a more precise and objective categorization of individuals using detailed analysis of the origin and affinities of individuals. It suggests that we will need extensive epidemiological studies to assess genetic disease risk in many human populations. The resulting genotypic categories will become increasingly valuable as ongoing population mixture decreases the reliability of current phenotypic classifications.

GENE GEOGRAPHY AND DISEASE PREVALENCE

The frequency of many diseases varies markedly between populations. This may be explained by genetic or environmental factors or, frequently, by an interaction of the two. Here we consider how well-known geographical distributions of genetic variation explain global patterns of disease.

Monogenic diseases: three examples

Far more progress has been made in analysing the molecular genetic basis of monogenic than of complex diseases. For monogenic disorders, there is often a correlation between the prevalence of particular diseases or disease alleles and the genetic classification of individuals. For complex diseases, it is much harder to assess how much variation in disease symptoms is caused by genetic variation; some approaches are suggested (see also Chapter 20).

Cystic fibrosis

One cystic fibrosis mutation (DF-508) accounts for about 70 per cent of the cases observed in central and northern Europe. The distribution of this allele in Europe is approximately clinal, with frequencies decreasing from the north-west to the south-east (Devoto 1991).

The pattern of DF-508 frequencies resembles the gradients of allele frequencies for other genes described over much of Europe (Cavalli-Sforza *et al*. 1994) and is attributed to the effects of a population expansion from the Near East that accompanied the neolithic spread of farming and animal breeding (Menozzi *et al*. 1978; Sokal *et al*. 1991). That expansion carried genes of Near Eastern origin to distant locations, with larger contributions of neolithic farmers to local gene pools closer to the Near East. This suggests that the DF-508 mutation was present in the preneolithic European population and originated in preneolithic times. Morral *et al*. (1994) estimated its probable age from the amount of microsatellite polymorphism at three intronic sites. Their estimate, 55 000 years, agrees well with the scenario described above.

Globin gene variants

Extensive studies of the thalassaemias have produced a detailed picture of the molecular basis of these globin gene disorders. Mutations in the α- and β-globin genes that produce thalassaemia syndromes of varying severity are found at high frequencies in most tropical and subtropical countries, mirroring the historical distribution of the selective agent, *Plasmodium falciparum* malaria. It is quite striking that largely different sets of thalassaemia alleles are found in Mediterranean, Oriental, Asian Indian, and African populations, consistent with a relatively recent arrival of malaria in these regions and selection of a different repertoire of variants in each place (Flint *et al*. 1993*a*). Presumably the selection of different sets of mutations in each region reflects stochastic effects determining which allele or mutation was in the population at low frequency when malaria was introduced. Molecular evidence consistent with this scenario comes from the pattern of molecular markers on the background haplotypes flanking these globin gene mutations. A single predominant background haplotype is associated with most point mutations, again suggesting their relatively recent origin.

Globin variants and haplotypes have also served as useful population markers. Analysis of background haplotypes has often been used to assess the age of some of these mutations in particular populations. For example, a particular thalassaemia mutation found only in Pacific Islanders provided early evidence of genetic affinity between Melanesians and Polynesians (Hill *et al*. 1989). This α-globin haplotype has a more recent mutation in its coding sequence producing a population-specific haemoglobin variant, haemoglobin J Tongariki. The sequence of mutations and the relative age of these molecular variants can be roughly estimated from the diversity of flanking genetic markers.

Triplet expansion disorder genes

Many diseases caused by the expansion of nucleotide repeats show uneven geographical distributions. They include some well known, although relatively uncommon, degenerative diseases of the central nervous and neuromuscular systems, such as fragile X syndrome, myotonic dystrophy, and Huntington's disease. All triplet expansion diseases are single gene disorders, frequently dominant, and of high

Box 6.1: DNA triplet repeat disorders

Triplet expansion diseases have attracted great interest because of their underlying genetic mechanism (for review see Gusella and MacDonald 1996). Triplet expansion is a new mechanism of mutation which has not been associated with genetic diseases in other species. In all of these diseases the function of the gene is disrupted by the enlargement of a tandemly repeated trinucleotide motif somewhere within it. The exact nature of the expanded repeat varies between diseases and is correlated with its genic location but, irrespective of these details, the molecular aetiology of the diseases involves instability of the repeat sequences in cells that undergo meiosis during gametogenesis, and depends on variations of repeat lengths within populations. Although the exact lengths of repeats that give rise to disease differ depending on genic location, repeat instability in all cases only occurs when the repeat in an allele exceeds a threshold length. Above that threshold, repeats show very high mutation rates (approaching 1) and increasing length in successive generations. Increasing lengths of repeats are associated with increasing severity of disease, and earlier age of onset in successive generations. This is known as genetic anticipation.

Population analyses of repeat length variation in the different genes commonly show a bimodal distribution (trimodal in myotonic dystrophy (Zerylnick *et al.* 1995)). The lower part of the distribution shows a modal repeat length well below the threshold for meiotic instability, whereas the higher part of the distribution contains alleles close to the instability threshold (premutational alleles). These longer alleles are thought to represent a pool that can approach the threshold, by undergoing mutation by strand slippage during replication, to replace alleles lost from the population because of disease.

penetrance (see Box 6.1 for background details).

Comparisons of allele frequency distributions in these genes have revealed considerable population heterogeneity at several levels, suggesting fairly recent origins of premutational alleles. Haplotype analysis of the Machado–Joseph gene in affected individuals shows almost all cases world-wide are derived from two Azorean families of Portuguese origin (Gaspar *et al.* 1996). The frequency of dentatorubral pallidoluysian atrophy is much higher in Japan than elsewhere; haplotype analysis suggests a founder effect during the peopling of Japan (Yanagisawa *et al.* 1996). Huntington's disease has a prevalence of 1 in 10 000 in Western Europe compared with 1 in a million in Japan, and haplotype analysis suggests a northwestern European origin for most disease chromosomes (Rubinsztein *et al.* 1994; Squitieri *et al.* 1994). Premutational alleles for myotonic dystrophy are rarely found in native African populations (Zerylnick *et al.* 1995; Deka *et al.* 1996); the exceptions appear to have been introduced by Europeans during colonization (Goldman *et al.* 1996). The fragile X repeat shows greater variation in African than other populations (Kunst *et al.* 1996), consistent with an ancient origin.

Knowledge of the population distribution of premutation length alleles of different triplet expansion disease genes may be helpful in rapid diagnosis of the diseases, some of which share common clinical features, by allowing selection of the most appropriate tests based on the ethnic affinities of patients. This may speed both diagnosis, especially where the costs of polymerase chain reaction (PCR)-based tests are high, and the initiation of appropriate therapy, important because these are progressive degenerative disorders. Detailed population analyses will be needed for these genes, for some premutational alleles are known to have arisen independently in distinct populations (e.g. Burke *et al.* 1994).

Complex diseases

How well can genetic differentiation between human populations explain global differences in disease frequency? Currently, any general conclusion is difficult for two reasons.

First, we lack high quality epidemiological data using standardized methodology for many diseases in many countries.

When such data are available, the contribution of population differentiation to global differences in disease frequency can be demonstrated, as in the WHO Multinational Project for Childhood Diabetes, known as the DIAMOND Project (WHO DIAMOND Project Group 1990). The large variation in insulin-dependent diabetes mellitus (IDDM) incidence across countries was used to evaluate the contribution of genetics to differences in disease frequency among countries with IDDM registries. DNA sequences coding for an amino acid other than aspartate in position 57 of the HLA-DQ β-chains were strongly associated with IDDM susceptibility (Todd *et al.* 1987), as was the presence of arginine in position 52 of the DQ α-chain (Gutierrez-Lopez *et al.* 1992). The heterodimers formed by the combination of the two alleles appeared to have the strongest quantitative diabetogenic effect. Much of the geographical variation in IDDM could be explained by the distribution of non-Asp-57 alleles (Dorman *et al.* 1990). Thus molecular genetics can be used to determine genotype-specific incidence of disease that may explain population-based differences in disease patterns.

The second difficulty in assessing the role of genetics in the determination of global patterns of disease is distinguishing between genetic and environmental influences on disease. Environmental differences can explain changes in disease frequencies between populations with the same genetic background, for example coronary artery disease frequencies in migrants: Irish and Japanese men who emigrated to the United States experienced a higher incidence of coronary heart disease (Kushi *et al.* 1985; Robertson *et al.* 1977).

At least three types of study could provide relevant data on the overall impact of human genetic variation on disease risk. Global comparisons of gene frequency distributions with disease maps may suggest correlations, but allowing for the confounding effects of environment will be difficult. Comparisons of disease frequency in ecologically similar regions containing genetically very different populations would be informative. Perhaps most productively, different immigrant populations can be compared in the same locality after correcting for confounding environmental factors.

THE ROLE OF INFECTIOUS DISEASES

Selection by infectious pathogens has probably had significant impact on human genetic diversity. Evidence that host genetic factors partly determine susceptibility to fatal infectious diseases comes from twin and adoptee studies (Comstock 1978; Sorensen *et al.* 1988), segregation analysis (Abel and Demenais 1988), and from studies of genes or genetic regions in individuals and families with particular infectious diseases (Hill 1996b). Associations between several blood groups and infectious diseases have been known for some time; more recently, both genetic linkage and association studies have suggested a role for infections diseases in explaining genetic variation in the MHC. Studies of genes that might affect disease resistance have been particularly successful in malaria, where 12 genes are now implicated (Hill 1992). How well currently recognized genetic polymorphisms explain the overall genetic component of variable susceptibility remains uncertain, for most infectious disease susceptibility genes have probably not yet been identified.

With the exception of malaria, little is known about the genetic component of host susceptibility to other major selective agents such as measles, tuberculosis, pneumococcal disease, and diarrhoeal infections. To identify genes involved in susceptibility to these diseases using whole-genome linkage methods, families with affected sib-pairs are required. Recruiting such families is difficult for acute infectious diseases, particularly in developing countries. Recently, genome scans have been undertaken for chronic infections such as schistosomiasis (Marquet *et al.* 1996) and tuberculosis (Bellamy *et al.*, manuscript submitted). Tuberculosis is one of the largest infectious causes of global mortality, and locating its resistance and susceptibility genes (e.g. Bellamy *et al.* 1998) will be of particular interest.

Another strategy of genome-based association studies would increase sample sizes by analysing large numbers of unrelated families with only a single affected case. Using data from unrelated families has important consequences. Whereas 300 evenly spaced microsatellite markers may be sufficient to scan the genome in families, far more—at least 30 000 —informative markers are needed for genome scanning in association studies of unrelated cases. The technical capacity required, though daunting, is foreseeable (Risch and Merikangas 1996). Association studies using unrelated cases and controls also require detailed knowledge of the ethnic composition of the population to avoid false-positive and false-negative results due to population stratification. Increasing knowledge of the population specificity of molecular markers should improve our ability to match cases and controls properly.

High frequencies of several genetic diseases and disorders in some populations are the result of natural selection by infectious pathogens (Motulsky 1960). This is well established for some haemoglobinopathies and glucose-6-phosphate dehydrogenase deficiency. For other genetic diseases, such as cystic fibrosis, selection by an infectious pathogen has been suggested but is far from proved. If the relevant selective pathogen is now extinct, direct proof of a protective effect may not be possible. For example, the high prevalence of the CCR5 chemokine receptor mutation that confers resistance to HIV infection in Europeans (Samson *et al.* 1996) might result from selection by an ancient epidemic of a pathogen that no longer exists.

Genes involved in defence against infectious pathogens evolve at a higher rate than most other genes (Murphy 1993), are more often polymorphic in human populations, and are good candidates for determining variable susceptibility to infectious diseases. When variable host molecules interact with polymorphic components of infectious pathogens, the polymorphism in the pathogen may result from selection pressure to escape recognition by the host defence system. It follows that associations may be observed between variable host genes and strains of an infectious pathogen defined by some polymorphic pathogen sequence. Such strain-specific associations have only occasionally (e.g. Apple *et al.* 1994, Gilbert *et al.* 1998) been searched for in infectious disease studies but might be relatively frequent. As in some simpler host–parasite systems, epistatic interactions may be common in susceptibility genes for human infectious disease, particularly when complex immune defence mechanisms are involved. The likely consequence of such epistasis will be population heterogeneity in immunogenetic associations with infectious disease, as found in several HLA studies. Geographical heterogeneity in allele frequencies of a pathogen probably also underlies some variable immunogenetic associations and produces different selection pressures on host variants in particular regions.

Another route to identifying susceptibility genes for human infectious disease is through genetic analysis of laboratory animals, usually mouse models of the infection. Several such genes, for example Mx, an influenza resistance gene (Staeheli *et al.* 1986), and Nramp, a leishmanial/mycobacterial resistance gene (Vidal *et al.* 1995), have been identified and human homologues characterized. However, an evolutionary perspective suggests that polymorphism in inbred mice will often not be carried through to polymorphism in humans. Functional polymorphisms that affect resistance to human infectious disease have only recently been identified in the human homologues of genes shown to play a major role in susceptibility in mouse strains (Bellamy *et al.* 1998). Nonetheless, genes identified in animal models may characterize important molecular pathways of resistance and can suggest candidate genes.

Whether selection by infectious pathogens can explain the prevalence of genes predisposing to common degenerative diseases remains of great interest in evolutionary medicine. Selection of sickle-cell anaemia by malaria remains the classic example, but there are some new possibilities. Polymorphism in the vitamin D receptor is associated with the common bone disease osteoporosis in several population studies and the same vitamin D receptor genotype was associated with resistance to tuberculosis and other infectious diseases in recent studies (Chapter 5). A gene or genes on a segment of chromosome 5 appear to contribute to susceptibility to atopy/asthma, and resistance to schistosomiasis has recently been mapped to the same region (Marquet *et al.* 1996). Many autoimmune and some infectious diseases have HLA associations, but the alleles involved usually vary. The mycobacterial diseases, leprosy and tuberculosis, have been repeatedly associated with HLA-DR2 in Asia (for review see Hill 1998); this HLA type has also been associated with several autoimmune diseases.

As new genetic variants are identified that predispose to the major degenerative diseases of modern societies it will be interesting to see whether those alleles affect infectious disease susceptibility. To do so we will need DNA samples from many well-characterized families and cases with infectious diseases of potential evolutionary significance. Such host genetic studies should be integrated with long-term epidemiological studies of the disease and infectious pathogen in the same ecological setting.

ECOGENETICS AND PHARMACOGENETICS

Ecogenetics studies the role of genetic variation in response to any kind of environmental agent with particular reference to deleterious factors (see Chapter 4). Infectious agents are described above; here non-infectious agents and factors are discussed.

Nutrition

Genetic variation affects nutritional requirements and nutritional metabolism. The rare inborn error of metabolism (phenylketonuria) is a cardinal example of how genetic inability to metabolize phenylalanine from food causes high phenylalanine levels that produce mental retardation by direct brain injury. Marked reduction of phenylalanine in the diet prevents mental retardation. All newborn infants in developed countries are screened for this condition, and with a restricted phenylalanine intake the genetic defect becomes innocuous. Lesser degrees of nutritionally related enzyme defects—as seen in heterozygous carriers for certain inborn errors—could affect nutritional metabolism less seriously, particularly under conditions of infection or rapid growth. Since there is ubiquitous polymorphism affecting many enzymes involved in nutrition, requirements for a variety of nutrients probably vary between individuals. More work in this area needs to be done.

A well-studied ecogenetic polymorphism affects absorption of lactose (Simopoulos and Childs 1990; Flatz 1992; Motulsky 1996*a*). All mammals including humans can absorb lactose at birth from the intestine using the enzyme lactase that hydrolyses lactose to glucose and galactose. In most populations this enzyme stops functioning after weaning: individuals become lactose intolerant and develop increased peristalsis and diarrhoea on drinking moderate amounts of milk. This condition illustrates how the definition of 'normality' and 'abnormality' clearly depends upon population origin. A mutation that produces persistence of lactase production presumably gave a selective advantage to its carriers. Both heterozygotes and homozygotes for lactose absorption are lactose tolerant and do absorb lactose during childhood and adult life. This trait is predominant in populations (Western and central Europeans, certain groups in Arabia and Africa) that depend upon animal milk for protein nutrition, where persistence of lactase production beyond the weaning period probably conferred a selective advantage.

A recently discovered polymorphism in an enzyme involved in folic acid metabolism illustrates the interaction of diet and genetic variation. About 10–15 per cent of populations of European origin are homozygotes for a thermolabile form of methylene tetrahydrofolate reductase (MTHFR) (Motulsky 1996b). Such homozygotes have elevated homocysteine levels with suboptimal folic acid blood levels. Homocysteinaemia is associated with coronary artery disease, strokes, and peripheral vascular disease as an independent risk factor. The frequency of the MTHFR polymorphism appears higher in patients with premature coronary artery disease, few other risk factors, and presumably, poor folate nutrition. Various studies are in progress (Motulsky 1996a).

That obesity has genetic determinants has been shown by family, twin, and adoption studies (Bray 1992). Obesity-related genes include leptin and the leptin receptor. Future work will assess the role of polymorphic variation at these and other obesity-related loci in predisposing certain individuals to obesity under conditions of identical caloric intake.

Another interesting ecogenetic trait is the acetaldehyde dehydrogenase polymorphism common in Japanese and Chinese populations. This variant causes less effective degradation of acetaldehyde—the first metabolic product of alcohol. Higher acetaldehyde levels cause flushing of skin and an unpleasant feeling that apparently deters gene carriers from excessive drinking. The frequency of this variant is lower among Oriental alcoholics than matched controls (2 compared with 50 per cent): it appears to act as an 'anti-alcoholism gene' in Oriental populations (Higuchi et al. 1995). Gene carriers are less likely to become chronic alcoholics.

Hypertension does not occur in traditional societies with a very low salt intake (Burke and Motulsky 1992), but when such people move to western environments with a high salt intake, their average blood pressure increases and some develop hypertension. The role of genes that may play a part in determining salt-sensitive hypertension is under study. Genes for salt sensitivity may have been the original state, and the failure of salt to raise blood pressure may result from a relatively new mutation (Inoue et al. 1997). Salt sensitivity is currently tested by placing subjects on a metabolic ward for lengthy investigations; work on the angiotensinogen polymorphism may lead to simpler tests.

Many genes appear to be involved in predisposition to coronary artery disease in addition to environmental factors such as smoking and high lipid diets (Motulsky and Brunzell 1992). The 'garden variety' of coronary artery disease appears to be multifactorial and polygenic. An increased susceptibility to this disease can be detected by measurement of total cholesterol, low-density lipoprotein (LDL) cholesterol, high-density lipoprotein (HDL) cholesterol, and lipoprotein (a). This lipid profile results from interactions of genetic and dietary factors. Specific gene–gene and gene–environment interaction of the various lipid and other genes involved in coronary artery disease need further study with large population samples. Searches for those that are at high risk is medically worthwhile since current lipid-lowering therapy is effective in reducing morbidity and mortality. A population approach that advises reduction of lipid intake for everyone also has much to recommend it even though certain individuals will derive little or no benefit. Searching for those at high risk and reducing lipid intake in the whole population are not mutually exclusive.

Environmental carcinogens

In the unusual monogenic cancer syndromes, affected individuals vary greatly in age of onset, even carriers of the same molecular mutation. This adds to the evidence that any cancer is the consequence of a complex interaction between a given genotype and exposure to endogenous and exogenous carcinogens. Undefined modifying genes may also play a role. Known genetic factors include mutations and dysregulation of oncogenes, tumour suppressor genes, and variable inflammatory and immune responses.

Many natural and man-made chemicals must be metabolized by detoxifying enzymes before becoming effective carcinogens. The metabolic systems involved vary genetically both within and among populations, and many studies have investigated the associations between specific genotypes and susceptibility to particular cancers (Smith et al. 1995a). In some cases a clear relationship has been established, as for bladder cancer in relation to the slow/fast acetylator phenotype (Risch et al. 1995a). In others the evidence remains equivocal (Nebert et al. 1996).

Importantly, a given genotype may be at an increased risk only in the presence of a particular environmental stimulus. Consider hepato-cellular carcinoma susceptibility in the presence of endemic infection by hepatitis B virus (HBV) and food contamination by the mycotoxin aflatoxin in South-East Asia. The ability to detoxify aflatoxin differs between genotypes at the EPHX and GSTM1 loci, and the proportion of hepatocellular carcinoma cases also differs between genotypes. A person infected by HBV with at least one copy of a mutant EPHX allele has a 77-fold increase of hepatocellular carcinoma risk compared with individuals with neither risk factor, whereas the increase of risk due to HBV infection alone (computed in non-susceptible genotypes) is 15-fold (McGlynn *et al*. 1995). The incidence of hepatocellular carcinoma in China is manyfold higher than in other areas (Africa) with similar food contamination by aflatoxin and chronic HBV infection, probably as a consequence of the observed variation of gene frequency at the detoxifying loci. Another example of genotype–environment interaction comes from a study on postmenopausal women (Ambrosone *et al*. 1996), in which the polymorphism in the *N*-acetyltransferase 2 gene, resulting in a slow acetylation phenotype, was associated with a fourfold increase in breast cancer risk in smokers but not in non-smokers.

Various polymorphisms of cytochrome P450 enzymes and of glutathione transferases (GSTs), all showing different population frequencies, have been associated with cancer risk. For example, homozygotes for a deletion in GSTM1 or in GSTT1 are at higher risk of developing several cancers (Ketterer *et al*. 1992; but see Kempkes *et al*. 1996). More work on the interaction of gene carriers for several polymorphisms involved in carcinogen detoxification is needed to assess the role of this type of genetic variation in carcinogenesis (Brock-moller *et al*. 1996).

Because individual differences in cancer susceptibility may not be very large for many loci when considered independently of all others, future research should aim to develop well-defined markers of genetic susceptibility to help assess the combined effect of multiple genetic risk factors. An analysis of individual risk should then include environmental exposures such as nutrition and smoking habits. This should lead to better identification of individuals at increased risk, with benefits for public health and cancer prevention.

Drugs

Whereas genetic variation of drug kinetics is established as a major cause of interindividual differences in drug responses, the study of genetic variation in receptor function is just beginning. A growing number of genetic polymorphisms of drug- (and xenobiotic-) metabolizing enzymes have been characterized; their frequencies in different populations differ significantly. The number of isozymes and the high incidence of mutations in these genes are consistent with selection, initially probably during 'plant–animal warfare' followed by continuous generation of new chemical and biological toxins. The frequency of some phenotypes is strikingly high in some populations. For example, 90 per cent of individuals in some North African populations are slow acetylators due to homozygosity for decreased function alleles of *N*-acetyltransferase 2 (NAT2). This compares with a 10 per cent incidence of slow acetylators in Oriental populations and 40–70 per cent slow acetylators in European populations (Meyer and Zanger 1997). The variable expression of particular alleles of these enzymes is reflected by differences in food preference (Britto *et al*. 1991) and the marked difference in genes for drug-metabolizing enzymes in carnivores and herbivores (Smith 1991).

Polymorphisms of drug metabolism are of medical interest, for they are associated with drug toxicity when normal doses of drugs are given to 'poor' or 'inefficient' metabolizers. In contrast, when an active metabolite is responsible for the drug effect, a poor metabolizer cannot form this metabolite and lacks drug response. For example, the analgesic effect of codeine depends on metabolism to morphine. In poor metabolizers of debrisoquine (homozygous for loss-of-function alleles of cytochrome P450 CYP2D6), which occur with a frequency of 5–10 per cent in most populations, no analgesic effect of codeine is demonstrable (Sindrup and Brøsen 1995).

Genetically based variation of drug metabolism and its reflection in population differences is now routinely assessed in the pharmaceutical industry for preclinical and increasingly for clinical development of drugs for global therapeutic application. However, its impact on drug prescription by physicians has been minor. One reason is the lack of prospective studies in large populations with different frequencies of these genetic traits. Future improvement in pharmacogenetic DNA tests should help to provide these data.

THE HUMAN MAJOR HISTOCOMPATIBILITY COMPLEX

Selection at the MHC

The human major histocompatibility complex (MHC) encodes HLA (human leucocyte antigens) as well as many other proteins and is one of the most extensively studied regions of the human genome. The function of HLA molecules is to present self and non-self peptides to T lymphocytes and thus to mediate the adaptive immune response. HLA loci are highly polymorphic. This extensive polymorphism is maintained by balancing selection acting on sites that encode the peptide-binding amino acid residues of the MHC molecules (for reviews see Klein 1986, Klein and Klein 1991; Klein *et al.* 1993).

Studies of the human MHC suggest that natural selection has acted and continues to act to maintain polymorphism (see Box 6.2). However, the difficulty of detecting selection in human populations should not be underestimated. Although the effects of several mechanisms of natural selection on the evolution of the polymorphism are well understood in populations at equilibrium, real populations typically live in a fluctuating environment. Moreover, large effects may result from the accumulation of small selective effects over many generations. It may thus be difficult to detect evidence of selection and to determine whether departures from neutrality are caused by other factors (such as genetic drift, bottlenecks, admixture) or by disease associations.

Different mechanisms have been proposed to underlie the selective maintenance of MHC polymorphism. MHC gene products play a major role in various aspects of the immune

---

Box 6.2: Selection and the MHC

Evidence for selection at human MHC loci comes from four different sources.

1. Genetic association and linkage has been found between HLA types and several infectious diseases in field studies (Hill *et al.* 1991, 1992; Thurz *et al.* 1995; de Vries *et al.* 1976), implying selection of particular HLA alleles by infectious pathogens.

2. A higher rate of non-synonymous (coding) than synonymous (silent) mutations is found in the genes encoding class I and class II molecules, and most of the non-synonymous changes occur at peptide-binding sites (Hughes and Nei 1988, 1989; Hughes and Hughes 1995).

3. Phylogenetic analyses have revealed that some HLA alleles are more similar to MHC alleles found in the chimpanzees than to each other (Klein 1987; Figueroa *et al.* 1988; Lawlor *et al.* 1988; Mayer *et al.* 1988; Fan *et al.* 1989), indicating the persistence of lineages older than at least 5 millions years in our species. Such old lineages are not expected for neutral loci, but can be maintained in a population due to balancing selection (Takahata and Nei 1990; Takahata 1991; Satta *et al.* 1994). Moreover, the presence of several shared alleles between chimpanzees and humans suggests that there has been no severe bottleneck throughout the history of our species.

4. The distribution of allele frequencies at loci HLA-A, -B, and -DRB1 is not compatible with expectations from the neutral theory (Klitz *et al.* 1986; Tiercy *et al.* 1992; Sanchez-Mazas *et al.* 1995), in the sense that there is an excess of intermediate frequency haplotypes, leading to the rejection of the Ewens–Watterson selective neutrality tests (Watterson 1978, 1986) in many populations of various origins.

response and are in direct contact with environmental agents. MHC molecules play a critical part in the initiation of the adaptive immune response, and the polymorphic sites overlap to a large extent with sites encoding the peptide-binding region. This strongly suggests that the selection pressure is exerted by infectious agents. Heterozygote individuals, capable of responding to a broader range of pathogens, would hold an advantage over homozygote individuals. Frequency-dependent selection has also been suggested. Rare alleles would be at an advantage because parasites would not have had time to adapt, whereas old alleles that had been in the population for a longer time would already have elicited an evolutionary response from the parasites, and would thus be at a disadvantage. Another mechanism, fluctuating selection, involving a turnover of advantageous alleles in response to epidemics (repetitive episodes of directional selection) has also been advanced. Finally, the MHC may be associated with a system responsible for mate choice and assortative mating (Chapter 9), promoting the choice of sexual partners with disparate MHC alleles and leading to increased heterozygosity.

There is no need to assume that only one particular mechanism is correct, for all could have been acting simultaneously or at different times to promote the genetic diversity of human MHC loci.

MHC polymorphism and population genetic affinities

In spite of evidence for selection, MHC polymorphism is considered an excellent genetic marker for population studies. Such studies have been included for 25 years in successive HLA workshops where an increasing effort has been focused on the analysis of well-defined populations from all over the world using DNA typing technology (Bodmer *et al.* 1997). These data were designed both to investigate the genetic affinities of humans in relation to their historical relationships and, more controversially, to constitute a databank of 'control' samples for disease association studies.

In most cases the frequency distributions of MHC alleles and haplotypes can be explained by events linked to the history and demography of populations. For example, some HLA haplotypes (like A1-B8-DR3) exhibit clinal frequency variation from south-eastern to north-western Europe that correlates well with frequency clines observed for many other genetic markers. These clines have been attributed to the demic diffusion of neolithic farmers from the Near East to Europe, some 10 000 years ago (Menozzi *et al.* 1978; Ammerman and Cavalli-Sforza 1984; Sokal *et al.* 1991; Cavalli-Sforza *et al.* 1994). Similarly, HLA frequency patterns fit apparent routes of migration suggested by archaeological and linguistic studies, for example in sub-Saharan Africa (Excoffier *et al.* 1987, 1991; Dard *et al.* 1992; Tiercy *et al.* 1992) or the Pacific (Serjeantson 1989; Serjeantson and Gao 1995). A broad correlation is observed for HLA between genetic distances and the geographical distribution of populations (Sanchez-Mazas and Pellegrini 1990; Tiercy *et al.* 1992; Cavalli-Sforza *et al.* 1994; Grundschober *et al.* 1994). Differentiation resulting from isolation-by-distance is thus arguably a major cause of HLA variation among present populations.

Hence, we are left with an apparent paradox. For reliable inference of history, the markers used should be neutral. MHC polymorphism is certainly affected by selection, yet HLA frequencies suggest a plausible and internally consistent reconstruction of some major episodes in human history. One possible explanation is that the overall selection detected on this system is weak. However, even weak selection pressures are cumulative. Thus the good correlation between population affinities determined by putatively neutral markers and by HLA types suggests that each population may have had a unique and characteristic history of encounters with infectious diseases.

The MHC and non-infectious diseases

There is much interest in medicine regarding the presence of various inherited HLA gene constellations as predisposing genetic factors for autoimmune diseases. Thus in insulin-dependent diabetes, HLA class II alleles are the principal genetic determinants in addition to other poorly understood loci elsewhere on the genome. The exact mechanisms by which autoantibodies are formed and destroy pancreatic islet cells remain unknown but probably involve viral injury of the pancreas. Similar mechanisms are observed in other autoimmune diseases such as rheumatoid arthritis, Graves' disease (affecting the thyroid), and Sjögren's syndrome (various secretory glands). Very striking associations with HLA types are seen with ankylosing spondylitis, HLA-B27, and narcolepsy (100 per cent of patients have HLA-DR2) and remain unexplained.

SOME RECOMMENDATIONS

The Human Genome Project and the Human Genome Diversity Project

The remarkable progress of the Human Genome Project (HGP) in mapping and sequencing the genome and the potential impact of the proposed Human Genome Diversity Project (HGDP) are important considerations for research strategies in human diversity, evolution, and medicine (Cavalli-Sforza et al. 1991; Collins and Galas 1993). World-wide collaboration among scientists in the use of a well-characterized core set of reference families, made available by the Centre d'Etude du Polymorphisme Humain (CEPH) in Paris, has helped the human genome project achieve one of its first 5-year goals, the construction of high resolution (i.e. 2–5 cM) genetic maps (Dausset et al. 1990). Such genetic maps with polymorphic markers have been extremely useful in identifying the genes for such important monogenic diseases as cystic fibrosis, Huntington's disease, polycystic kidney disease, amyotrophic lateral sclerosis, neurofibromatosis, and myotonic dystrophy. They have also facilitated mapping several genes underlying susceptibility to common diseases such as breast and colon cancer, diabetes, hypertension, and Alzheimer's disease.

In addition to high-resolution genetic maps, the HGP is making impressive progress towards completion of a 100-kb resolution sequence-tagged-site map by the end of this century; a

reference human genome sequence should be completed by 2004. This will provide evolutionary biologists, anthropologists, and human population geneticists with an unprecedented resource for tracing the origin, history, and spread of mutations causing diseases.

Let us consider the benefits of international collaboration in using a core set of reference CEPH families to construct high-resolution genetic and physical maps. A comparable resource of human DNA collected world-wide from human population groups has been proposed in the human genome diversity project to facilitate studies of human genome variation. Although the diversity project involves controversial ethical, legal, and social issues, the proposed world-wide collection of DNA from human population groups (Lehrman 1996; Macilwain 1996) would be invaluable for human evolutionary studies. A generally available set of DNAs representative of the global human population would be a natural complement to the human genome project. The reference human genome sequence (i.e. completion of the HGP) will make it possible to design primers for amplifying and analysing variations in any segment of genomic DNA. When combined with global epidemiological data, DNA sequences from geographically defined human populations would significantly improve our knowledge of population-based differences in disease profiles and the evolution of health and disease.

Thus the objective of the HGDP is commendable. The sampling should be carried out in full collaboration with local scientists. A broader medical view of this ambitious programme would correctly emphasize the medical utility of the gene frequency data, facilitate sampling through the involvement of established professional networks and facilities in countries world-wide, such as those already linked by the WHO, and could also provide an immediate dividend to local participants by allowing an offer of enhanced health care.

Population affinities and movements

Often the effect of one or more major genes of medical interest is modulated by interactions with other, often unidentified, components of the genome. Related individuals will share alleles to an extent that can be estimated from their pedigrees. Because populations, like individuals, are interrelated, the closer the evolutionary relationships between populations, the more similar are their pools of genes (e.g. see Nei and Roychoudhury 1993). Therefore, patterns of population relatedness may give us insight into phenomena such as different response to drugs, different disease prevalences, and different characteristics of the same disease in different populations. Such a research programme complements the study of the selective mechanisms which also play a role in producing and maintaining genetic diversity.

Patterns of gene flow and barriers to individual dispersal suggest levels of population subdivision that may reflect medically relevant genetic differentiation. Groups that are or have long been isolated are likely to show the reduced genetic variation expected after founder effects and population bottlenecks. Geographically distant populations between which gene flow has been demonstrated may share a greater-than-expected fraction of their gene pool, and therefore may resemble each other in the incidence of some disorders.

The history of human settlement is informative *per se* on many phenomena of medical relevance. However, a closer interaction between clinical and population geneticists would be facilitated by including in comparative population studies genetic polymorphisms associated with diseases as well as apparently neutral highly polymorphic markers in close linkage with disease genes.

SUMMARY

Molecular and genetic approaches now provide powerful tools for investigating the origin of human populations and the evolution of genes affecting both complex and monogenic traits. This allows a biologically appropriate classification of individuals and population groups based on genotypes and gene frequencies rather than appearance; this information will facilitate disease risk assessment. Many of the observed differences in gene frequencies between human populations may be accounted for by population movements, and it is difficult to assess how large a role natural selection has played in population differentiation. For several disease genes there is evidence that environmental agents, such as infectious pathogens, dietary factors, and environmental toxins, have been responsible for an increased frequency in some populations. Further examples of such selection may emerge from current efforts to define the genetic basis of many polygenic common diseases. We need more information on molecular genetic differences between human populations; we could obtain that information from the successful completion of the Human Genome Project and the proposed Human Genome Diversity Project.

Part III: Natural selection, conflicts, and constraints

7

GENETIC CONFLICTS OF PREGNANCY AND CHILDHOOD

David Haig

Haldane (1932*a*) warned against a fallacy which 'has been responsible for a good deal of the poisonous nonsense which has been written on ethics in Darwin's name'. This was the common assumption that natural selection always makes an organism fitter in its struggle with the environment. Haldane recognized that competition within a species can favour characters that confer a competitive advantage but that are otherwise maladaptive. The competitive advantage disappears once competitors also possess the character, but the fitness costs remain. His examples of natural selection with potentially maladaptive outcomes included prenatal competition among the members of a litter for limited maternal care. Trivers (1974) recognized that sibling rivalry is also present when offspring are produced one at a time. That is, parental expenditure of time and resources on one offspring has an opportunity cost that means less time and resources are available for future offspring (or for older offspring still receiving parental care). For this reason, he argued, there will be an evolutionary conflict between parents and offspring, with offspring selected to acquire more parental investment than parents are selected to supply. Haig (1993, 1996*a–c*) applied Trivers' theory to interactions between human mothers and fetuses during pregnancy. I proposed that evolutionary conflicts between maternal and fetal genes were responsible for some otherwise inexplicable features of pregnancy and could help to explain why pregnancy so often entails costs to a mother's health.

The evolutionary view of parent–offspring relations I wish to foster is not one of unmitigated struggle but of a subtle interplay between conflicting and mutual interests. Co-operation is possible, but the form of co-operation when parties also have conflicting interests differs from that form when their interests are identical. In the latter case, there is no need for compromise (problems of co-ordination may still arise).

For many purposes, the best model for thinking about the prenatal relation between mother and fetus is the postnatal relation between the same genetic individuals. Mother and child have many interests in common, but their relationship is not without conflict. When conflicts arise, sometimes the mother gets her way, sometimes the child gets its way, and sometimes a compromise is negotiated (see Godfray 1995 for a discussion of conflict resolution in models of parent–offspring relations). Some relationships proceed smoothly with little overt conflict whereas others are tempestuous, and similar contrasts may exist within the same relationship at different times.

The first four sections of this chapter extend —and in some places correct—my previous work on genetic conflicts during pregnancy: **Parental justice** reviews the kinds of conflicts that can occur among the autosomal genes of mother and offspring; **Pregnancy termination** considers the function of early pregnancy losses, the evolution of menstruation, and the control of parturition; **Maternal–fetal communication** discusses how genetic conflicts impede the exchange of information between mother and fetus, and modifies my previous views on the role of human placental lactogen; **Maternal circulation** presents a simple model of conflict over the distribution of maternal blood during pregnancy. The penultimate section—**Growth** —considers ways in which intergenerational conflicts may have influenced the evolution of human postnatal growth. The final section— **Evolution and medicine**—addresses some common anxieties about the application of evolutionary ideas to medicine.

PARENTAL JUSTICE

How can there be a genetic conflict between parents and offspring if the genes of the offspring come from the parents? There are several ways to answer this question, but they do not always convince sceptics. Here is an attempt that borrows ideas from political science and emphasizes the information available to genes. Consider a family of full-sibs. All of the genes of the offspring are present in one or other of the parents, but genes in the parents usually have no way to tell which offspring have inherited their copies. Parental genes therefore make decisions about the allocation of resources to individual offspring behind a meiotic 'veil of ignorance' that ensures each gene has the same probability of being present in each offspring and that each gene's interests are best served by choosing the allocation that maximizes the expected number of surviving offspring (cf. Harsanyi 1953; the metaphor of the veil is borrowed from Rawls 1971). However, the veil of ignorance is partially lifted once genes find themselves in offspring, because a gene is definitely present in the offspring in which it is expressed and can take actions that benefit these offspring at the expense of their siblings. Thus, if the genes of offspring, rather than the genes of parents, control the allocation of parental investment they are subject to an evolutionary 'tragedy of the commons' (Hardin 1968). All genes will be worse off if all pursue their own interest. On the other hand, unilateral restraint will be exploited. As a consequence, genes expressed in offspring have been selected to obtain higher levels of parental investment than is favoured by genes expressed in parents. The

allocation of resources favoured by parental genes is 'just' in the sense that it maximizes the aggregate fitness of offspring considered as a group.

A closely related conflict is possible within the genomes of mothers. If some maternal genes are able to peak behind the veil of ignorance and learn which offspring receive their copies, such genes will be selected to favour offspring with their copies in preference to offspring without. For example, a maternal haplotype could cause the abortion of embryos that did not inherit its copies. Such favouritism will usually be opposed by other genes that segregate independently of the nepotists (Haig 1996a).

A parent's reproductive value can be divided into the contribution of a particular offspring and the contribution of all other offspring. Parent–offspring conflict arises because genes expressed in offspring will evolve to discount costs and benefits to a parent's residual reproductive value relative to costs and benefits to the offspring's reproductive value. Whenever mating systems depart from strict lifetime monogamy (i.e. whenever females have offspring by more than one father), physiological costs imposed on a mother by an offspring will usually have a greater impact on the mother's residual reproductive value than on the father's residual reproductive value. Therefore, conflict can exist between the genes that an offspring gets from its mother and the genes it gets from its father. Expression of this conflict requires that mechanisms of genomic imprinting allow the relevant genes to be differentially expressed on the basis of their parental origin. If so, the paternal genes of offspring will have been selected to impose greater physiological costs on mothers than will the offspring's maternal genes (Haig and Westoby 1989; Moore and Haig 1991).

PREGNANCY TERMINATION

Selective abortion

Many organisms initiate more offspring than ever complete development. For example, opossum mothers produce more young than they have teats for attachment (Harder *et al.* 1993); the elephant shrew *Elephantulus myurus* releases about 50 ova from each ovary but has only a single implantation site per uterine horn (van der Horst and Gillman 1941); and intra-uterine siblicide is a regular feature of pregnancy in the pronghorn *Antilocapra americana* (O'Gara 1969). Stearns (1987*b*), among others, has suggested that the overproduction of offspring may be adaptive if it allows the selective elimination of offspring of low quality. In this view, parental fitness is enhanced by the early death of a subset of offspring if death occurs before major parental investment and if the subset of survivors are of higher quality (on average) than the subset of non-survivors. The lower the cost of abortion relative to the cost of raising an offspring until independence, the more selective a parent can afford to be (Kozlowski and Stearns 1989; Haig 1990). Such models can be rendered more sophisticated by allowing parental selectivity to vary in response to inputs of information about how favourable current circumstances are for reproduction. This section will discuss the extent of pregnancy loss in humans and consider whether some of these losses may be adaptive.

Unless an implanting embryo produces sufficient chorionic gonadotrophin (hCG) to block regression of the corpus luteum, the lining of the human uterus is shed about 14 days after ovulation as concentrations of luteal progesterone fall. The proportion of fertilized ova that fail to block menstruation is controversial, but it is none the less clear that the highest rate of attrition at any stage of the human life cycle occurs during these two short weeks. Most of these losses occur without a woman being aware she is pregnant. Estimates of the monthly rate of clinically recognized pregnancy in sexually active, non-lactating women who are not using contraception are typically 25–35 per cent (Tietze 1959, 1968; Suchindran and Lachenbruch 1975). Such estimates suggest an upper bound of at most 75 per cent for the rate of early pregnancy loss (because this figure assumes that a conception occurs in every cycle), whereas sensitive assays of hCG in a woman's urine during the late luteal phase suggest a lower bound of at least 30 per cent for the proportion of human conceptions lost before term (Wilcox *et al.* 1988; Ellish *et al.* 1996; Zinaman *et al.* 1996). These assays miss an unknown number of conceptions that fail to implant or whose hCG levels do not reach the critical value on the days sampled.

Two caveats are important. First, the data of the previous paragraph all come from North American populations and may not be typical of other ecological contexts. Second, the estimates are average values and hide considerable variation within populations. The probability per cycle of a clinically recognized pregnancy varies with a woman's age, body composition, energy expenditure, and energy balance, both within and across populations (for review see Ellison *et al.* 1993). Without endocrinological data, it is usually impossible to separate the effects of reduced fertility due to anovulatory cycles from effects due to ovulatory cycles with 'luteal insufficiency' (i.e. reduced activity of the corpus luteum). The latter (but not the former) would be associated with an increased frequency of early pregnancy losses. Maternal malnutrition appears to have little effect on fetal mortality rates once a pregnancy is clinically recognized (Bongaarts 1980).

About 10–20 per cent of clinically recognized pregnancies miscarry in the first trimester and the majority of these have some form of chromosomal abnormality (Hassold 1986). For this reason, the idea that most clinical miscarriages reflect the adaptive functioning of an evolved mechanism of quality control is now widely accepted. By contrast, very little is known about the genetic properties of earlier losses. A simple interpretation would be that embryos lost in the first few weeks are even more severely compromised than later losses. Clearly, some early losses must have severe abnormalities, but it seems unlikely that all early losses are of this kind because fecundability is responsive to environmental factors (see above) and there seems no reason to suppose that these factors should cause a change in the frequency of defective genotypes.

An alternative hypothesis would view mothers' bodies as making an early decision whether or not to carry an embryo and would view clinical miscarriages as 'mistakes' that got through the initial screen. Because the cost of a 1-month delay in reproduction is so much less than the cost of raising an offspring of low expected fitness, maternal physiology may have evolved to select among embryos on the basis of small differences that need be only weakly correlated with subsequent vigour. In a study of women attempting to conceive, Wilcox *et al.* (1988) observed that a third of early pregnancy losses were followed by a conception in the next cycle and that women with an early loss were more likely than others eventually to have a clinically recognized pregnancy. Thus, early losses may have minimal costs for a woman's lifetime reproductive success.

The metaphor of offspring as examination candidates (or job applicants) and the mother as examiner is useful for thinking about the action of natural selection in systems of selective abortion (Haig 1987). An examiner's aim is to design tests that provide accurate information as cheaply as possible. The examinee wishes to pass. If the examinee is indeed of high quality, she and the examiner have a common interest in conveying this information accurately, but, if the examinee is of poor quality, her best interest may be to dissemble. How then can a mother obtain useful information about embryo quality? A requirement for the final product of a synthetic pathway tests all the steps of the pathway. Thus, the ability of a small embryo to make large quantities of hCG demonstrates the efficient functioning of the embryo's machinery of transcription, translation, and glycosylation. If mothers require more than some threshold level of hCG to block menstruation, this will constitute a crude screen on embryo quality. If the threshold is adjustable, mothers could raise the required level of hCG in ecological circumstances that are less favourable for reproduction. Alternatively, the threshold could be kept constant but embryos could be handicapped when conditions are less favourable.

Any testing procedure selects for individuals that are good at passing tests. But there is rarely a perfect correlation between test scores and competence in the task for which the test is designed. (A test would be abandoned if there were no correlation.) Therefore, early pregnancy losses can favour genes with negative effects later in life either because the genes are better than average at avoiding early losses or because they can bias the testing procedure to handicap offspring that do not inherit their copies. The first possibility would be associated with higher than average fecundability (because the distorter causes embryos that would otherwise have failed to pass) and the second with lower than average fecundability (because the distorter causes embryos that would otherwise have passed to fail). Both possibilities would result in segregation distortion in the successful pregnancies of heterozygous mothers (Haig 1996a). Thus, prenatal advantages could explain how some alleles with deleterious postnatal effects are maintained at high frequency by natural selection (Diamond 1987), and segregation distortion (defined to include biased early pregnancy losses) could explain the persistence of common medical conditions that are associated with either reduced or increased fecundability (and should be considered as a possibility in such cases).

I have presented two perspectives on the evolution of chorionic gonadotrophins: that they evolved in primates (and equids) as fetal attempts to evade maternal mechanisms of pregnancy termination (Haig 1993), and that hCG conveys useful information to the mother about embryo quality. The two perspectives address different questions. The first addresses the origin of chorionic gonadotrophins: why embryos first produced the hormone. The second addresses the maintenance of the system: why mothers continue to respond. Because embryos sometimes have an incentive to deceive mothers about their true quality, credible signals of quality must be difficult to fake (Zahavi 1981; Grafen 1990; Johnstone 1997). The credibility requirement is reflected in the high levels of hCG produced. If the purpose of hCG were *solely* to signal the presence of an embryo of whatever quality, the signal could be sent cheaply, because signals of presence are always credible ('I speak therefore I am').

Menstruation

Shedding of the lining of the uterus with visible loss of blood is rare among mammals: it is well documented only in some haplorhine primates, a few bats, and elephant shrews. Interest in the evolution of menstruation has been revived by Profet's (1993) provocative hypothesis that it cleanses the female reproductive tract of sperm-borne pathogens. Profet argued that menstruation occurs at least covertly in most, if not all, mammalian species. However, her hypothesis has been strongly criticized by Clarke (1994), Finn (1994), and Strassmann (1996*b*) who all emphasize the restricted taxonomic distribution of menstruation, among other reasons for rejecting Profet's hypothesis. Adaptive modifications of the menstrual process to reduce the risk of infection would not be surprising (given that menstruation occurs), but defence against pathogens does not seem to be its *raison d'être*.

Finn (1987, 1994, 1996) argues that menstruation is a consequence of the initiation of the decidual changes of the endometrium independently of whether or not an embryo implants, where these changes function as a maternal defence against excessive invasion by trophoblast. 'The advantage would seem to be in providing protection for the uterus in anticipation of the presence of a blastocyst rather than in response to it' (Finn 1996). Strassmann (1996*b*) sees menstruation as a side-effect of the cyclical regression and proliferation of the endometrium that occurs in most mammals. In her view, the function of regression is to economize on the energy used to maintain a metabolically active endometrium. Regression usually occurs without blood loss but 'the unusually profuse bleeding in humans and chimps may be due to the large size of their uteri relative to adult female body size and to the design of the microvasculature in these and other catarrhines'. Thus, for both Finn and Strassmann, menstruation does not have a function (in and of itself) but is a side-effect of

other processes. However, given that menstruation occurs, the process can be co-opted or modified to serve other functions.

The evidence of early pregnancy losses suggests that a significant proportion of menses may have been associated with the loss of an implanted embryo. Therefore, Clarke (1994) proposed that the adaptive significance of menstruation may be explained as 'a way of eliminating defective embryos before a pregnancy has proceeded very far'. Menstruation could also function to eliminate normal embryos when social or ecological conditions are unsuitable for pregnancy. If a species has invasive implantation, elimination of an embryo requires simultaneous shedding of the attached endometrium, but why should this shedding be general rather than localized? One consequence of generalized shedding of a partially decidualized endometrium is that this obviates 'stealth strategies' in which embryos hide out in the endometrium without signalling their presence. Sloughing the endometrium is an effective means of eliminating a single embryo but would be an indiscriminate form of embryo selection if it resulted in the loss of an entire litter. Menstruating species usually produce singletons. (Elephant shrews are an exception and produce twins: one from each uterine horn.) Many polytokous eutherians are able to resorb some members of a litter while leaving other embryos intact (Morton *et al.* 1982), perhaps serving an analogous function of selective embryo abortion.

Menstruation provides other members of society with information about a woman's reproductive status and can be used by these individuals to enhance their own reproductive interests, even if these interests are opposed to those of the woman. Who controls menstrual information thus becomes an important issue (Strassmann 1992). Many human societies enforce cultural rules which ensure that a woman's menstrual status is public knowledge. Strassmann (1992) has proposed that such practices among the Dogon of Mali primarily serve the interests of married men and their patrilineages. If other individuals observe whether or not a woman menstruates and interpret this information with fitness consequences for the woman, then the interpretations become a selective factor in shaping how and when menstruation is expressed. Many women experience a series of menstrual cycles with a low probability of pregnancy, both at the beginning and end of their reproductive years. It would be interesting to know whether this is simply the warming-up and running-down of the female reproductive system or whether non-fertile cycles have been retained by natural selection because of how menstruation is interpreted by other individuals.

Gestation length

Obstetricians have long recognized a continuum from shedding of the menstrual decidua, through early miscarriages and premature deliveries, to shedding of the decidua with the placenta at a term birth (Tyler Smith 1856). The normal duration of pregnancy is considered to be 9 months (40 weeks from end of last menses), but infants born up to 3 months earlier had some chance of survival before the modern medical era. In nineteenth-century Britain, expulsion of the contents of the uterus during the first 6 months of gestation was classified as an abortion because the fetus was considered inviable prior to 28 weeks gestation but viable thereafter (Tyler Smith 1856). Fetuses can also survive in the uterus for considerable periods beyond normal term. Exceptionally prolonged gestations have been reported for some anencephalic fetuses (Anderson *et al.* 1969; Milic and Adamsons 1969), including one case in which the fetus was still alive after an estimated gestation of 389 days (Higgins 1954).

If one assumes that an offspring's contribution to its mother's reproductive value declines for gestation lengths longer or shorter than some optimal duration and that the cost to the mother's residual reproductive value is an increasing function of gestation length, then the length of gestation favoured by genes expressed in fetuses will exceed that favoured by genes expressed in mothers (Fig. 7.1). Who then controls the decision when to end a particular pregnancy? Mothers probably have

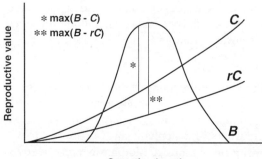

Fig. 7.1 A mother's reproductive value is increased by a benefit (B) that is a hump-shaped function of gestation length. Her residual reproductive value is decreased by a cost (C) that is an increasing function of gestation length. Genes expressed in mothers are selected to maximize ($B - C$) whereas genes expressed in offspring are selected to maximize ($B - rC$) where r is a measure of the degree to which genes of offspring discount costs to a mother's residual reproductive value. Fetal genes will favour longer gestations than maternal genes, although the difference will be slight if the 'hump' of the benefit function is narrow and the rise in the cost function gentle.

effective control during early gestation, but the relative power of maternal and fetal genes shifts as the conceptus grows larger. At later stages, the human fetus appears to control when it comes out. This is suggested by cases of prolonged gestation associated with anencephalic fetuses and by cases of superfetation in which twins of different ages are delivered weeks apart (Dorgan and Clarke 1956; Abrams 1957). Experimental evidence shows that the duration of pregnancy in sheep is controlled by the fetus and the same is believed to be true of human pregnancies, although the physiological mechanisms are different and poorly understood in humans (Nathanielsz 1996).

If gestation length is determined by fetal interests (adjusted for the interest of fetal genes in the mother's residual reproductive value), the human fetus will have evolved to stay inside its mother's body for as long as intrauterine existence is more desirable than life on the outside. For this reason, one would expect longer gestations when intrauterine conditions are relatively favourable but rapid delivery if the uterus becomes unsafe (after extrauterine survival is possible). Intrauterine infection is a major cause of premature delivery (Fields *et al.* 1996), and infants that are small for gestational age are more likely to be delivered early than appropriately grown infants (Ott 1993).

In normal singleton pregnancies, fetal growth velocity (grams/day) falters at about the 36th week of gestation but then accelerates after delivery (Forbes 1987). This has been interpreted as evidence that delivery occurs when the placenta has reached its capacity for supplying nutrients. However, weight changes do not fully represent the caloric cost of fetal growth because body composition is also changing as fetuses deposit an increasing proportion of energy-rich fat as term approaches.

Once it has survived the first trimester, a fetus is usually the sole occupant of the uterus with a low risk of death and first call on its mother's nutrient reserves. The fetus's placental supply line is already established and has considerable reserve capacity, at least until the final stages of pregnancy. Therefore, if the fetus controls when it leaves the uterus, there seems no particular reason why it should grow at the fastest rate possible if there are advantages of taking its time, either in terms of the quality of construction of its body or in reducing the demands on its mother's health. By contrast, in species that produce litters, competition within the litter will favour rapid development (Haldane 1932*b*).

MATERNAL–FETAL COMMUNICATION

Credibility problems

Communication between parties with conflicting interests is more difficult than communication between parties with identical interests. If conflicts of interest are possible, messages cannot always be trusted and untrustworthy messages are often best ignored (Johnstone 1997). These problems are exacerbated in maternal–fetal relations because there will often be ambiguity about whether maternal or fetal cells are sending a particular message.

One consequence is that maternal physiology during pregnancy will lack many of the checks, balances, and feedback controls that are present in the non-pregnant state. Some complications of pregnancy that endanger both mother and fetus may be the result of the inability of a physiologically threatened mother to communicate her dire state credibly to a demanding fetus. Messages between a mother and fetus can be credible under some circumstances, but these messages are likely to be simpler, and much less detailed, than the messages that can be conveyed between the genetically identical cells of a single body. Placental hormones may provide mothers with general information about offspring size and vigour—for this purpose one hormone is as good as another—but their production does not show the kind of temporal fluctuation that could communicate moment-to-moment variation in fetal need (cf. Chard 1993).

Communication is problematic, not only between maternal and fetal cells, but also between cells of the same genetic individual because a cell may be uncertain about the genotype (fetal or maternal) of other cells it encounters and about the origin and trustworthiness of incoming signals. For this reason, endocrine communication within the mother's body is compromised by potential disinformation from placental hormones, whereas the fetus's internal lines of communication are relatively secure because maternal hormones do not have equivalent access to the fetal circulation. The complexities of maternal–fetal communication are discussed at greater length by Haig (1996*b*, *c*).

Placental growth hormones

Growth hormone (GH) and prolactin (PRL) are structurally related hormones with structurally related receptors. The growth hormone gene cluster on human chromosome 17q22–q24 contains a single gene expressed in the anterior pituitary (*hGH-N*) plus four genes expressed in the placenta (*hCS-A, hCS-B, hCS-L, hGH-V*). All five genes are derived from the duplication of an ancestral gene that encoded a pituitary growth hormone: whether *hCS-L* encodes a functional product is unknown; *hGH-N* encodes human growth hormone (hGH); *hCS-A* and *hCS-B* encode human placental lactogen (hPL); and *hGH-V* encodes human placental growth hormone (hPGH). The hormones have different relative affinities for GH and PRL receptors: hGH binds strongly to both GH and PRL receptors; hPL has high affinity for PRL receptors but negligible affinity for GH receptors; and hPGH has high affinity for GH receptors (Chen *et al.* 1989; Strauss *et al.* 1995).

During human pregnancy, the placenta secretes into the maternal blood a ligand of maternal GH receptors (hPGH) and a ligand of maternal PRL receptors (hPL). The production of both ligands increases as pregnancy progresses. At term, 1–3 g of hPL per day are released into the maternal circulation, accounting for roughly 10 per cent of all placental protein production and reaching concentrations of 5–15 μg/ml (Strauss *et al.* 1995). To put this in perspective, a near-term fetus accumulates about 4 g of protein per day, with a somewhat larger daily increment of fat (Ziegler *et al.* 1976; Widdowson 1980). hPGH is produced in much smaller, although still substantial, quantities. Its concentration in maternal serum reaches 20 ng/ml in the latter stages of pregnancy. Near term, the maternal pituitary has almost ceased producing hGH but produces hPRL at higher levels than in the non-pregnant state. Thus, maternal GH receptors encounter only their placental ligand but PRL receptors encounter high levels of both maternal and placental ligands (Stefaneanu *et al.* 1992; Mirlesse *et al.* 1993).

Haig (1993) proposed that the function of hPL was to activate maternal PRL receptors for fetal benefit. At that time, I had not recognized the significance of elevated production of maternal hPRL during pregnancy, nor had I considered the possibility that an abundant placental hormone could act as an antagonist of maternal receptors. If maternal PRL receptors respond in the same way to hPL and hPRL, the genetic-conflict hypothesis predicts that maternal production of hPRL should decrease in response to increasing placental production of hPL, contrary to observations (Haig 1996b). However, if one of the hormones functions as an agonist and the other as an antagonist (or if hPL and hPRL have different effects via maternal PRL receptors), the hypothesis is compatible with elevated production of both hormones. At present, the interactions of hPRL and hPL with human PRL receptors are poorly understood and the possibility of antagonistic effects remains an untested hypothesis.

MATERNAL CIRCULATION

A simple model of the maternal circulation during pregnancy is presented in Fig. 7.2. Cardiac output from the left side of the heart is shared between two subcirculations arranged in parallel. The uteroplacental subcirculation (with resistance R_p) represents all maternal blood diverted through the intervillous space of the placenta, whereas the non-placental subcirculation (with resistance R_m) represents the systemic blood supply to maternal tissues. The fetal share of maternal cardiac output is given by the ratio of non-placental resistance to total systemic resistance, namely

$$\text{Fetal share} = \frac{R_m}{R_m + R_p}$$

The conflict hypothesis predicts that fetal genes have been selected to appropriate a greater share of maternal blood than maternal genes have been selected to supply. Therefore, placental factors are predicted to increase R_m and decrease R_p, whereas maternal factors are predicted to decrease R_m and increase R_p. Conflict over R_p is largely played out during the first trimester of pregnancy. The maternal arterioles that supply the endometrium increase rapidly in length during the first weeks after ovulation, becoming highly convoluted as they outpace the growth in thickness of the endometrium. For this reason, these vessels are known as spiral arteries. Their increase in length (and many twists and turns) increase R_p. This effect is countered by trophoblast which invades and remodels the spiral arteries, destroying the vessels' muscular lining and

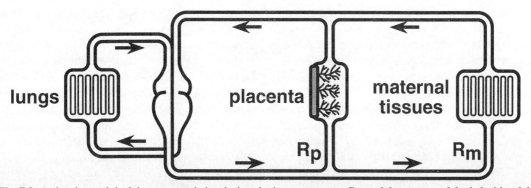

Fig. 7.2 A simple model of the maternal circulation during pregnancy. Part of the output of the left side of the mother's heart is diverted through the intervillous space of the placenta. The proportion of maternal cardiac output that is available to the fetus is determined by the ratio of the resistance in the non-placental circulation (R_m) to the total systemic resistance ($R_m + R_p$).

greatly expanding their radius. In this manner, trophoblast reduces R_p and destroys the mother's ability to constrict her spiral arteries, with the consequence that R_p effectively becomes a fixed quantity. Subsequent conflict over the fetal share of maternal systemic circulation therefore focuses on control of R_m.

Conflict over R_m intensifies as pregnancy progresses because the nutritional requirements of the fetus steadily increase and because growth of the placenta confers greater power on the conceptus to influence maternal physiology. Early in pregnancy there is a decrease in maternal peripheral resistance (Duvekot *et al.* 1993) and a concomitant decrease in maternal arterial pressure (Redman 1989). The conflict hypothesis interprets these changes as a maternal adaptation to reduce the fetal share of cardiac output because increased blood pressure is associated with increased blood flow to the placenta (for given R_p). Maternal blood pressure reaches its nadir in the second trimester before a progressive rise towards term (Redman 1989). The conflict hypothesis interprets this increase in maternal blood pressure during the third trimester as a fetal adaptation to direct extra maternal blood to the placenta.

For a subset of pregnant women, the pregnancy-associated rise in blood pressure is sufficient for a diagnosis of pregnancy-induced hypertension. A subset of this subset develop a far more serious condition known as pre-eclampsia in which high blood pressure is associated with widespread endothelial damage to maternal vessels. Haig (1993) interpreted pre-eclampsia as the outcome of an adaptation of poorly nourished fetuses to increase utero-placental blood flow by causing vasoconstriction of the non-placental circulation. This hypothesis has a strong and a weak version. In the strong version, pre-eclampsia itself has been associated during human evolution with benefits to fetuses relative to otherwise-equivalent fetuses that did not cause pre-eclampsia. In the weak version, pre-eclampsia is the occasional non-adaptive byproduct of a fetal adaptation that operates in non-pre-eclamptic pregnancies. If the symptoms of pre-eclampsia are caused by an evolved fetal response to inadequate nutrition, identical symptoms could be precipitated by multiple causes (anoxia at high altitude, competition with a twin, defects of the placenta, etc.), just as a raised heart rate can have many different antecedents.

GROWTH

The total physiological cost of pregnancy is probably larger for human mothers than for our closest living relatives. Allometric comparisons among primates reveal that human pregnancies are prolonged and human neonates are large relative to maternal body size (Huggett and Widdas 1951; Leutenegger 1974). When humans and chimpanzees are directly compared, maternal body weights are similar but human gestation is about 5 weeks longer (Gavan 1953) with humans growing faster than chimpanzees after 27 weeks gestation (Schultz 1940). As a result, birth weights average about 3.5 kg for humans and 2.0 kg for chimpanzees (offspring of well-nourished mothers: Smith *et al.* 1975; Arbuckle *et al.* 1992). By contrast, postnatal growth is substantially slower for humans than for chimpanzees (Schultz 1936). The physiological cost of human pregnancy is exacerbated by the exceptional adiposity of human babies. The next section discusses the development of human adipose tissue. The subsequent two sections discuss reasons for our slow postnatal maturation.

Fat

Human fetuses deposit large amounts of fat during the final weeks of pregnancy. The 'reference fetus' of Ziegler *et al.* (1976) is only 0.1 per cent lipid by weight at 24 weeks but 11.2 per cent lipid at 40 weeks. Other studies report somewhat higher levels of fat at term. Six human neonates varied from 11 to 28 per cent fat (average 16 per cent; Widdowson 1950), and the 'reference boy' and 'reference girl' of Fomon *et al.* (1982) are born with 13.7 and 14.9 per cent fat, respectively. Over half of the energetic cost of the later stages of human pregnancy may be expended in the storage of fetal fat (Southgate and Hey 1976). Ziegler *et al.*'s 'reference fetus' adds 5.4 g of fat per day from 36 to 40 weeks gestation, whereas Widdowson (1980) calculated that fetuses deposit 8.8 g of fat per day over the same 4-week period. The fat percentage of human infants continues to increase after birth, reaching 25 per cent at 6 months before declining (Fomon *et al.* 1982; de Bruin *et al.* 1996).

Most fat is laid down after the fetus is sufficiently mature to have a chance of survival if it is delivered early. Survival in the event of premature delivery thus appears to be the fetus's first priority, but, after this priority is met, a major proportion of the fetus's nutrient intake is stockpiled as fat rather than lean tissue. As might be expected, fat is the most variable component of birth weight. Fat averaged 14 per cent of birth weight but accounted for 46 per cent of the variance in a study of middle-class American mothers (Catalano *et al.* 1992) and for 70 per cent of the reduction in birth weight of babies whose mothers maintained intense physical exercise throughout pregnancy (Clapp and Capeless 1990). Other evidence is suggestive: birth length decreased less than birth weight for babies exposed during third trimester to the Dutch famine of 1944–1945 (Stein *et al.* 1975), and babies of poor Indian mothers are leaner (11 per cent fat) than babies of Western mothers (Apte and Iyengar 1972).

Comparative data are limited but suggest human babies are fatter than neonates of most other mammals. Widdowson (1950) reported 1–2 per cent fat in newborn pigs, cats, rabbits, mice, and rats. Only neonatal guinea pigs, with 10 per cent fat, approached the 16 per cent of newborn humans. Alexander (1962) reported 2–3 per cent fat in newborn lambs, depending on the level of maternal nutrition. Grey seals and Antarctic fur seals are 6 per cent fat at birth (Iverson *et al.* 1993; Arnould *et al.* 1996), whereas hooded seals—which are weaned after 4 days—are 14 per cent fat at birth (Oftedal *et al.* 1993). Anecdotal remarks suggest that non-human primates are lean at birth (Schultz 1969: 152), but I have found little quantitative data. Neonatal fat has been reported for two New World monkeys: 3 per cent in squirrel monkeys (Russo *et al.* 1980) and 5 per cent in cebid monkeys (mature stillbirths: Ausman *et al.* 1982).

Human adults are also remarkably fat by primate standards. Wood Jones (1929) claimed that our layer of subcutaneous fat was a distinctive feature of humans among primates, but more recent studies suggest the distribution of fat in obese non-human primates resembles that of humans (Pond and Mattacks 1987; Pereira and Pond 1995). Subcutaneous fat depots are among the last to be filled, and the prominence of the subcutaneous layer in humans reflects the exceptional adiposity of humans rather than a qualitatively new structure (Pond 1968; Pond and Mattacks 1987). Human infants may be an exception to this generalization because they have well-developed subcutaneous fat depots but little perirenal and omental fat (Knittle 1978). The subcutaneous fat of humans has been suggested to substitute for the insulation provided by hair in other primates, although Pond (1968, 1997) questions the strength of evidence for this popular hypothesis. Morgan (1982) takes the unconventional view that subcutaneous fat was an adaptation for a semi-aquatic phase in our past.

Why should human infants be so much fatter than the infants of other primates? Fat is insurance against future food shortage. There are at least three kinds of reasons why natural selection might favour increased insurance. First, extra insurance would be favoured if the risk of food shortages is higher. For example, some human societies exposed newborn infants to the elements to test their viability. If similar tests of quality were part of our shared evolutionary past, selection to survive these tests could have favoured infants with larger energy reserves. Hrdy (1998) has explored the related hypothesis that chubbiness communicates offspring quality to human mothers. Second, extra insurance would be favoured if the cost of insurance was reduced (cheaper premiums). The depletion of maternal reserves would contribute to the cost of fetal insurance if the mother were thereby less able to provide care in the future. It could be argued that the cost of insurance is reduced for human infants because their mothers have more fat to spare or because alternative care-givers can substitute for a depleted mother. Third, extra insurance would be favoured if food deprivation was associated with higher costs for the uninsured. Human brains continue to grow rapidly after birth and account for more than half of basal metabolism during the first postnatal year (Holliday 1978). If brain development were particularly vulnerable to malnutrition, mobilization of fatty acids could conserve glucose for use by the brain.

Brains and bodies

The allometric relationship between human brain and body size closely follows the pattern of other primates during fetal development (Martin 1990). Thus, among primates, human brain size at birth is not exceptional when compared with neonatal body size. Newborn squirrel monkeys, for example, have almost twice the relative cranial capacity of human neonates (Leutenegger 1974) and, among our closer relatives, newborn gibbons and orang-utans have proportionally larger heads than newborn humans (Schultz 1926). However, human babies do have large heads relative to the size of their mothers because human babies are allometrically large relative to maternal size (Leutenegger 1974). Although its prenatal growth is not unusual, the human brain uniquely maintains fetal growth rates for the first year of postnatal life (Martin 1990). Brain growth then steadily declines, with minimal increases in weight after 6 years (Laird 1967; Dekaban and Sadowsky 1978; Cabana et al. 1993).

The bodies (and brains) of most mammals exhibit a simple decelerating growth curve with an asymptotic approach to final size. The human body shows such a growth curve during its first few years, more or less coincident with the period of brain growth. However, this initial phase is followed by a protracted period of slow growth until puberty, when there is a dramatic acceleration in linear growth (the adolescent growth spurt) before the final asymptotic approach to adult height (Bogin 1988). The idiosyncratic trajectory of human body growth contrasts with a conventional decelerating growth curve for the human brain. The distinctive feature of human brain growth is its unusual postnatal prolongation of fetal growth rates, but in other respects the growth curve resembles growth curves for the brains of other primates.

If a prepubertal child's growth falters because of malnutrition or temporary illness, the child grows at a faster rate than is typical for its age upon recovery until it has regained its former growth trajectory (Tanner 1973). This phenomenon of 'catch-up growth', together with the adolescent growth spurt, suggests that prepubertal children have considerable reserve capacity for faster growth. The implication seems to be that nutrient intake is not *directly* limiting growth and that human children have evolved to delay growth and remain small. A consequence of the conservatism of brain growth relative to body growth 'is that the primate, and particularly man, spends a considerable part of his growth period with a more or less full size brain, but with a body much smaller than that of the adults around him' (Laird 1967).

Intergenerational conflicts

Eshel and Feldman (1991) presented an elegant model of parent–offspring relations that may be relevant to the evolution of the human growth curve. They showed that an action, taken by an offspring, that handicapped the offspring relative to its siblings could be favoured by natural selection if parents respond to the handicap by providing extra care. Their model can be interpreted as a model of parent–offspring conflict in the sense that the parent's fitness is reduced, and the offspring's fitness enhanced, as a result of the offspring's unilateral action. But it is not a model of parent–offspring conflict in the sense that the parent takes an action to maximize its own fitness (given the offspring's *fait accompli*) and this action benefits the offspring.

If hominid mothers possessed a flexible behavioural repertoire that responded adaptively to variation in infant need, this flexibility could have been exploited by offspring that prolonged their period of helplessness. An infant that delayed brain maturation would be helpless relative to earlier-maturing siblings but, if its mother responded with increased attention and prolonged care, the slow-developers might have gained the competitive advantage of a more highly developed brain. A newly arisen gene for prolonged helplessness would thus tend to increase interbirth intervals and reduce maternal fitness but, in what almost seems an injustice, mothers would value their slower-maturing infants (with their better prospects) more highly than their less costly siblings. Models of this kind may explain how hominid infants were able to shift much of the cost of neural development and learning on to other individuals, particularly mothers. Young children

can gain experience about the world in relative safety because they are able to exploit the sensory and muscular systems of their parents to stay out of trouble.

A distinctive feature of the human life cycle is that offspring are weaned and mothers conceive again before older offspring are nutritionally independent. Thus, human infants often grow up with (and share maternal care with) younger or older siblings (Sulloway 1996). This enables human mothers to have a shorter duration of lactation and shorter interbirth intervals than chimpanzee mothers even though each offspring is dependent for a longer period (Bogin 1996). It is generally believed that hominid mothers were able to care for more than one offspring at a time because other members of their social group assisted with child care. Potential helpers include fathers, current sexual partners, grandmothers, and older children, but their relative contributions to allomaternal care is a subject of debate (Hawkes *et al.* 1997).

Genes expressed in mothers will favour shorter interbirth intervals than will genes expressed in offspring. Therefore, the overlap in the care of successive offspring can be considered a maternal adaptation to the increased period of infant dependency caused by delayed maturation. Whether intergenerational conflicts also played a role in the delayed onset of puberty is unclear because this requires knowledge of when children ceased to be a net cost to their parents and how these costs changed with sexual maturity. Parent–offspring conflict would have been present if progressive delays of puberty had costs for the residual reproductive value of parents but benefits for offspring.

If one assumes that the timing of puberty was adaptive in ancestral environments, one can infer that each extra year of experience prior to puberty made a greater expected contribution to a child's fitness than an extra year of reproduction. The fact that children grow at far below their evolutionary potential, and that puberty coincides with a dramatic acceleration of growth, suggests that children gained strong advantages from remaining small while they were non-reproductive because it would have been theoretically possible for puberty to have occurred at the same age but for children to have reached final adult size several years earlier. Three advantages of small size can be hypothesized: small size may have reduced maintenance costs; small size may have minimized dangers of being misperceived as an adult; and small size may have aided in extracting extra parental investment. Bogin (1994) discusses these and other evolutionary questions raised by the extended juvenile period of humans and the adolescent growth spurt.

EVOLUTION AND MEDICINE

Health, fitness, and the pursuit of happiness

Shorter interbirth intervals are associated with increased childhood mortality. Nevertheless, Hobcraft *et al.* (1983) observed: 'For what it is worth, we note that any family trying to achieve maximal numbers of surviving children at any cost would, in the light of these results, continue to bear children at the most rapid rate possible. The dramatic excess mortality is not enough to negate the extra births. However, it is hard to recommend a pattern with such disastrous human consequences.' This quotation illustrates two important distinctions. First, maximizing the fitness of a parent need not maximize the fitness of individual offspring. Second, health and fitness are not synonyms when fitness is understood in its genetic sense. Where there is a conflict between the self-defined interests of human individuals and the interests of their genes, medicine should serve the former. However, what individuals will choose for themselves does not bear any simple relation to health or fitness. Our choices sometimes promote health over fitness and sometimes fitness over health. When a woman chooses to be pregnant, she takes an action that enhances her fitness but has risks for her health. When she uses contraception, her choice may be good for her health but reduces her fitness. These effects are consequences of her choice but need not be the reasons for the choice.

Human diversity

Medicine and evolutionary biology have different approaches to variation. Medicine tends to be normative: some states (health) are better than other states (disease). Evolutionary biology is similarly concerned with the causes and consequences of variation, but particular states are not intrinsically more valuable or desirable than others. Differential reproduction is a consequence of interest but not a measure of value. Despite a common misconception, evolutionary biology is concerned with environmental as well as genetic sources of variation. A central evolutionary question is the extent to which the plastic human response to different environments enhances genetic fitness, but whether a particular response is adaptive or non-adaptive (in the evolutionary sense) says nothing about the desirability of the response. The idea that some variation is 'normal' and some 'abnormal' (or 'natural' and 'unnatural') has no place within evolutionary theory (for a discussion of the many senses of 'normal' used in medicine, see Murphy 1972; Davis and Bradley 1996).

Normative questions are appropriate in some contexts but inappropriate in others; this is not the province of evolutionary biology. If it were convincingly shown that some men have a genetic predisposition to homosexuality (encoded at Xq28 or wherever), the discovery would raise interesting evolutionary questions, but there seems no reason to treat sexual orientation as a medical problem, just as few people would now see left-handedness as a problem needing correction. On the other hand, if it could be shown that variation in growth between human populations is an adaptive response to their different levels of nutrition (Seckler 1980), the answer would also be of evolutionary interest but would not absolve us of asking why some people should have more food than others. I raise these issues because critics often object to the application of evolutionary theory to our own species because they fear that the theory has normative implications, or will be perceived as having such implications.

Evolutionary biology is not going to provide easy answers to medical dilemmas, nor provide a simple guide for intervention, but a dialogue between evolutionary biology and medicine should nevertheless be of benefit to both disciplines. Most immediately, the vast database of medicine provides unparalleled opportunities to test evolutionary theory and suggest new avenues of evolutionary research. I hope that evolutionary biology will be able to repay some of this debt by providing medicine with new insights into old questions.

ACKNOWLEDGEMENTS

This work has benefited from discussions with Nadine Allal, David Barker, Roger Gosden, Sarah Hrdy, Colin Finn, Manfred Milinski, Elaine Morgan, Pierre-Yves Robillard, Steve Stearns, Beverly Strassmann, David Thaler, and Claus Wedekind.

8

HUMAN EVOLUTION AND DISEASE: PUTTING THE STONE AGE IN PERSPECTIVE

Beverly I. Strassmann and Robin I.M. Dunbar

The search for single gene mutations is one of the best funded areas of medical research budgets, yet these mutations may account for a small fraction of all diseases. The true culprits are often wild-type genes that were adaptive in past environments. For example, less than 2 per cent of all breast cancers can be attributed to the BRCA mutations and other susceptibility genes (Peto *et al.* 1996); 98 per cent of malignancies are probably caused by normal genes in the face of novel life-history patterns (Short 1976; Eaton *et al.* 1994).

The change in the environment of the genes has been one of the centrepieces of the new evolutionary medicine (Eaton *et al.* 1988; Williams and Nesse 1991; Nesse and Williams 1994). In particular, it has been argued that human biology is adapted to the hunter–gatherer lifestyle of the 'Stone Age'. From a genetic standpoint, modern humans have been viewed as Upper Palaeolithic preagricultural foragers (Eaton *et al.* 1988). To situate these arguments in time, we note that the Old Stone Age or Palaeolithic is the period from about 2.5 million years ago to 10 000 years ago (Lewin 1993). Anatomically modern *Homo sapiens* first appeared in Africa approximately 130 000 years ago (Foley and Lahr 1992). Hunting and gathering predominated over scavenging by 55 000 years ago (Steiner and Kuhn 1992), and the technological revolution of the Upper Palaeolithic began about 40 000 years ago (Lewin 1993; Jurmain *et al.* 1997). Those who view contemporary humans as 'Stone Agers in the fast lane' (Eaton *et al.* 1988) assume that the 'environment of evolutionary adaptedness' ended 10 000 years ago (Symons 1979; Tooby and Cosmides 1989; Barkow *et al.* 1992).

Ten thousand years is long enough for roughly 400 human generations, implying that we are genetically similar, but not identical, to our ancestors of the Upper Palaeolithic. The forces of evolution (natural selection, gene flow, mutation, and drift) continue to act on human populations and have demonstrably altered allele frequencies since the origin of agriculture (Durham 1991). The best documented examples of natural selection in modern human populations are the evolution of malaria resistance (Neel 1947; Livingstone 1958, 1984; Cavalli-Sforza and Bodmer 1971; Durham 1991) and lactose tolerance (Simoons 1978; Durham 1991), both of which are thought to have originated within the past few thousand years. In other species, natural selection has been observed over just a few generations (Grant 1986), reminding us that it is imprudent to underestimate the importance of micro-evolution in contemporary human populations. None the less, it is unlikely that a large percentage of the 100 000 odd genes in the human genome are more recent than the Stone Age (see Sokal *et al.* 1991; Neel 1994).

If our genes have changed relatively little over the past 10 000 years, does it follow that we have mostly Stone Age genes? The answer must surely be that 2.5 million years is a flash in the pan compared with the nearly 4 billion years of evolution behind *Homo sapiens*. The molecular evidence shows that most of our genes are of far greater antiquity than the Pleistocene. For example, we continue to use the same genetic code as all other species on Earth including bacteria (Ridley 1993). Many of our proteins, such as the histones that package our DNA, are virtually identical to those of other organisms ranging from yeast, to sea urchins, to marigolds, to mice (van Holde 1989). From a genetic standpoint, the Stone Age may have no greater significance than any other period of our evolutionary past (see Foley 1995/1996).

Although there is no genetic reason for emphasizing the Stone Age, perhaps there is an ecological reason. Was the Palaeolithic/ Neolithic boundary and the rise of agriculture the most important transition in the aetiology of modern disease? The view that agriculture was the chief novelty that selection has not caught up with is implicit in much of the work on Darwinian medicine (e.g. Eaton *et al.* 1988, 1994). It seems to derive from the broader supposition that human behaviour is rarely adaptive except among populations that pursue a lifestyle similar to ancestral hunter–gatherers (e.g. Barkow *et al.* 1992). Recently settled foragers under the influence of missionaries are sometimes thought to provide better data on human adaptation than traditional agriculturists who do not use contraception and have practised essentially the same mode of subsistence for centuries, if not millennia (e.g. Eaton *et al.* 1994).

In this chapter we will attempt to put the Stone Age in perspective by demonstrating that the neolithic revolution was just one of many transitions relevant to human disease. Different medical problems have their origins in different transitions, and it is helpful to be judicious in deciding which problems can, in fact, be ascribed to agriculture. Thus, rather than search for Stone Age legacies, we will discuss a variety of disease-provoking transitions over prehistoric and historic time.

BIPEDALISM

Disease-provoking transitions both predate and postdate the palaeolithic. We begin with bipedalism, the transition that distinguishes

hominids from other primates. Bipedalism evolved in the Pliocene over 4 million years ago (White *et al.* 1994; Wolpoff 1995), but the oldest clear-cut evidence is provided by footprints in 3.5-million-year-old volcanic tuff at Laetoli in northern Tanzania (Leakey and Hay 1979). Although bipedalism is of ancient origin, some features of the human skeleton remain poorly adapted to this locomotory style. For example, the lower back and knee joints are plagued by osteological malfunction, with lower back pain among the foremost causes of lost working days. These instabilities reflect the need for joint surfaces to carry weights for which they were never originally designed: the body's full weight is now borne on two legs rather than four. Physical design constraints have apparently made it difficult to modify the joints to the standards required to maintain full anatomical stability without rendering them so large as to impede locomotion. Vertebral arthritis is found even in prehistoric hunter–gatherers, and is not merely a pathology of modern lifestyles. The lower spine bears the most weight and is usually the most affected (Bridges 1994).

During the shift to bipedalism, the long narrow pelvis characteristic of the great apes was shortened and broadened, although the bowl shape characteristic of modern humans was not achieved until around 2 million years ago (Aiello and Dean 1990). This remodelling provided a stable base for support of the trunk and gut during bipedal locomotion. It also narrowed the birth canal through which the baby has to pass. Another major change in hominid evolution was a dramatic increase in brain size. Among the Australopithecines, the encephalization quotient (ratio of the species' actual brain size to that expected for a primate of that species' body mass) was 2.5, whereas in common chimpanzees (*Pan troglodytes*) the ratio is 2.0. The ratio grew to 3.1 in early *Homo* and to 5.8 in modern humans (Lewin 1993).

The passage of a fetus at birth depends on the size of the infant's head relative to the size of the maternal birth canal. Therefore, brain evolution was only possible in the human lineage through alteration of the typical primate trajectory of brain:body growth. In other primates, brain mass increases less rapidly in relation to body mass from birth onwards (Martin 1989). In humans, however, the rapid brain growth characteristic of fetuses continues for about 12 months after birth. Specifically, from birth to the completion of brain growth, the brain increases in size by a factor of nearly 3.5 in humans, compared with a factor of only 2.3 in other primates (Martin 1989). Human brains are thus small at birth relative to adult size. None the less, the head size of human infants at birth is scarcely smaller than adult female pelvic diameter (Smith and Tompkins 1995). This is due not only to a narrow pelvis, but also to the fact that human neonates are heavier than other primates when maternal weight is controlled. In contrast with the situation in humans, the pelvic outlet in apes permits comparatively easy passage of the newborn infant. It is not surprising, therefore, that birth in humans is fraught with more obstetric complications than in other primates (Smith and Tompkins 1995). In summary, the shift to bipedalism coupled with large brain size contributed to medical problems ranging from back ache and knee trouble to obstetric and perinatal complications.

The significance of all perinatal complications relative to other health problems was assessed in the Global Burden of Disease Study. The purpose of this study was to derive a measure, called the disability-adjusted life year (DALY), that permits the severity of different conditions to be compared (Murray and Lopez 1996). DALYs are calculated from: (i) the number of years of life lost due to premature mortality, and (ii) the number of years of life lived with disability (adjusted for the severity of the disability). Thus, they reflect the mortality and morbidity associated with each health problem. In 1990, conditions arising during the perinatal period (low birth weight and birth asphyxia or birth trauma) ranked third among the leading causes of DALYs world-wide (Table 8.1).

Table 8.1 The leading causes of morbidity and mortality as measured by disability-adjusted life years (DALYs)

Disease or injury	1990 Rank	Percentage of total DALYs
Lower respiratory infections	1	8.2
Diarrhoeal diseases	2	7.2
Conditions arising during the perinatal period	3	6.7
Unipolar major depression	4	3.7
Ischaemic heart disease	5	3.4
Cerebrovascular disease	6	2.8
Tuberculosis	7	2.8
Measles	8	2.6
Road traffic accidents	9	2.5
Congenital anomalies	10	2.4
Malaria	11	2.3
Chronic obstructive pulmonary disease	12	2.1
Falls	13	1.9
Iron-deficiency anaemia	14	1.8
Protein–energy malnutrition	15	1.5
War	16	1.5
Self-inflicted injuries	17	1.4
Violence	19	1.3
HIV	28	0.8
Trachea, bronchus, and lung cancers	33	0.6

From Murray and Lopez (1996).

AGRICULTURE

We now take a long step from the origin of bipedalism to the advent of agriculture early in the Holocene. The health implications of this transition were examined at a symposium on palaeopathology at the origins of agriculture (Cohen and Armelagos 1984a). We briefly summarize the conclusions of this symposium, then discuss recent criticisms of these conclusions (Wood et al. 1992). Skeletal evidence suggested that Palaeolithic foragers experienced seasonal and periodic physiological stress, but not the severe or chronic stress found in Neolithic farming populations (Roosevelt 1984). During the Mesolithic, a transitional period of intensive hunting–gathering and incipient agriculture, the proportion of starch in the diet rose, producing various dental diseases, but dietary quality remained high (Cohen and Armelagos 1984b). With the intensification of agriculture in the neolithic, skeletal evidence suggested that malnutrition and disease had become widespread. Compared with the skulls of Palaeolithic foragers, those of Neolithic farmers had a high frequency of tooth enamel hypoplasias indicative of nutritional deficiency or childhood illness (e.g. measles) (Roberts and Manchester 1995). The Neolithic material also displayed a high incidence of porosities of the skull (porotic hyperostosis) and orbits (cribra orbitalia) associated with anaemia. Infectious disease lesions were reported to be much more numerous among farmers than among their hunting–gathering predecessors at the same locales (Cohen and Armelagos 1984b). Ten of 13 palaeodemographic sequences implied a decline in mean age at death with the adoption of agriculture and, in some cases, a downward trend in life expectancy for adults (Cohen and Armelagos 1984; Goodman et al. 1984b).

In their critique, Wood et al. (1992) argue that it is difficult to tell, for any locale, whether health got better or worse with the transition to agriculture. They point out that skeletal lesions (e.g. enamel hypoplasias, porotic hyperostosis, cribra orbitalis) occur only in individuals who were sufficiently healthy to survive the illness that caused the lesions. When the illness is fatal the skeleton is unaffected. Therefore, the higher frequencies of skeletal lesions in neolithic compared with Palaeolithic samples could reflect an enhanced ability to survive episodes of illness and stress. They might also be due to a decline in other, competing causes of death that leave no mark on the skeleton. For these reasons, Wood et al. suggest that improved health may lead to worse skeletons, not better ones as previously supposed. (To explore this idea, it would be helpful to compare the skeletons of elites and commoners in stratified societies.) The decline in mean age at death is also hard to interpret because it may reflect an increase in fertility as well as mortality (Milner et al. 1989; Wood et al. 1992). Until the relationship between the health of a living community and that of the skeletons it buries is better understood, it will be difficult to draw firm conclusions about the epidemiology of the Neolithic.

None the less, if we suppose that the spread of agriculture produced waves of disease and mortality (Cohen and Armelagos 1984), this might seem to support the caricature that there was a qualitative leap from adaptation to maladaptation from the Palaeolithic to the Neolithic (see Barkow et al. 1992). The alternative hypothesis, however, is that agriculture became an increasingly viable mode of subsistence as the returns from foraging diminished (Cohen 1977; Smith et al. 1984). The Pleistocene big game of Europe had vanished or was on the wane, and Palaeolithic population growth had spurred competition for wild food resources. In areas where wild grains could be efficiently cultivated, the shift to agriculture made more calories of food available (Flannery 1973).

In the Near East, agriculture originated about 9000 years ago. By 5000 years before present it had spread to almost all suitable areas of Europe (Ammerman and Cavalli-Sforza 1984; Sokal et al. 1991). The radial rate of advance averaged about 1 km per year (Menozzi et al. 1978). Using genetic evidence, it is possible to determine whether the mechanism of transmission was cultural diffusion, with few genetic consequences, or demic diffusion, defined as the geographical expansion of a population whose size is increasing. The evidence supports the demic diffusion hypothesis. Agriculture spread because the farmers outreproduced the hunter–gatherers (Menozzi *et al.* 1978; Sokal *et al.* 1991). This view is reinforced by palaeodemographic data. Population growth had begun among the foragers, but was greatly accelerated among the farmers (Cohen and Armelagos 1984*b*).

The higher fitness of the agriculturists is at odds with the hypothesis that the transition from foraging to farming was such a dramatic cultural change that it rendered human behaviour maladaptive (e.g. Barkow *et al.* 1992). The mere fact that humans passed more generations as hunter–gatherers than as farmers does not establish the significance of the transition. The advent of agriculture was not the first change in mode of subsistence over human evolution, nor was it even the first change in which culture played an important role. It would be just as persuasive to argue that the behaviour of Upper Palaeolithic hunter–gatherers was out of joint with its selective background because most of the ancestors of these foragers were vegetarians. Technological change during the Upper Palaeolithic, as documented by the explosion in artefact types, was already monumental (e.g. Butzer 1977; Jurmain *et al.* 1997).

We do not dispute the lag periods that inevitably separate environmental change from adaptation, nor do we dispute the argument that when environmental change is particularly rapid, maladaptation will tend to be more pronounced. We do, however, question the tendency to view the Palaeolithic/Neolithic boundary as a definitive cut-off that separates the human lineage into: (i) inclusive fitness-maximizing hunter–gatherers, and (ii) maladapted cultural beings. A growing body of field and historical studies suggests that reproductive striving continues to translate into reproductive success in traditional, kin-based societies that have not undergone the demographic transition of the past century. The mode of subsistence is variable across these societies and does not appear to be a crucial parameter with respect to the adaptiveness of behaviour (see Alexander 1979, 1987; Irons 1979; Dickemann 1981; Hartung 1982; Borgerhoff-Mulder 1987, 1991; Chagnon 1988; Cronk 1991; Dunbar 1991; Smith 1992*a*, *b*; Dunbar *et al.* 1995; Betzig 1997).

Although the Palaeolithic/Neolithic boundary was probably not a watershed between adaptation and maladaptation, it did change the kinds of diseases affecting human population. Palaeolithic hunter–gatherers fought a limited range of infectious diseases because their small population sizes afforded limited opportunities for infection to pass from host to host (Cohen 1989). Among the most common afflictions of hunter–gatherer bands were zoonoses and chronic infections that survive for a long time in a single host and can therefore be reliably transmitted. Examples include bacterial infections (e.g. staphylococcus, streptococcus), various intestinal protozoans such as amoebas, and possibly the herpes virus (Cohen 1989). Many of the epidemics of recent history are caused by pathogens that survive only in large populations in which new hosts are continuously produced through birth and immigration. Measles, which may have originated as a canine or bovine virus, is a good example (Cohen 1989). The measles virus will die out at any one locale unless a fresh supply of new victims arrives as quickly as the old hosts are used up. Black (1975, 1980) has documented the disappearance of the measles virus from isolated island populations and makes a convincing case that it could not persist before the emergence of large population centres.

Protection from infection in small, isolated groups is also well illustrated by the plague epidemic in France in 1720–1722. Eighty-eight per cent of villages with fewer than 100 people were spared, but all cities with more than 10000 inhabitants were severely afflicted (Biraben 1968). The fates of towns of intermediate size were closely proportional to population size.

Although the agricultural revolution brought an upsurge in population, the increase in numbers was small compared with the acceleration in population growth after the Industrial Revolution (Roberts and Manchester 1995). Limited population growth occurred even among prehistoric hunter–gatherers and may have been one impetus for the adoption of agriculture (Cohen and Armelagos 1984*b*; Roosevelt 1984). Thus, although population growth has been one of the most important factors impinging on human health, it has occurred at many times and places and was not unique to the Neolithic.

In addition to spurring population growth, the neolithic revolution caused people to settle down in villages. This change probably resulted in local accumulation of human wastes and the proliferation of diseases transmitted through the faecal–oral route. Although plumbing and sewerage systems now help substantially where available, diarrhoeal diseases continue to be a major public health problem. The Global Burden of Disease Study ranked diarrhoeal disease as the second leading cause of morbidity and mortality in 1990 (Murray and Lopez 1996).

Compared with foragers, sedentary populations tend to build substantial houses and spend more hours indoors in enclosed air spaces. The result is increased exposure to the many pathogens that do not survive in sunlight, such as the influenza virus. Permanent dwellings are prone to infestation by vermin such as rats and fleas (Cohen 1989; Dyer 1989). Urban yellow fever is hosted by a mosquito that breeds almost exclusively in water stored in human habitations (Johnson 1975).

Sedentism also increases the transmission of schistosomiasis, malaria, and other diseases that must somehow be passed from host to host despite dilution in water or air. After being excreted into fresh water by a human host, the fluke that causes schistosomiasis passes a developmental stage in a freshwater snail before it can again be transmitted. If the human population has moved away from the area, transmission is blocked. But if the population has stayed put, transmission is more likely (Cohen 1989). Malaria was favoured by sedentism because an infectious mosquito must bite two or more people within a short time-span for transmission to occur (Livingstone 1958, 1984).

Although sedentism promoted some diseases, such as malaria, studies of modern populations during the transition to sedentism usually report a reduction in mortality. For example, among the Ache of Paraguay, mortality rates on reservations over the past 15 years are demonstrably much lower than those among nomadic forest Ache prior to contact (Hill and Hurtado 1996). Roth (1985) reported lower mortality in sedentary than in nomadic groups in six populations, with no difference in two other populations. The health advantages of sedentism include enhanced ability to care for the diseased and infirm, and opportunities to stockpile surplus food in granaries. Sedentism also enables people to develop some immunity to the parasites of a given region, at least until new parasites are introduced from outside. Even today, traveller's diarrhoea is the unpleasant result of encounters with new pathogens for which prior immunity has not been established.

Subsequent to the agricultural revolution, novel means of transportation (shipping, railroads, and air travel) have been major forces in the spread of infectious diseases (Cohen 1989). Although the waves of disease have been numerous, three major transitions stand out: (i) the linking of China, India, and the Mediterranean by land and sea; (ii) the spread of the Mongol empire in the thirteenth century; and (iii) the advent of European exploration in the fifteenth century (McNeil 1976, 1980). The Old World epidemics of measles, malaria, and smallpox were probably introduced into the Americas around the time of Columbus (Merbs 1992). Bubonic plague had a limited geographical distribution in Asia until it was spread by human transportation (Davis *et al.* 1975; McNeil 1980). In the twentieth century, the rapid intercontinental spread of AIDS can be attributed to air travel.

MODERN AFFLUENT DIETS

Population growth, sedentism, and mobility have challenged human health through the spread of infectious disease. Other changes have led to epidemics of chronic disease, including ischaemic heart disease, the killer ranked fifth in the Global Burden of Disease Study (Table 8.1). The chronic diseases are often called 'diseases of affluence' because they were uncommon among both hunter–gatherers and peasants. Diseases of affluence owe their origin and frequency to lack of exercise as well as to rich and unbalanced diets, especially diets that are low in fibre and high in fats, sweets, and red meat (Widdowson 1991). For example, studies of Australian Aboriginals have shown that populations following ancestral lifestyles are significantly less prone to chronic conditions such as obesity, hypertension, diabetes mellitus, coronary heart disease, and insulin resistance (O'Dea 1991). In contrast, westernized Aboriginals living settled urban lives often exhibit these lifestyle diseases at elevated frequencies compared with their European counterparts. Even a temporary reversion to a traditional lifestyle can dramatically reverse these effects (O'Dea 1984).

O'Dea (1991) has commented on the 'feast-and-famine' dietary patterns characteristic of hunter–gatherer Aboriginals, especially with respect to fatty foods. It is not unheard of for an Aboriginal to consume 2 kg of meat at a sitting, and such meat is likely to be taken from the fattest parts of the carcass. However, most of the wild-caught animals eaten by Aboriginals are leaner than domestic stock (especially those from intensive farming systems), and their fat contains significantly more polyunsaturated fats (principally those of the n-3 series) than that of modern farm breeds. When body fat reserves have been depleted by seasonal food stress, selection of fat-rich animals is actually healthier than a diet of lean meat, in part because lean meat requires more energy to digest (Speth 1987). Well-fed urbanized individuals who pursue a high fat diet are prone to diabetes and other chronic diseases (O'Dea 1991). The attractiveness of sweets and fats is particularly dangerous when combined with an equally natural inclination to minimize energy expenditure (an adaptive strategy when food is scarce). These two factors together are largely responsible for the rapidity with which Aboriginals succumb to diseases of affluence when they settle in urban environments.

An additional feature of the diets of hunter–gatherers is their sheer breadth (Flannery 1973). Hunter–gatherers commonly eat in excess of 100 different plant species, whereas people in most traditional agricultural societies typically only eat about 10 to 15. This in itself has been responsible for a wide range of dietary deficiencies and their associated medical conditions, including kwashiorkor, xerophthalmia, and conditions associated with low body mass (Widdowson 1991). Eaton *et al.* (1996) (see also Chapter 22) demonstrate numerous differences between putative Pleistocene hunter–gatherer diets (based on the diets of modern foragers) and modern affluent diets. For example, the ancestral diet was probably significantly lower in the fatty acids that raise serum cholesterol, but higher in protein from lean meat animals. It was also lower in sodium and higher in fibre. In particular, modern Americans may consume only one-fifth as much fibre as foraging peoples, and perhaps only one-tenth as much as free-living chimpanzees (Eaton *et al.* 1996).

Traditional agricultural populations lack the dietary breadth of hunter–gatherers but continue to eat a high fibre diet. They suffer a concomitantly lower incidence of appendicitis, diverticulosis coli, gallbladder disease, and tumours of the colon (Burkitt *et al.* 1972), all of which were uncommon in England until the end of the nineteenth century or later and may be associated with improved milling techniques. Central African populations eat a high fibre diet and remain relatively free of these digestive system diseases; stool weight in these populations is up to four times that of the British. Fibre acts to increase the rate of transit in the large intestine, and Burkitt *et al.* (1972) suggested that this helps to reduce the intestinal muscosa's exposure to passing oncogens. The highly processed diets of contemporary industrial societies allow more complete extraction of nutrients, thereby exacerbating problems of obesity, and reduce the rate of faecal passage, thereby increasing exposure to oncogens.

THE DEMOGRAPHIC TRANSITION

Eaton *et al.* (1994) have shown that the reproductive experiences of women in affluent Western nations differ greatly from those of contemporary forager women, whom they refer to as the best available surrogates for discerning reproductive patterns prior to the origin of agriculture. Compared with women in eight forager populations, they show that American women experience early menarche (12.8 versus 16.1 years), late first birth (26.0 versus 19.5 years), nursing of short intensity and duration (3 months versus 30 months), and low final parity (1.8 children versus 5.0). Epidemiological evidence has linked each of these changes to increased risk for reproductive cancers (Apter and Vihko 1982; Brinton *et al.* 1988; Layde *et al.* 1989). The primary mechanism underlying these changes is an increase in the proportion of the lifespan spent in menstrual cycling (Short 1976; Henderson *et al.* 1985, 1993). During menstrual cycling, the ovarian hormones stimulate cell division in breast and uterine epithelium, enhancing the risk for the DNA-copying errors responsible for neoplastic transformation (Henderson *et al.* 1993; Spicer and Pike 1993).

Eaton *et al.* conclude: '... the reproductive risk factors for women's cancers suggested by epidemiological investigations are largely, though not exclusively, a recapitulation of reproductive differences between women in affluent 20th century nations and women living before the advent of agriculture'. Thus, Eaton *et al.* imply that cultural changes dating all the way back to the origin of agriculture are relevant to reproductive cancers.

However, the reproductive patterns that Eaton *et al.* report for foragers continue to be widespread among traditional, agricultural populations in the Third World (Campbell and Wood 1989). For example, among the Dogon, who are millet farmers in Mali, West Africa, the median age at menarche is 16 years, first birth occurs at about age 19, postpartum amenorrhoea lasts a median of 20 months, and the total fertility rate is 8.6 live births (Strassmann 1992, 1997). Menstruation was monitored among Dogon women via a census of women's visits to menstrual huts and the results were corroborated by hormonal data from urine collections (Strassmann 1996*a*). In the study population, the median number of menses per lifetime was estimated at 109 (Strassmann 1997). This value is much lower than the approximately 400 menses experienced by Western women, but is not strikingly different from the estimate of 160 ovulations per lifetime among foragers (Eaton *et al.* 1994). The apparent difference may reflect the fact that the estimate for foragers was primarily based on information on nursing and interbirth interval length, rather than on data on menstruation or ovulation.

Both traditional agriculturists and contemporary foragers lack the reproductive parameters (early menarche, late first birth, bottle-feeding, and low parity) that are risk factors for reproductive cancers. They practise natural, rather than target, fertility, which means that reproduction is not controlled in a parity-dependent fashion (Henry 1961). The implication is that the demographic transition, not the origin of agriculture, played a major role in the aetiology of women's reproductive cancers. In industrialized nations, this transition occurred as recently as the past century. Adiposity may be another risk factor (Ziegler *et al.* 1996), but the evidence is less conclusive: for adiposity, we can thank the modern affluent diet.

AIR POLLUTION

Indoor air pollution may be as ancient as the control of fire by *Homo erectus* but continues to be a major cause of lower respiratory tract disease in developing countries. In the Global Burden of Disease Study, lower respiratory infections were the number one cause of morbidity and mortality (Table 8.1). Young infants who inhale wood smoke from cooking fires while being carried on their mothers' backs are at particularly high risk (Pio *et al.* 1985; Collings *et al.* 1990; Morris *et al.* 1990; Brauer *et al.* 1996). Outdoor air pollution is an environmental novelty that dates at least as far back as the prohibition on burning soft coal in England in 1273. Other early evidence includes a treatise on the problem of smoke published in 1684 (Shprentz 1996). At least as far back as the thirteenth century, English royalty escaped polluted air by extended visits to the country-side. The problem of low air quality is worth examining in some detail because it has become so pervasive.

The Harvard Six City Study (Dockery et al. 1993) is one of the most comprehensive recent studies of the health effects of air pollution. The researchers established air quality moni-toring stations in six American cities and fol-lowed the health of 8000 adults over a 16-year period. Confounding factors such as age, sex, smoking, education, occupational exposures, and body-mass index were statistically con-trolled. After adjusting for these variables, the risk of mortality associated with exposure to fine particles was approximately 26 per cent higher for the residents of the most polluted city—Steubenville, Ohio—compared with the cleanest city—Portage, Wisconsin (odds ratio = 1.26; 95% confidence interval = 1.08–1.47). The researchers concluded that this increase in relative risk shortens life expectancy by 1 to 2 years for the Ohio residents compared with the Wisconsin residents. Many of the premature deaths due to air pollution appear to be caused by aggravation of cardiopulmonary disease (e.g. hypertension, ischaemic heart disease, atherosclerosis, bronchitis, and pneumonia). The risk of death from cardiopulmonary dis-ease was approximately 37 per cent higher (odds ratio = 1.37; 95% confidence interval = 1.11–1.68) for the residents of the most pol-luted city compared with the least polluted city.

None the less, the risk of death from air pol-lution was not as high as the risk from smoking. After adjustment for confounding factors, the relative risk of mortality per annum among smokers was 2.0 compared with non-smokers. The risk of death from lung cancer was 8.4 times greater among smokers than non-smok-ers; the risk from cardiopulmonary disease was 2.3 times greater among smokers.

A study by the American Cancer Society (ACS) reached similar conclusions (Pope et al. 1995). This study linked monitoring data on fine particles in 50 American cities with the health of nearly 300 000 people who were prospectively studied from 1982 to 1989. Resi-dents of the most polluted city had an approxi-mately 17 per cent greater risk of mortality (odds ratio = 1.17; 95% confidence interval = 1.09–1.26) than residents of the cleanest city. The ACS study, like the Harvard Six Cities Study, concluded that modest air pollution exposures are shortening the lives of Ameri-cans by 1 to 2 years in the most polluted areas. The Natural Resources Defense Council (NRDC) (Shprentz 1996) extrapolated the more conservative findings of the ACS study to 239 American cities for which data on air qual-ity were available. Assuming a linear relation-ship between particle concentration levels and risk of premature mortality, the NRDC esti-mates that 64 000 (range: 38 000–88 000) Ameri-cans die prematurely each year on account of particulate air pollution in the 239 urban areas.

The magnitude of the air pollution problem becomes evident when deaths due to air pollu-tion are compared with deaths from other causes. In the 239 urban areas in 1989, a total of 29 000 deaths were caused by automobile accidents, less than half the mortality associ-ated with air pollution (Shprentz 1996); 19 000 deaths were attributed to homicide. Deaths from breast cancer and AIDS for the 239 cities are not available, but nation-wide there were 16 000 deaths from AIDS and 43 000 deaths from breast cancer. Thus, if the NRDC is cor-rect, more deaths are caused by particulate air pollution than breast cancer and AIDS com-bined. Mortality attributable to air pollution pales only in comparison with mortality from smoking: 419 000 American deaths in 1990 (Shprentz 1996).

LOSS OF KIN SUPPORT

In the Global Burden of Disease Study, unipolar major depression ranked fourth among health problems in 1990 (Table 8.1) and was projected to rise to second place by 2020. The projected increase does not reflect age-specific changes in depression, but rather an increase in the proportion of young adults in the population (Murray and Lopez 1996). Few studies have tackled the causes of this global epidemic of depression, but a contributing factor may be the breakdown of kin support networks and the attendant loss of psychological and material security. In traditional subsistence economies, the introduction of cash crops has been the major cause of the dissolution of extended families. Examples include coconut trees in Fiji and onion gardens in West Africa, both of which provoked competition among relatives and reduced the number of kin who work together in collaborative units. In the West, the breakdown of kin support networks can probably be traced to urbanization, occupational specialization, and mobility subsequent to the Industrial Revolution.

Evidence that kin support protects health and survivorship derives from several studies. One of the most powerful is a recent study of helping behaviour in the Israeli city of Haifa (Shavit *et al.* 1994). The study took place during the 1991 Gulf War, when Iraq launched 40 Scud missiles against coastal Israeli cities. The investigators were interested in how Israelis used their social networks during this mortal threat. The study design included two surveys of a stratified random sample of residents of Haifa, the first during the Gulf War and the other a year later. Respondents to both surveys indicated that they were more solicitous of kin than non-kin. For example, they were more likely to provide wartime shelter to kin than to friends. During and after missile attacks, they preferentially made phone calls to check up on kin to find out if assistance was needed. Less costly forms of aid, such as advice (e.g. how to seal a room against gas), were more likely to be exchanged between friends than kin (Shavit *et al.* 1994). Because the researchers were unaware of inclusive fitness theory (Hamilton 1964), their findings were not prejudiced by theoretical expectations.

The results of the Israeli study echo the finding that Americans tend to rely on non-kin for companionship but borrow money from kin (Fischer 1982). They also corroborate a growing body of evidence that help is given more freely to kin without corresponding demands, whereas with non-kin strict reciprocation is the rule (e.g. Dunbar *et al.* 1995). Although the Israeli study revealed the potential importance of kin support, a historical study of Plymouth Colony provided tentative evidence that kin support may actually translate into higher survivorship (McCullough and York Barton 1990). After the arrival of the Mayflower at Plymouth Rock in December, 1620, 52 per cent of the population died in the first winter from disease, malnutrition, and lack of preparedness. However, individuals with more relatives in the population had lower mortality. Similarly, an epidemiological study in Newcastle, England, demonstrated that kin support can reduce morbidity and mortality under chronic poverty (Spence 1954). On the Caribbean island of Dominica, children raised in household environments in which kin support was lacking were at risk for abnormal cortisol levels and a high frequency of illness (Flinn and England 1995).

These observations have important implications for the health and well-being of individuals living in modern large-scale polities. Although many individuals work hard to maintain their extended kinship networks (Dunbar and Spoor 1995), the physical distances that often separate family members make it difficult to do so. Where kin networks are not readily available to provide support, we should expect morbidity to be higher. Even the social support offered by a pet has been reported to improve survivorship in nursing homes. Measures that help to re-establish kinship networks, and their surrogates, may contribute positively to public health.

SUMMARY

In evolutionary medicine, environmental novelty has often been linked to the end of the Pleistocene 10 000 years ago. Modern health problems have been viewed as the outcome of differences between current lifestyles and the behaviour of ancestral hunter–gatherers. The implicit assumption is that unhealthy lifestyles originated with agriculture. This view is vulnerable in two respects. First, the palaeodemographic data do not show, unequivocally, that health and survivorship actually deteriorated from the Palaeolithic to the Neolithic (Wood *et al.* 1992). Second, the transition to agriculture is just one of many candidates for disease-inducing transitions in the human past. In this chapter we emphasized both the shift to bipedalism, which goes all the way back to the Pliocene, and more recent changes, such as the demographic transition, air pollution, and the loss of kin networks.

Changes that promoted infectious disease include population growth and the expansion of transportation and trade networks. Although both occurred during the neolithic, the salient episodes of growth and expansion are considerably more recent. Epidemics of most chronic degenerative diseases (e.g. osteoporosis, diabetes, coronary heart disease, reproductive and digestive cancers) are associated with affluence and continue to be relatively uncommon in contemporary agrarian societies. Short life expectancy is the main difference: people die of other causes before the diseases of senescence set in. The lower incidence of degenerative disease in traditional agrarian societies is also due to lifestyle differences such as ample exercise, diets low in fat and high in fibre, and infrequent ovarian cycling (see Burkitt *et al.* 1972; Gray *et al.* 1979; Panterbrick 1993; Strassmann 1997). The problem of dental caries is found in both affluent and agrarian societies and is one of the few chronic health conditions that is firmly rooted in the transition to agriculture (Roosevelt 1984).

The Global Burden of Disease Study (Murray and Lopez 1996) permits comparison of the relative importance of different health problems. The five leading causes of death and disability in 1990, as estimated in this study, were as follows: (i) infection of the lower respiratory tract, (ii) diarrhoeal disease, (iii) perinatal problems, (iv) unipolar major depression, and (v) ischaemic heart disease. These afflictions have multiple causes that are considered in other chapters of this volume. Here we focused specifically on the role of environmental novelty. Rather than equate environmental novelty to the Holocene (the past 10 000 years), we tried to clarify some of the novel circumstances relevant to each health problem. This approach draws attention to immediate and correctable concerns such as smoking, air pollution, and traffic accidents.

The alternative view is that human biology is adapted to a specific period of our evolutionary past, the so-called 'environment of evolutionary adaptedness'. This period is usually matched to the upper palaeolithic (40 000 to 10 000 years ago) or to the entire Pleistocene (1.8 million years ago to 10 000 years ago). The major weakness of this concept is that it ignores the fact that human evolution has been mosaic in form: different components of our biology evolved at different stages. Moreover, it discards human evolution before and after two arbitrary cut-off points, the second of which corresponds to the rise of agriculture. Our analysis of the transition to agriculture uncovered no empirical evidence that it was a watershed between adaptation and maladaptation. If we were forced to make a simple dichotomy of this sort, the demographic transition of the past century would be a stronger candidate.

9

PATHOGEN-DRIVEN SEXUAL SELECTION AND THE EVOLUTION OF HEALTH

Claus Wedekind

Conflicts are intrinsic to sexual reproduction. Normally, males and females differ greatly in their parental investment (Trivers 1972), and males usually have a much higher potential rate of reproduction than females. Both cause conflicts between and within the sexes (Clutton-Brock and Parker 1992). These conflicts give rise to a new and important kind of selection. If they are to propagate their genes, individuals must not only survive the threats posed by harsh climates, predators, pathogens, and competitors, they must also find a mate (intersexual selection) and withstand competition from rivals of the same sex for access to mates (intrasexual selection) (Andersson 1994).

Sexual selection raises the costs of several aspects of reproduction, including the cost of resources invested in being attractive (e.g. the peacock's tail) and the risks of injury, death, or distraction by predator surveillance during intrasexual struggles and courtship. The transmission and growth of many pathogens are also increased during the courtship of their host, either directly by sexual behaviour itself (transmission of ectoparasites or sexually transmitted microparasites) or indirectly by a reduction of host immunocompetence (e.g. Grossman 1985; Folstad and Karter 1992) that may result from adaptive resource reallocation during courtship (Wedekind and Folstad 1994), mating, and reproduction (Gustaffson *et al.* 1995; Oppliger *et al.* 1996).

In this chapter I develop an evolutionary view of the implications of sex and sexual selection for the impairment of health. First I outline the evolutionary conflict between pathogens and their hosts. Then I list some

arguments why sex itself, and sexual selection in particular, could strongly influence the coevolutionary dynamics of parasite–host systems and the health of hosts. I distinguish several forms of parasite-driven sexual selection, some currently not well supported by evidence, to stimulate research.

EVOLUTIONARY CONFLICTS BETWEEN HOSTS AND PATHOGENS

The most important aspect of the pathogen–host interaction is the harm the pathogen causes its host (virulence). However, virulence is not a characteristic of pathogens but the result of the interaction between pathogen and host. Both parties normally suffer from the effects of virulence (except when a specific aspect of virulence enhances pathogen transmission). Damage to the host is obvious, but damage to the pathogen requires some explanation (see Frank 1996 or Chapter 14 for details).

Pathogens evolve to maximize their fitness —as do hosts. What counts to the pathogen is therefore the transmission of its own genes into the future, and it has two ways to do so: its own survival and its own reproduction. Survival in a host will normally involve countering the host's immune defence, which in itself may damage the host. To reproduce the pathogen must extract resources from its host—the more resources extracted, the greater the damage to the host. Furthermore, transmission of offspring is often improved if the host is damaged to a certain degree (Ewald 1994).

From the perspective of the pathogen, an optimal level of virulence may, but need not, exist. For example, non-linear or non-continuous relationships between factors important for the evolution of virulence could exclude the possibility of optima (Frank 1996), and competition between individuals or clones in multiple infections can shift an existing optimum towards greater virulence depending on the genetic relatedness of competitors within the host (Nowak and May 1994, 1995; Frank 1996).

For simplicity, let us assume that there is an optimal level of virulence from the viewpoint of the pathogen—nothing is driving the pathogens to evolve ever increasing virulence. (The following argument does not depend on this simplification, but the simplification clarifies the arguments.) The parasite's optimum must be above zero, while the optimal virulence for the host is, of course, zero. This creates a conflict that drives the coevolving system of pathogens and hosts towards an observed level of virulence somewhere between the optima for the pathogens and the hosts. This level of virulence, determined by several factors, may vary in time. In the following I discuss some selective mechanisms that have the potential to increase or decrease virulence.

SEX, SEXUAL SELECTION, AND LEVELS OF VIRULENCE: THE HOSTS' VIEW

What factors keep the level of virulence below the pathogen's optimum?

First, the ability of the host to reproduce sexually plays a significant role (Ebert and Hamilton 1996). This follows from the Red Queen hypothesis (Jaenike 1978; Hamilton *et al.* 1990; Howard and Lively 1994), which proposes that pathogen–host coevolution plays a crucial role in the maintenance of sexual reproduction itself (Michod and Levin 1987; Stearns 1987*a*). Sexual reproduction (outcrossing) results in novel rearrangements of host

genes in the host offspring. Pathogen populations that have adapted to one host genotype have to readapt to new offspring genotypes. Hence, host sexual reproduction reduces the adaptation of pathogens to their hosts. In most cases this means that the pathogens are less virulent than would be expected if they could optimize their fitness (Ebert and Hamilton 1996). Point 1 in Fig. 9.1A shows this reduction.

This chapter discusses other factors, consequences of active mate choice, that may further reduce the level of virulence in locally adapted host–pathogen systems. To relate mate choice to the evolution of virulence, we must know what criteria determine mate choice and identify their relative importance.

The literature usually groups such criteria into three classes (e.g. Andersson 1994): (i) criteria that result in direct benefits, such as good parental care, nuptial gifts; (ii) so-called 'Fisher-traits', that is, criteria that are attractive to members of the other sex and that do not reveal anything apart from that (Fisher 1930; Lande 1981); and (iii) criteria that reveal 'good genes' (Zahavi 1975; Hamilton and Zuk 1982; Grafen 1990; Wedekind 1994; Johnstone 1995). 'Good genes' are of special interest because some of them are advantageous in the coevolution between pathogens and their hosts (Hamilton and Zuk 1982). Mate choice for good genes may therefore contribute to determining the level of virulence in natural pathogen–host interactions.

In their original hypothesis, Hamilton and Zuk (1982) suggested that individuals in good health and vigour are preferred in mate choice because they probably have heritable resistance to the predominant pathogens. Preferring healthy mates should yield more resistant progeny and result in populations of hosts less susceptible to the local pathogens and in a lower average level of virulence (point 2 in Fig. 9.1A) (see also Grahn *et al.* in press).

However, the mechanisms Hamilton and Zuk (1982) suggested do not appear to be universal, for they lead to populations where all individuals of one sex have the same mate preference, and less attractive individuals would only be taken as mates if the more attrac-

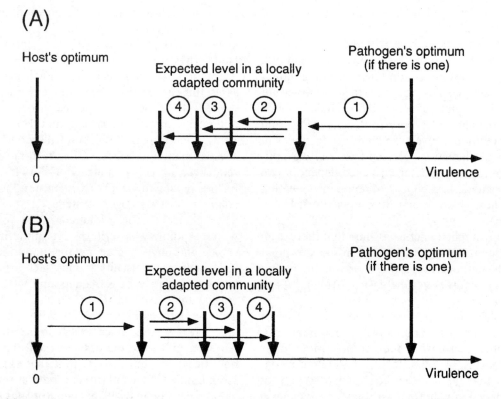

Fig. 9.1 (A) The potential influence of sex and different forms of parasite-driven sexual selection in the host on the observable level of virulence in a locally adapted host–pathogen community. The horizontal arrows indicate the possible influence of : (1) sexual reproduction without any mate choice or with a preference for outcrossing, (2) a preference for healthy and vigorous mates, (3) a preference for mates with complementary resistance genes, and (4) a conditional preference for mates with complementary resistance genes which takes into account the present pathogen populations.

(B) The potential influence of sex and sexual selection in the pathogen on the observable level of virulence in a locally adapted host–pathogen community. The horizontal arrows indicate analogously to the ones in (A) the possible influence of: (1) sexual reproduction without any mate choice or with a preference for outcrossing, (2) a preference for vigorous mates, (3) a preference for mates with complementary virulence genes, and (4) a conditional preference for mates with complementary virulence genes which takes into account the present host genotypes.

tive ones are not available. In some species, this prediction does not appear to be fulfilled (see literature cited below). Furthermore, an offspring's level of resistance depends on both its mother's and its father's genetic contribution. At loci important for the parasite–host interaction (e.g. immunogenes), certain combinations of alleles may be more beneficial than others. If individuals choose mates to produce such beneficial allele combinations, their preferences would have to depend on their own genotypes as well as their partners. As a conse-

quence, individuals with different resistance genes would show different mate (or sperm) preferences, and there would be no universally valid order of sexual attractiveness with respect to signals that reveal heritable disease resistance (or immunogenes).

Preferences for mates or for sperm of genetically dissimilar types have been observed in several species. Olsson *et al.* (1996) found in sand lizards (*Lacerta agilis*), where most females mate with more than one male, that genetically more dissimilar males sire more offspring, both in the field and in the laboratory. They concluded that female reproductive tissue actively selects genetically dissimilar sperm. The ascidian *Diplosoma listeranum* is a colonial, sessile, marine filter-feeder that disperses sperm into surrounding water. Sperm are collected by the feeding apparatus and pass through the oviduct to reach oocytes within the ovary. Autoradiography of labelled sperm revealed that sperm from the same clone were normally stopped in the oviduct while sperm from other clones progressed to the ovary (Bishop 1996). A weak negative correlation was also found between the mating success of pairs and their overall genetic similarity (Bishop *et al.* 1996). While in these species the loci that determine reproductive compatibility are not yet known, they are known in other examples: pollen growth in some plant species is influenced differentially by the stigma tissue —the combination of male and female alleles at the self-incompatibility locus is the decisive factor (e.g. Franklin-Tong and Franklin 1993). Similar observations have been made in the tunicate *Botryllus* in which the chances of fertilization are greater when eggs and sperm carry dissimilar alleles at the fusibility locus (Scofield *et al.* 1982).

The best studied example of mate preferences that depend on the chooser's own genotype occur in the mouse. The most important genes are those in the major histocompatibility complex (MHC) (Klein 1986). Mice base their mate choice to a large extent on odours. These odours reveal some of the allelic specificity of the MHC (Yamazaki *et al.* 1979, 1983), and this information is used by males and females (Yamazaki *et al.* 1976; Egid and Brown 1989; Potts *et al.* 1991) to choose mates according to their own MHC types, apparently to achieve or avoid certain allele combinations in the progeny. Human noses also appear to be capable of discriminating odours of two mouse strains that are congenic except for their MHC (Gilbert *et al.* 1986)—probably because some similarities remain between murine and human MHC antigens. The MHC also correlates with human male and female preferences for human body odours (Wedekind *et al.* 1995;

Wedekind and Füri 1997). Preferences for human body odours correlated with actual mate choice in two independent test series: odours of MHC-dissimilar persons reminded the test subjects more often than expected by chance of their current or former mate (Wedekind *et al.* 1995; Wedekind and Füri 1997). Recently, Ober *et al.* (1997) supported these findings in a study on American Hutterites: the MHC types of 411 couples differed more often from each other than would be expected if matings were random (their calculation of the null expectancies was controlled for non-random mating with respect to colony lineage and kinship).

The MHC may simply serve as a marker of kinship to avoid inbreeding (Potts and Wakeland 1993; Brown and Eklund 1997), or MHC-correlated mate preferences may optimize the offspring's immunogenetics. If they do, host defence against pathogens would be improved and the level of virulence would be further reduced (point 3 in Fig. 9.1A). Further effects on virulence depend on what kind of MHC complementarity is preferred. Do individuals simply prefer other types to ensure a higher proportion of MHC-heterozygous offspring (Brown 1997) because heterozygosity at the MHC is beneficial on average (Doherty and Zinkernagel 1975; Hedrick and Thomson 1983), or does mate choice aim for specific allele combinations that are more beneficial under some environmental conditions than others (Wedekind 1994)?

There is not much evidence that beneficial allele combinations exist for whatever reason. Hirayama *et al.* (1987) found that a schistosome evoked epistatic effects in at least two human MHC antigens. The strong linkage disequilibrium observed between some alleles of the MHC could be explained by long-term epistatic fitness effects (Klein 1986; Maynard Smith 1989). More research is needed on the fitness of homo- or heterozygous combinations of MHC haplotypes under different pathogen pressures.

Mice and humans appear to prefer dissimilar MHC types. If the MHC is not just a marker for kinship, the evidence on MHC-based mate choice has an alternative explanation: a general preference for heterozygous offspring. In a recent study in which male and female students scored the odours of T-shirts worn by other students, Wedekind and Füri (1997) tested whether MHC-correlated odour preferences aim for specific allele combinations or simply heterozygosity. They found evidence for the latter but not for the former, confirming an earlier study (Wedekind *et al.* 1995). This result does not rule out such preferences in populations that are under stronger pathogen pressure than Swiss students.

Choice for complementary alleles would yield the highest fitness return if individuals could choose mates conditionally. If condition-dependent mate selection has evolved in some hosts, it would take into account the present pathogen pressure and would promote allele combinations in offspring that defend against these specific pathogens. Such choice requires mechanisms which have not yet been demonstrated.

A study on mice suggests that mate choice takes into account the male and female MHC genotypes and is conditional in that it depends on at least one external factor that varies over time. In an *in vitro* experiment with two congenic mouse strains, Wedekind *et al.* (1996) tested whether the MHC plays a role after mates have been chosen, that is, whether (i) eggs select for sperm and whether (ii) the second meiotic division of the egg is influenced by the genotype of the fertilizing sperm. They found that neither egg–sperm fusion nor the second meiotic division is random; both processes depend on the MHC of both the egg and the sperm. However, non-random fertilization did not simply select for heterozygous MHC combinations (as had been expected), but eggs appeared to vary from day to day in their preference for homozygous and heterozygous combinations. This heterogeneity between different days of the same experiment was statistically significant and suggested that an external factor that varied over time had influenced the experiment. Wedekind *et al.* (1996) proposed that this external factor was an epidemic of mouse hepatitis virus (MHV) that occurred

during the experiment. The presence of MHV appeared to stimulate a preference for heterozygous combinations; when it was absent, the mice seemed to prefer homozygous variants. Rülicke *et al.* (1989) tested this hypothesis by experimentally infecting some mice with MHV while sham-infecting others. Infected mice produced significantly more MHC-heterozygous embryos than sham-infected mice.

It is not yet known whether MHC-heterozygous offspring of the mice strains resist MHV infection better than homozygous variants, nor is it known whether homozygous offspring survive better in the absence of MHV. If they do, non-random fusion of egg and sperm with regard to their respective MHC and to the presence or absence of MHV would improve the health of the progeny: it would further decrease the observed level of virulence in a locally adapted host–pathogen system (point 4 in Fig. 9.1A).

SEX, SEXUAL SELECTION, AND LEVELS OF VIRULENCE: THE PATHOGENS' VIEW

Sexual reproduction, which results in novel combinations of alleles in offspring, may also help pathogens adapt rapidly to new host genotypes (Read and Viney 1996). Recently, Gemmill *et al.* (1997) found an interesting connection between the degree of sexuality of a parasitic nematode and its hosts' immune system: the nematode *Strongyloides ratti*, a parasite of rats, can have a direct life cycle with clonal reproduction or an indirect life cycle with free-living sexual adults. Gemmill and his colleagues demonstrated that larvae from hosts that have acquired immune protection are more likely to develop into sexual adults than larvae from hosts that have an experimentally compromised immune status. This suggests that the hosts' immune response selects either for sex, for departure from immunologically competent hosts to a free-living stage, or for both in this nematode.

Sex in the pathogen population, followed by selection, can increase virulence (point 1 in

Fig. 9.1B); sex in the host can decrease it. As in the host, non-random mating that improves the virulence genetics of the next generation of a pathogen may benefit the parasite more than random mating or a simple preference for out-crossing (points 2–4 in Fig. 9.1B). In pathogens where mate choice is not too costly, it may even have evolved to conditional choice for complementary virulence genes (point 4 in Fig. 9.1B). However, whether pathogens have any kind of mate preference that improves the virulence of the next generation remains to be shown. The few studies that address mate choice in parasites include Lawlor *et al.* (1990) and Tchuem Tchuenté *et al.* (1995, 1996).

CONCLUSIONS

Natural mate choice probably affects the health of host populations, enabling them to react in evolutionary time to coevolving pathogens so that offspring suffer less from the virulence caused by pathogens. This could have implica-tions for breeding programmes of endangered species, of farm animals, and of artificially stocked wild populations, such as fish species in which ripe females are stripped and the eggs are fertilized by sperm of any available male, not necessarily one of the males the female would have chosen. It could also have implications for all procedures in human reproductive tech-nology that bypass natural mate choice.

Mate choice in parasites deserves more atten-tion, for free mate choice based on genetic virulence criteria could increase the virulence of a parasite population.

ACKNOWLEDGEMENTS

I have profited from discussions with many friends and colleagues. I especially thank Manfred Milinski for his encouragement and support and Theo C.M. Bakker, Dieter Ebert, David Haig, Ed LeGrand, Lukas Schärer, Steve Stearns, and George Williams for helpful comments on the manuscript.

10

EVOLUTIONARY INTERPRETATIONS OF THE DIVERSITY OF REPRODUCTIVE HEALTH AND DISEASE

*Roger G. Gosden (rapporteur), Robin I.M. Dunbar, David Haig, Evelyn Heyer,
Ruth Mace, Manfred Milinski, Gilles Pichon, Heinz Richner, Beverly I. Strassmann,
David Thaler, Claus Wedekind, and Stephen C. Stearns (chair)*

Evolutionary theory poses testable hypotheses and addresses key questions about human reproductive behaviour and physiology. The questions considered in this chapter include:

1. What factors guide decisions about family size, sex composition, contraception, and abortion?

2. Why are patterns of fertility and child care so diverse?

3. Can reproductive responses to migration, changes in wealth and disease, and other environmental changes be predicted?

4. Can evolutionary homologies reveal facts of practical importance about the origins and significance of reproductive phenomena?

5. What is the adaptive significance, if any, of menstruation and menopause?

6. Does assisted reproductive technology sometimes conflict with evolved characteristics and tendencies?

Our group was represented by a wide range of scientific disciplines—biology, psychology, anthropology, and demography. These distinctive perspectives and traditions confronted us with the dilemma of how to analyse genetic and non-genetic transmission of information between generations, and their interactions. Genes contain only some of the information that is transmitted from one generation to the next. Other sources of information include genomic imprinting (Haig and Graham 1991), modifications of the environment, and culture.

The biological and behavioural characteristics of individuals develop as plastic responses of genes interacting with the overall environment. In humans, the cultural environment—the mating system, rules of inheritance, and nature of resource acquisition—is particularly important.

Phenotypic traits that favour reproductive success—or to use the succinct evolutionary expression, 'fitness'—are expected to be maximized by natural selection, given enough time and variation. Other forces of biological change, such as drift or pleiotropy (where a heritable effect is a side-effect of an adaptive trait), sometimes outweigh natural selection. Phenotypes (by which we imply physiology and/or behaviour) become fixed in the population if they produce the highest contribution to the future gene pool under the prevailing constraints. The contribution to this gene pool can be either in terms of direct reproduction or by helping kin that share genes (Hamilton 1964).

Differences in phenotype often represent adaptive responses that individuals make when they experience changing circumstances, such as climatic change or population pressure. In such cases, it is not the phenotypes themselves that have been genetically coded, but rather the ability to respond appropriately to changing circumstances that has evolved by natural selection. Human cultural variation can be adaptive—a fact that most early critics of human evolutionary ecology failed to appreciate. Some branches of evolutionary anthropology have treated cultural traits as separate entities (sometimes called 'memes', after Dawkins 1976) and have sought theoretical examples to show that transmission properties of memes could lead to genetically maladaptive outcomes (e.g. Boyd and Richerson 1985). However, most evolutionary anthropologists regard culture and behaviour as inseparable in humans. As Williams (1966) pointed out, it is not necessary to consider the mechanism of gene (or meme) transmission to study the adaptiveness of the phenotype. A large body of evolutionary theory (including life history theory, game theory, and optimality theory) investigates function by exploring the consequences of phenotypic variation for reproductive success. Most of the evolutionary hypotheses have been tested with experiments on animals. For obvious reasons, experiments are not usually possible when dealing with humans, but statistical and epidemiological methods have been designed to investigate natural variation and test hypotheses.

How well adaptive models are likely to apply to humans was controversial within the group—a controversy that will only be resolved with data. We did agree that evolutionary theory is a useful tool for examining patterns and gaining useful insights in human demography, health, and physiology. The scope of this subject encompasses: child health, disease, and mortality; family size and contraception; fertility and infertility. This chapter therefore moves from evolutionary anthropology to demography and finally to reproductive physiology and technology.

CHILDHOOD MORTALITY AND MORBIDITY

That parents sometimes manipulate the survival prospects of their children is known from the high rates of female infanticide and abandonment observed on the Indian subcontinent and in China (notably since the implementation of the one-child family policy) (Dickeman 1979). That there is an evolutionary explanation for such behaviour may not yet be widely familiar to medical and health professionals.

Evolutionary theory assumes that individuals act to maximize their genetic fitness—represented by the number of offspring produced in a lifetime. However, in our species, close and prolonged parental care is required for successful reproduction, and individuals can maximize their fitness by producing an intermediate number of offspring and investing more heavily in each of them. This well-established principle of evolutionary theory was first demonstrated empirically in birds by Lack (1947).

Variation in sex ratio

An important implication is that parents may manage family size and the ratio of sons and daughters in order to make most effective use of their resources, especially in societies where wealth is inherited. In such cases, parents may be able to increase the reproductive success of their children, thus maximizing the number of potential grandchildren, by investing prudently in their offspring. They can do this by producing a completed family size that is optimal for their resources or by preferring one sex over the other (Trivers and Willard 1973). The original Trivers–Willard hypothesis stated that parents would adjust the sex ratio at birth, raising either more sons or daughters according to which sex promised the greater dividend in fitness for themselves. While there is good evidence of this from several animal species (Clutton Brock 1982), it is unlikely that humans could vary natal sex ratios so readily. However, the principle applies to any form of postnatal investment, and we may expect parents to adjust their investment in offspring in an analogous way.

In humans, wealth is probably the single most important factor influencing unequal investment in sons and daughters. Wealth allows sons to have higher rates of reproductive success because they can use it to acquire more mates—legally or otherwise—whereas it has little impact on the more limited rate at which daughters can reproduce. Hence, we can expect wealthier families to favour sons and poorer families to favour daughters. We begin by considering the question of whether parents invest equally in their offspring in other respects, then review the evidence for financially based favouritism.

Infanticide

The effects of parental decisions about investment are most conspicuous when they result in the death of children, and infanticide is the most dramatic evidence of differential investment. Infanticide is widespread among animal species. For example, it is sometimes used by a new male to return lactating females to breeding condition so that the females can raise his offspring instead (e.g. lions and langur monkeys, Hrdy 1979).

The most compelling evidence in humans was obtained by Daly and Wilson (1988) in a study of childhood homicide rates in contemporary Canada. They found that the risk of mortality for children under the age of 2 years was 65 times higher for those living with one step-parent and one biological parent than for children living with both biological parents. They have subsequently shown the same statistical effect on child abuse and neglect in the United Kingdom, the United States, and New Zealand.

There are ethnographic reports of infanticide in almost every traditional society (Daly and Wilson 1988) and, when the people involved are interviewed, four common reasons are given: (i) inappropriate paternity (fathers are unwilling to support a child fathered by another male), (ii) such limited resources that parents cannot afford to rear another child (e.g. the frequency with which children were abandoned or given to orphanages in eighteenth-century France correlated with the price of barley—an index of starvation), (iii) a poor prognosis for child survival (e.g. due to early signs of physical or mental disability, Bugos and McCarthy 1984), and (iv) a short interval between successive births, which increases the likelihood that offspring die in most premodern societies (Lee 1979).

Among the Ache hunter–gatherers of Paraguay, males often deliberately kill the children of women whom they 'marry' after the death or emigration of her previous partner, openly declaring their unwillingness to rear another man's offspring (Hill and Kaplan 1988). A child's chances of dying before the age of 15 years is 15 times higher if the father leaves the group or dies. But it is not always the male that kills offspring. The Ache childen who die between birth and the age of 2 years are more likely to be killed by their mothers (usually because they exhibit signs of disability or the birth interval is too short), whereas children dying between the ages of 2 and 15 years are more likely to die at the hands of males. Maternal infanticide (usually within hours of giving birth) has been documented among peoples as diverse as the !Kung San of southern Africa (Lee 1979), the Ayoreo Indians of Bolivia (Bugos and McCarthy 1984), the Eipo of New Guinea (Schiefenhövel 1989), and the historical Inuit of northern Canada (Irwin 1989).

In Ostfriesland (Germany) in the eighteenth and nineteenth centuries, a young widow's chances of remarrying were significantly higher if her child died than if it survived, whereas a widower's offspring were significantly more likely to die after he remarried (a point reflected in the widespread occurrence of wicked stepmothers in European folk tales) (Voland 1988). The apparent contrast to the high frequencies of wicked stepfathers in contemporary Western societies (as documented by Daly and Wilson 1988) probably reflects the contrasting patterns of remarriage. In premodern societies, high maternal mortality during childbirth often meant that the subsequent incomer into the family was a woman. Improved health care has, of course, dramatically reduced maternal mortality, but the high frequency of divorce in contemporary society, combined with the fact that children typically remain with the mother, means that the new member of the family is usually a stepfather.

Birth order

Among the Ostfriesland petty farmers of the eighteenth and nineteenth centuries, a son's chances of surviving the first year of life (and his chances of reaching adulthood and marrying) decreased with the numbers of brothers (Voland and Dunbar 1995). This phenomenon seems to be a consequence of families attempting to ensure the economic viability of their estates (the most important factor influencing their long-term reproductive success). In effect, they tried to ensure that they had one son to inherit and one son as a back-up in case the first son died. Male heirs had to buy off their younger brothers (often by selling part of the estate), which was a drain on family resources, whereas the smaller dowries of daughters made extra females a less serious problem.

Boone (1989) demonstrated the same effect among the late medieval Portuguese nobility: once pressure on land became serious, the nobility shifted from partible inheritance (equal inheritance by all sons) to primogeniture (eldest son inheriting all). Younger sons were encouraged to seek their fortunes abroad (where many died), whilst elder sons were kept at home. Similarly, Mace (1996) has found that Gabbra pastoralists from northern Kenya prefer their eldest sons to their younger brothers: the fewer older brothers a man has, the more likely he is to marry (and at a younger age), the more animals he receives at marriage from the family herd, and the more children he is likely to have compared with other men of the same age. These birth order effects were not found among Gabbra women, whose costs of marriage are far less.

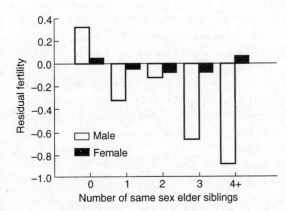

Fig. 10.1 Reproductive success and bith order in a traditional African population (the Gabbra) contrasting the strong effect of number of older brothers with little effect of sisters (adapted from Mace 1996).

Sex-biased preferences

Other recent studies have reported striking examples of differential investment in one sex over the other. Gaulin and Robbins (1991) studied babies born in Philadelphia over a 6-month period and found that, for several indices of parental investment (including the amount of breastfeeding and the length of the subsequent birth interval), families earning more than US$ 60 000 per year invested more heavily in sons than in daughters, whereas the reverse was true in families earning less than $10 000 per year.

Likewise, Bereczkei and Dunbar (1997) found that contemporary Hungarian gypsies invested significantly more heavily in daughters than in sons, as measured by the duration of breastfeeding, the length of the subsequent birth interval, and the length of secondary education. On the other hand, the relatively wealthier native Hungarians exhibited the reverse pattern. More significantly, the relative investment patterns in four populations (two of each ethnic group) very closely matched the relative numbers of grandchildren gained through both sexes of offspring. Each group appeared to be acting in ways that maximized the future reproductive opportunities of the two sexes of offspring in their particular socio-economic context.

Such reproductive decisions can be quite flexible, changing with the economic circumstances of the parents from one generation to the next. Voland et al. (1997) found that, in six north German peasant communities from the mid-nineteenth century, the preference for sons compared with daughters as measured by their respective mortality rates during the first year of life varied in direct proportion to the growth rate of the population. When populations were able to expand into virgin land, sons were preferred, but when populations were at saturation levels and there was little opportunity to acquire new land, daughters were preferred. There appeared to be about a 30-year or one-generation lag between the environmental stimulus and the corresponding behavioural response. The early French-Canadian populations of Quebec exhibited the same lag in adjusting their fertility patterns to the new conditions they encountered after migrating from France (see below).

These results do not necessarily imply that people deliberately set about killing their children, although it is clear that they sometimes do. In many cases, the death is brought about by neglect or other less obvious forms of physical abuse and psychological or economic discrimination. We would not expect such behaviour to be necessarily genetically determined in any simple sense. Selective forces are weak or absent when circumstances vary considerably between individuals in the same population and over time. Hence, at the motivational level, an individual's beliefs about his or her genetic relatedness to a child may be as important in influencing the way he or she treats that child as the fact of being related.

Public health workers and doctors need to be aware that evolution does not seem to have led our species to become unconditionally attached to children. Our willingness to bond is a flexible strategy that allows parents to manage the way they allocate their resources to their offspring. The spectacular nature of some of the examples discussed should not blind us to the fact that sometimes these strategies can be extremely subtle and complex. Infanticide is the tip of a very large iceberg that extends to physical and psychological abuse and to willingness to invest in the health and education of offspring. The well-publicized sex-specific abortion and fertilization rates made possible by prenatal screening and the new reproductive technologies are consistent with the evolutionary theory of parental investment described.

FAMILY SIZE, CONTRACEPTION, AND THE DEMOGRAPHIC TRANSITION

Recent history

Before considering insights from life-history theory, we begin by reviewing recent demographic history. Populations began to grow rapidly in most parts of the world in the eighteenth century, but growth rates are now decelerating at variable rates almost everywhere. Most Western populations went through a demographic transition in the nineteenth and twentieth centuries in which there was a dramatic decline in child mortality followed by a reduction in fertility. Asian, South American, and lastly African populations have followed the patterns of Europe and North America, but the proximate determinants of these changes remain controversial. For several decades, demographers have sought a universal explanation for the transition, but this quest has been fruitless, for a web of socio-economic and cultural factors are responsible (Cleland 1995). When mortality is declining and population density increasing, economic and ecological circumstances are also in flux, making it difficult to assess the contribution of each factor.

Constraints on fertility

Before considering the adaptationist view, it is pertinent to consider traditional societies in which family size is not planned. For example, the rates of prevalence of sexually transmitted diseases (STDs) are thought to have been a major determinant of variation in fertility in some African populations. When colonial administrations in southern Africa distributed antibiotics to combat yaws and other diseases, fertility increased rapidly. The total number of births per woman in the !Kung and Herero peoples of Botswana doubled from around three to six (Pennington 1992), suggesting that fertility had been constrained by STDs. Where this was obviously not the case, a range of proximate mechanisms has been proposed to contribute to fertility variation—from coital frequency to the duration of postpartum amenorrhoea (Bongaarts and Potter 1983). Populations that do not appear to aim for a target family size are described as showing 'natural fertility'.

The evolutionary view

The evolutionary view is that decisions that influence family size have evolved to maximize biological fitness. Family size is thus considered to be an adaptation to local environmental and economic circumstances. The theory helps to predict how changes in circumstances are associated with changes in family size, which may or may not involve demands for contraception. It does so by focusing on the causes of variation in the proximate determinants of family size, and looks for variation both within and between populations. The availability of resources is seen as a key factor. The precise nature of the resources varies with the society. In most cases, however, private wealth is an important determinant of access to limited resources such as food, and there are many examples of a positive correlation between wealth and reproductive success in traditional societies.

Where the costs of raising or marrying children of each sex are strikingly different, the number of children of a particular sex in the family influences interbirth intervals, even in societies without access to modern contraceptives (Arnold 1993; Mace and Sear 1997). When contraceptives are provided, their use depends on whether the mother already has the desired number of sons and daughters (Gadalla et al. 1985; Rahman and DaVanzo 1993).

One of the most puzzling paradoxes of human demography is that a decline in wealth is not a trigger for the contemporary decline in fertility. In fact, a general overview of variation in family size suggests that wealthy countries have the lowest fertility—a fact that has been used to argue against any evolutionary explanation for family size in modern populations (e.g. Vining 1986).

Evolutionary anthropologists have suggested several reasons why wealth and reproductive success may become decoupled in modern societies. Most of these explanations focus on the quantity/quality trade-off in offspring familiar in evolutionary ecology since Lack (1947). Kaplan (1995) has argued that humans maximize status and income-earning opportunities, both of which are enhanced by education. In our evolutionary past, these benefits might have been associated with increased fertility, and we continue to maximize these objectives even though they are no longer im-

proving our biological fitness. Rogers (1995a) argued that a motivation for wealth is as good a proxy for fitness in some circumstances as a motivation to have children, and thus we have evolved to aim to achieve both. The prediction of a positive correlation between wealth and fertility still holds true however.

Mace (1997) has modelled optimal family size in various environments where inherited wealth affects reproductive success and plays a key part in parental investment. She found that the populations with high levels of parental investment had the smallest family sizes and the greatest wealth, even though wealth within a population was positively correlated with family size. She argued that the puzzling negative relationship between wealth and reproductive success across large populations or countries appears because populations are not homogeneous but include groups with significantly different average levels of parental investment. When the groups are combined, the overall influence of wealth appears to be negative, but for any individual an increase in prosperity would be expected, on average, to increase his family size. Because parents tend to replace dead children with additional births, changing the child mortality rate alone will not have much impact on family size. The model also suggests that, when other things are equal, improved agricultural production only increases fertility. Thus, the most effective motivator for small families, and hence for contraception, is an environment in which the benefits—the number of grandchildren, among other things—that result from substantial levels of parental investment in each child are perceived to be large.

Environmental change and time lags

Age-specific mortality and fertility rates are not, of course, hard-wired characteristics of populations but vary temporally and spatially. One well-documented response to environmental changes was the demographic change that occurred among the founders of 'New-France' (Quebec) in the seventeenth century. Despite their common origins and social status, the fertility of the women who were born in Quebec was higher than others who migrated there or those who remained in France (Charbonneau *et al.* 1993). Even the pioneers who arrived as children had intermediate fertility rather than adopting that of their new country.

These differences reveal a striking response to environmental change and indicate that the conditions of the early years of life can set the pattern of future reproductive investment (see Chapter 21 for other connections between events early and late in life). The fast demographic changes of Canadian settlers clearly demonstrate the ability of human populations to respond rapidly and flexibly to environmental changes, even if there is a lag in some groups. Interpreted from an evolutionary standpoint, we may expect the migrants to exploit the opportunities of a new land by a rapid increase in fertility, although they may have paid a cost in terms of disease burden.

There was also an increased variation in family size of the settlers, which may explain the high frequency of inherited recessive disorders 10 generations later among contemporary Quebecois. Even a 'well-adapted' strategy can apparently carry unexpected costs later on, and it is difficult to identify the long-term adaptive strategy in complex environments. In some circumstances the best short-term strategy may not be best for the long term. Since human populations have been subjected throughout history to unpredictable fluctuations, selection (natural or cultural) for plastic behaviour that is adaptive in the short term is not necessarily always happening, particularly in populations where there is a high frequency of unpredictable events.

MENSTRUATION

Menstrual complaints, including premenstrual syndrome, amenorrhoea, and dysmenorrhoea, are among the most common reasons for women to consult their gynaecologists (Baird and Michie 1985). Research on these problems has illuminated many of the proximate mechanisms that trigger menstruation, but has largely ignored the functional question: why has menstruation evolved in the first place? Studies of the adaptive significance of traits such as menstruation are sometimes dismissed as 'teleological', a criticism that is unwarranted because ophthalmic and cardiac surgeons, for example, perform their work better by understanding the functions of the organs they are treating. Although the adaptive significance of menstruation is more obscure than that of either the eye or heart, it would be unwise to ignore it. At one time, vaginal bleeding was thought to be a beneficial way of eliminating toxins from the body, and amenorrhoea was assumed to be unhealthy. Another set of assumptions guides current treatment of menstrual problems, but if we knew why menstruation occurs, medical intervention might be different or in some cases be suspended.

Evolutionary biologists and anthropologists have generated several alternative hypotheses to explain the origins of menstruation. Profet (1993) suggested that it evolved to rid the uterus of sperm-borne pathogens, which leads to the logical extension that contraceptives suppressing menses may encourage uterine infection. However, this proposal, which was widely reported in the media, is incompatible with a substantial body of information (Clarke 1994; Finn 1994; Strassmann 1996*b*). Rather than quashing infection, menstrual blood in fact aggravates it by providing a rich culture medium of amino acids, proteins, and sugars in which bacteria can grow (Eschenbach 1976; Johnson *et al.* 1985).

Basically there are two phenomena to be explained: the cyclicity of the uterine lining or endometrium, and vaginal bleeding. During each menstrual cycle the endometrium proliferates and develops a glandular epithelium with a dense network of blood vessels under the actions of oestrogen and progesterone. However, the tissue regresses if implantation and pregnancy fail to occur because the corpus luteum lyses in the absence of the embryonic signal, human chorionic gonadotrophin (hCG). Vaginal bleeding is well documented only in Old World primates and shrews (Strassmann 1996*b*), though in virtually all mammals there is cyclical growth and retreat of the endometrium by tissue reabsorption (Nalbandov 1976; Johnson and Everitt 1988). Humans have unusually copious bleeding, but about two-thirds of the endometrial tissue is reabsorbed rather than shed (Johnson and Everitt 1988; Kaiserman-Abramof and Padykula 1989; Ferin *et al.* 1993).

Strassmann (1996*b*) proposed that the function of endometrial cyclicity is to conserve energy and that vaginal bleeding is a side-effect that arises when the capacity for re-absorption is exceeded. In all mammalian species, the endometrium is receptive for implantation for only a short time in the cycle coinciding with the appropriate stage of embryo development (blastocyst). In humans, the implantation window is just 3 days or less (Ferin *et al.* 1993) and, since suitable embryos are absent at other times, there is no selective advantage to extending the window. Given that endometrial receptivity need only be brief, the question arises: is it more costly to sustain the endometrium or to regenerate it in each cycle?

When the cost of tissue maintenance was measured in terms of oxygen uptake, it was found that endometrial energy consumption increased nearly sevenfold per milligram of protein per hour during the course of the menstrual cycle until the time of ovulation or implantation (Price *et al.* 1981). The endo-metrium consumed most energy when it was primed for implantation and least after it had regressed after menstruation. Thus, it is ener-getically more economical to regrow the endometrium during each cycle than to main-tain it continuously. The costs of preparing for implantation are not paid for by the endo-metrium alone because the cyclical actions of ovarian steroids on target tissues produce cycles in the metabolic rate of the whole body. In women, the metabolic rate is about 7 per cent lower during the follicular (preovulatory) phase than in the luteal (postovulatory) phase (e.g. Soloman *et al.* 1982; Bisdee *et al.* 1989). There is also a compensatory increase of up to 35 per cent in food intake (Dalvit 1981).

Because the costs of menstrual bleeding are more visible, the costs of maintaining the endo-metrium have been overlooked. However, put in perspective, less than 0.5 per cent of the body protein requirements for one cycle are lost during menstruation and, compared with faecal protein loss, this is trivial (Calloway and Kurzer 1982; Finn 1994). What is more, iron loss is slight except where there is unusually heavy and frequent menstrual bleeding. In natural-fertility populations in the past, as in the few remaining hunter–gatherers today, menstruation was a relatively rare event be-cause women of reproductive age were usually either pregnant or lactating (Short 1976; Strassmann 1997), and the diet was probably sufficient for replacing iron.

Fresh explanations about the origins and significance of menstruation are beginning to emerge, and it seems that endometrial re-gression is part of an evolved tendency in vertebrates for unused tissues to regress. This flexibility permits energy savings not only in the endometrium but also in the oviducts, gonads, and other reproductive tissues of many vertebrates both inside and outside the breed-ing season (Strassmann 1996*b*). Fluctuations in organ size have also been found in the small intestine, kidneys, liver, stomach, lungs, and heart of a variety of organisms, and energy spared at one site becomes available for others (Piersma and Lindstrom 1997). The Burmese python (*Python molurus*) is a spectacular example: within 2 days of consuming a large piece of meat, the intestinal mucosa doubles in mass, whilst intestinal absorption and meta-bolic rate increase sevenfold (Secor and Diamond 1995).

The energy-economy hypothesis carries medical implications because it implies that menstrual bleeding is not adaptive but a by-product of endometrial preparation for preg-nancy. It follows that when a woman does not seek to become pregnant, there is no biological need for waxing and waning of endometrial growth. Contrary to popular belief, a woman's body is not well adapted for repetitive cycles for 35 years, but for prolonged amenorrhoea (Chapter 8). Contraceptives that suppress menstrual cycles or reduce their frequency by mimicking pregnancy or lactational amenorrhoea need not therefore be regarded as 'unnatural', though they will not be widely acceptable if they cause weight gain. If a woman spends more time in amenorrhoea and less time in menstrual cycling, her risk of developing cancer of the reproductive system is diminished (see Short 1976; Henderson *et al.* 1993; Eaton *et al.* 1994). Contraceptive strategies that lower the risk for reproductive cancers (e.g. by mimicking the physiology of amenorrhoea), without increasing the risk for other health problems such as Alzheimer's disease and heart disease, could be a very positive development (Henderson *et al.* 1993; Spicer and Pike 1993).

MENOPAUSE

Ovarian failure

Menopause is the last menstrual cycle in the lifespan and is universal in women surviving to middle age. It is an inevitable consequence of the depletion of the store of about 1 million oocytes (or follicles) laid down at birth (Gosden and Faddy 1994), and such precocious senescence of an organ is unique and unparalleled in males. Women currently live one-third of their lives in a state of postmenopausal physiology and, because life expectancy continues to rise whereas menopausal age remains fixed, this fraction is set to lengthen. Menopause also has considerable medical significance since it raises the risks of cardiovascular disease, osteoporosis, and probably Alzheimer's disease. These major causes of morbidity and mortality in older women outweigh the risks of increasing breast cancer by taking hormone replacement therapy (HRT). Oestrogen deprivation in adult women is therefore an unhealthy state and justifies HRT (Gosden 1985).

Ovarian failure is rarely recorded in animals under natural conditions. Occasionally, a chimpanzee reaching a great age may become menopausal a few years before death, and cetaceans sometimes have a postreproductive phase of life (Austad 1994). Ovarian failure is otherwise only observed under the protected conditions of captivity. Human menopause is

therefore both exceptional and paradoxical. It is maladaptive because it triggers a decline in health and terminates the reproductive lifespan. We are now accustomed to regarding senescence and the diseases of old age as consequences of the evolution of adventitious, deleterious effects of trade-offs where the net benefit is for genes promoting health and fertility at young ages (Chapter 19). Has menopause emerged as a non-adaptive trait, or should it be regarded as biologically beneficial and better not thwarted? The answer to this question has a bearing on attitudes to HRT and the theoretical possibility of abolishing menopause by slowing down the rate of follicular wastage. A conclusive answer is difficult to reach in the absence of reliable data about the lifespan of early hominids, although some cautious suggestions can be made.

According to one hypothesis, menopause is a consequence of an increased lifespan without a commensurate increase in the size of the follicle store and/or more economical use of the follicles throughout life. Accordingly, menopause has emerged as a result of living in a more benign environment, which has occurred too recently to have had any selective effects. This hypothesis is consistent with comparative biology which provides no support for the idea that the human ovary is underendowed at birth or that follicles disappear more rapidly afterwards (Gosden and Telfer 1987). Menopause, from this viewpoint, is more an epiphenomenon of the evolution of an extended lifespan than an adaptive truncation of fertility.

Screening for egg quality

Another view is that menopause is a consequence of screening for egg quality. There is a huge excess of follicles in the newborn ovary, most of which never reach maturity and ovulate but are reabsorbed by the process of 'atresia'. A maximum of only 450 oocytes can be ovulated over the 35 reproductive years compared with the million at birth and almost 7 million germ cells formed initially in the fetus (Baker 1963). At least some of the defective germ cells forming during mitosis and meiotic pairing may be selectively eliminated prior to ovulation, although there is little indication of what the mechanism may be. Evidently, the screen is far from fully effective because chromosome anomalies are common in human oocytes and embryos, especially in the premenopausal years (Hassold *et al.* 1993). There is, of course, an opportunity to abort defective embryos or fetuses later on. The absence of any greater risk of birth defects among children conceived by standard *in vitro* fertilization treatment (IVF) could reflect the operation of this second screen, which is a safeguard when the normal process of follicular selection is abrogated by super-stimulating the ovaries with gonadotrophins.

It might be expected that the screening process would become more active with age. The rate of follicle attrition doubles in the human ovary after the age of 37 years, and thereby advances the age of menopause by some 20 years (Faddy *et al.* 1992). The facts are, however, equally consistent with accumulation of mutations and cytoplasmic damage, which are expected in any post-mitotic cell. We need not, therefore, invoke the hypothesis that adaptive programmed acceleration of ovarian ageing has evolved. Accordingly, a decline in the force of natural selection with age has not produced any greater investment in egg quality and menopause is only an adventitious product of evolving a screening mechanism.

The grandmother hypothesis

Finally, there is a sociobiological argument that late fertility may be sacrificed if it enables women to invest their energy in raising their kin, which is called 'the grandmother hypothesis'. For the female who survived to 50 years of age during hominid evolution, there would be the advantage of avoiding obstetric risks and the birth of a handicapped child if they affect the fitness of other children or grandchildren. Given all the relevant parameters, the hypothesis can be modelled and the supposed benefits and costs estimated, but no consensus has yet been reached, although there is some supporting evidence (see Chapter 19). What appears to be a relatively simple problem turns out to be complex, and reliable data are hard to obtain. Unless sufficient numbers

of women survived long enough, and in sufficient health to be useful nannies rather than burdens on the family group, menopause would not have evolved. Moreover, since grandmothers are genetically related by a factor of 0.25 to their grandchildren and by 0.5 to their sons and daughters, we might expect that the greater benefit of continuing reproduction would postpone the supposed advantage of grandmaternal investment. It may, in fact, be better for older women to lose their fertility for the sake of their own children, which we may call the 'mother hypothesis'. The frequency of births even in well-nourished natural-fertility populations such as the Hutterites declines after 30 years of age and half of the women are sterile by age 40 (Tietze 1957), which enables mothers to devote more care to existing children.

Menopause signals the end of reproduction and occurs when the prospects of raising another child drop to the point where it pays to invest only in the last child, rather than trying to have another one. At this point, parent and offspring are no longer in conflict as their interests coincide, and the parent switches off the possibility of further reproduction to invest everything in the last child. Thus, there is a parallel with the Trivers–Willard hypothesis for variation of the sex ratio, as the children may have moved the balance point to earlier in the lifespan by exercising conflict behaviour.

The costs of late parenting are potentially heavier in terms of miscarriage or delivery of a congenitally abnormal baby, whose prospects of reproduction would be low. Early ovarian failure could be advantageous in curbing the risks of trisomy. Axelrod and Hamilton (1981) have suggested that the increased frequency of Down syndrome (trisomy-21) as a mother ages could result from an intragenomic conflict: only one allele normally ends up in the oocyte and stands a chance of being inherited, but during the last ovulatory cycles the balanced segregation may give way to competition (Axelrod and Hamilton 1981).

There is no consensus about the origins of menopause, nor any agreement about whether early ovarian failure has any adaptive value. If it is adaptive, then programmed reproductive senescence stands out as a striking exception to life-history theory, which generally concludes that evolution has been blind to age changes. But even if menopause has evolved as a non-adaptive trait, there is still much to be explained about the genes that influence germ cell selection and the programme of follicular growth and atresia.

WHAT PRICE ASSISTED REPRODUCTIVE TECHNOLOGY?

Infertility

Humans are a very infertile species by the standards of domesticated and wild animals. Estimates vary, but even in the more affluent parts of the world between 1 in 6 and 1 in 10 couples seek medical advice at some stage for involuntary infertility. In some poorer areas, undernutrition, malnutrition, and a high prevalence of STDs add extra problems. Women and men each contribute approximately one-third of the total incidence of infertility, and the balance is due to mutual or unexplained causes (Mosher and Pratt 1990).

Assisted reproductive technology (ART)

In the past two decades, assisted reproductive technology (ART) has revolutionized the treatment of infertility. *In vitro* fertilization (IVF), which was originally developed to bypass fallopian tubes that became blocked after pelvic inflammatory disease, has been used to help a wide range of problems, including male subfertility due to low sperm count and quality (Table 10.1). Recently, intracytoplasmic sperm injection (ICSI) was introduced for treating oligospermic men (Palermo *et al.* 1992) and, as success rates rise, is set to become standard practice in IVF clinics for most patients. ART is now able to offer not only assistance with conception but the opportunity to screen pregnancy at the preimplantation stage for prospective parents who are carriers of genetic disease or are over 40 years old (Handyside *et al.* 1990). Preimplantation genetic diagnosis

Table 10.1 Assisted reproductive technologies

Gonadotrophic hormone superstimulation
In vitro fertilization (IVF)
Intracytoplasmic sperm injection (ICSI)
Gamete/zygote intrafallopian transfer
  (GIFT/ZIFT)
Donor insemination (AID)
Egg/embryo donation
Gamete/embryo cryopreservation
Tubal surgery
Ovarian diathermy for polycystic ovaries
Testicular/microepididymal sperm aspiration
  (TESA/MESA)
Preimplantation embryo diagnosis and other
  prenatal genetic screens

(PGD) avoids the risks of transferring abnormal embryos to the uterus, which would probably result in pregnancy loss or require medical termination after chorionic villus biopsy or amniocentesis. One or two blastomeres are removed from an eight-cell embryo for DNA amplification of target genes (e.g. cystic fibrosis gene) or fluorescent *in situ* hybridization (for sex-linked disease). It is a highly demanding technology and the number of babies born after PGD is still small, but rising.

The emergence of ART has coincided with legalized abortion in many countries and a consequent decline in the numbers of babies available for adoption. Most couples express strong preferences for genetically related children, but even some of those who would adopt are unable to because of an upper age limit set for foster parents (35 and 40 years in the United Kingdom and France, respectively). Delayed reproduction is further increasing the number of people seeking help from reproductive medicine.

Infertility annuls the choice of a reproductive partner made at marriage and, although ART is at best restorative, 'outbreeding' is a compromise that sometimes has to be struck. Where either partner is sterile, sperm or eggs or embryos are provided by young donors. In most cases, the donor is unrelated and anonymous, and in some countries such as the United Kingdom, financial incentives are prohibited. It would be interesting to know what impact, if any, gamete and embryo donation has on marriage unions and infant bonding, and whether the attitudes and responses of the father and mother are similar.

Prospective parents can only choose a few general characteristics of the donor(s), whose details are registered centrally in some countries (e.g. CECOS in France). But setting aside the donor issue, it is the clinical staff as much as the parents who make the crucial reproductive decisions in ART. They choose the embryos conceived by IVF that are transferred to the patient, and they decide which ones should be 'selectively reduced' if more implant than expected. Whilst ART overrides natural events and often abrogates reproductive choice, it can also increase choice by offering screening for genetic disease and sex—although access is restricted by economic factors.

Gamete selection and choice

The implications of bypassing mate choice in reproduction and normal processes of gamete selection in ART deserve mentioning since the female reproductive tract can exert a selective influence on sperm survival and the chances of fertilization. Parts of the tract are hostile to sperm (especially the vagina), and the sheer distance travelled to the site of fertilization ensures that only the most vigorous cells stand a chance of success. Sometimes immunological interactions are obstacles to sperm migration.

In principle, the likelihood of fertilization for a given spermatozoon is affected by three factors, namely: (i) sperm competition between different males, (ii) competition between the sperm of the same male, and (iii) female selection for certain sperm types (Overstreet 1983; Birkhead and Møller 1993; Birkhead et al. 1993; Wedekind 1994; Eberhard 1996). Sperm competition is potentially beneficial when the most viable and vigorous cells are selected (e.g. Birkhead and Møller 1992). However, a growing number of studies demonstrate that male strategies that combat sperm competition can lead to a conflict of interests between the sexes (see Stockley 1997). In the fruit fly (Drosophila melanogaster), for example, the seminal fluid contains products that support its native

sperm against foreign sperm whilst at the same time causing a significant reduction in female longevity (Chapman et al. 1995).

The possibility that the female reproductive tract can select particular sperm phenotypes has been less intensely studied. In sand lizards (Lacerta agilis), where most females mate with more than one male, the female reproductive system favours the success of genetically dissimilar sperm (Olsson et al. 1996). Likewise, in the ascidian Diplosoma listeranum, sperm are favoured if they are derived from other clones, and a weak negative correlation was found between a pair's mating success and their overall genetic similarity (Bishop 1996; Bishop et al. 1996). While the loci involved in these animals are not yet known, there are convincing examples in plants. Growth of the pollen tube in some plants is affected by the stigma and depends on the combination of male and female alleles on the self-incompatibility locus (Franklin-Tong and Franklin 1993). It would be interesting to know if immunogenes had equivalent effects in animals.

The genetic costs of ART

We must hope that ART, now responsible for tens of thousands of new births annually, does not reduce genetic fitness. We have already noted that the incidence of birth defects has not been noticeably elevated. Nor is it likely that there will be any impact on the frequencies of alleles of medical significance (such as those affecting disease resistance), given the relatively small proportion of the population undergoing treatment and the long intergenerational interval in our species. Although it is unlikely that there will be genetic implications of ART for the population, there could be significant genetic costs for certain individuals. For instance, ICSI enables almost every man to become a genetic parent because the technique can theoretically succeed even if only one sperm is available or if only immature, postmeiotic germ cells can be recovered from the testis or genital tract. Some oligospermic men are known to carry deletions on the Y chromosome (e.g. DAZ), and those with agenesis of the vas deferens have a high risk of carrying the cystic fibrosis gene. As a result of bypassing the genetic cul-de-sac of infertility, the children of these patients are likely to carry a larger burden of disease and infertility than normal.

SUMMARY AND CONCLUSIONS

We have discussed a few insights that can be gained by an evolutionary approach to human reproductive behaviour and physiology. In contrast to accounts of biological mechanisms which are the core of medical textbooks and lectures (so-called 'proximate causes'), this approach provides different levels of explanation for the origins of biological and biosocial phenomena, and indicates their adaptive significance ('ultimate causes', see Chapter 2). Decisions that are made by individuals about their family size, the preferred sex ratio, and the extent to which children are cherished can frequently be interpreted as being adaptive in so far as they affect long-term reproductive success ('fitness'). The value of this approach is not diminished because the driving forces are culturally transmitted. But the fact that reproductive phenomena have evolved does not necessarily imply that they are always adaptive. By becoming senescent in mid-life, human ovaries trigger maladaptive postmenopausal pathology; and menstrual cycles may have evolved only as a side-effect of another adaptation. Such examples should make us pause to reconsider assumptions about the 'normal' state of the body and what is optimal for health.

We recommend priority attention to and research on the following:

(1) factors affecting preferred family size and acceptability of contraception;

(2) contraceptive strategies that mimic the physiology of amenorrhoea;

(3) whether a mechanism exists for screening defective germ cells during gametogenesis;

(4) whether menopause has evolved as a beneficial adaptation;

(5) the impact of the new reproductive technologies on genetic and psychological well-being of children; and

(6) whether more patient choice should be offered in donor insemination programmes.

KEY REFERENCES

Boyd, R. and Richerson, P.J. 1985. *Culture and the evolutionary process.* University of Chicago Press.

Dunbar, R.I.M. (ed.). 1995. *Human reproductive decisions.* MacMillan, London.

Eaton. S.B., Pike M.C., Short R.V., *et al.* 1994. Women's reproductive cancers in evolutionary context. *Q. Rev. Biol.* **69**: 353–67.

Nesse, R.M. and Williams, G.C. 1994. *Why we get sick.* Times Books, New York, p. 290.

Short, R.V. 1976. The evolution of human reproduction. *Proc. Roy. Soc. London B* **195**: 3–24.

Trivers, R.L. and Willard, D.E. 1973. Natural selection of parental ability to vary the sex ratio of offspring. *Science* **179**: 90–2.

Williams, G.C. 1966. *Adaptation and natural selection.* Princeton University Press.

Box 10.1: Glasses for children—are they curing the wrong symptoms for the wrong reason?

Manfred Milinski

About a third of the world's people require spectacles because their eyes are too long to focus distant images on the retina (Wallman 1994). Parents who wear glasses themselves may worry about potential far- or near-sightedness in their young children. When an infant's eyes are hyperopic (far-sighted), it is tempting to provide one's child with glasses that correct this far-sightedness. Are glasses for children good advice?

In animals, as in humans, normal ocular development generally proceeds from an initial hyperopic state toward a state in which the axial length matches the focal length (Norton 1994). This may be achieved by a feedback mechanism that regulates the growth of the eye so that overall optical performance is optimized (Troilo and Judge 1993). The young eye can use visual information to determine whether to grow in the direction of myopia (that is, longer) or hyperopia (shorter), for if myopia or hyperopia is imposed on a chick or tree-shrew eye using spectacle lenses, the optical mismatch rapidly disappears (Irving *et al.* 1991; Siegwart and Norton 1993; Wallman 1994). Of course, if the lenses were removed, those eyes that had worn positive lenses would be quite hyperopic (far-sighted) and those that had worn negative lenses quite myopic (short-sighted). Because negative lenses are universally prescribed for myopic children, one hopes that this compensation is a peculiarity of birds and tree shrews, and that one need not fear that giving myopic children spectacles exacerbates their myopia (Wallman and McFadden 1995).

However, a study in infant rhesus monkeys showed that both positive and negative lenses produced compensating ocular growth that reduced the lens-induced refractive errors (Hung *et al.* 1995). In light of the close similarities between the sensory, oculomotor, and optical characteristics of the human and monkey visual systems, the ocular alterations produced by spectacle lenses in rhesus monkeys provide strong evidence for the idea that spectacle lenses worn by human infants can influence both absolute refractive-error development and the interocular balance of refractive errors. Full optical correction of refractive errors in very young children could preclude the normal reduction in refractive error that occurs during maturation (Hung *et al.* 1995). On the other hand, because lack of clear images leads to 'deprivation myopia,' not correcting myopia might also be ill advised (Troilo and Judge 1993; Wallman 1994). Hung *et al.*'s results provide analytical support for the partial correction strategy that has been advocated in recent clinical trials on the efficacy of spectacle lenses in the prevention of sensory vision disorders (Atkinson 1993). This is in line with the finding of several studies showing that the rate of myopia progression in children wearing lenses with a slight undercorrection of their myopia was significantly slower than in children wearing a full correction (for review see Goss 1994).

Yet another social influence that may interfere with the feedback mechanism that regulates the growth of the eye is early education. 'Near work' such as reading may cause a child's eyes to grow into focus at the distance of a page (whereas those of a person who largely lives outdoors may grow into focus at infinity). There is increasing correlational evidence supporting the association of myopia (short-sightedness) with increased education. The rapid rise in myopia may depend upon the introduction of schooling and modern life in many societies (Wallman 1994).

Can we think of an evolutionary advantage for the developmental plasticity of eye growth? It may be simply a means of adjusting for genetic variation that would otherwise produce either myopia or hyperopia. A more sophisticated explanation is that the sexes had different niches in our ancestral societies— such as hunting for men and collecting and preparing food for women. A feedback mechanism would be adaptive for the different occupations so that eyesight is well matched to its usual working distance in adults.

Part IV: Pathogens, drugs, and virulence

11

THE POPULATION BIOLOGY OF ANTI-INFECTIVE CHEMOTHERAPY AND THE EVOLUTION OF DRUG RESISTANCE: MORE QUESTIONS THAN ANSWERS

Bruce R. Levin and Roy M. Anderson

Primarily through declines in infectious disease mortality, during the course of this century the expected lifespan of a newborn infant in a developed country has risen by nearly 30 years. Diseases like diphtheria, typhoid, dysentery, typhus, polio, and tuberculosis have all but disappeared from the developed world, and the virus responsible for one previously major source of mortality and morbidity, smallpox, is now extinct in natural populations. Most of this decline in infectious disease mortality is due to changes in environmental conditions; improvements in sanitation, nutrition, the processing of food and water, and generally better living conditions (McKeown 1979). However, some of this decline, and most of our sense of well-being with respect to the threat of infectious diseases, can be attributed to preventive and interventive medicine.

Vaccines against the common viral diseases of childhood and antibiotics for bacterial infections have been extremely effective. Community-based vaccination programmes throughout the world have succeeded: the incidences of once common infections such as measles, whooping cough, polio, and diphtheria have dramatically decreased. With chemotherapy we can successfully treat most of the major lethal bacterial and protozoan infections of the past such as tuberculosis, pneumonia, dysentery, typhoid fever, meningitis, and malaria, as well as many of the normally non-lethal, debilitating infections such as gonorrhoea, cystitis, otitis media, and herpes that persist endemically in most countries. Even HIV is now viewed as a treatable infection for those who can afford the new triple therapies that appear to suppress viraemia to very low or even undetectable levels in patients who have had no prior experience of any single drug. Prophylaxis and/or treatment with antibiotics are also essential to the success of most cancer therapies, all organ and tissue transplants, and many of the other achievements of modern surgery.

Nevertheless, complacency about the threat of infectious disease in the developed world would be ill advised. Even excluding AIDS, the death rate from infectious diseases in the United States increased by approximately 20 per cent between 1980 and 1992. Recent outbreaks of *Escherichia coli* 0157, Hanta virus, and *Cryptosporidium*, the HIV/AIDS epidemic, and the resurgence of tuberculosis have called attention to our continuing battle to develop vaccines and treatments for infectious diseases. In part these emerging and re-emerging diseases arise as a consequence of the increase in the size of the human population, increased travel between major urban centres, and the increased concentration of the world's population in such centres. In part they result from the capacity of microparasites (viruses, bacteria, protozoa, and single-celled fungi) to evolve more rapidly than new vaccines and therapeutic agents can be developed and thus thwart our efforts to prevent or treat infections.

Because microbes evolve, anti-infective chemotherapy may be a short-lived achievement. Widespread antimicrobial chemotherapy has imposed a selection pressure of extraordinary magnitude, a pressure to which microbes have responded by rapidly evolving resistance. Since the first decade of this century, there has been evidence that resistance could evolve. The use of quinine for the treatment of malaria between 1910 and 1920 was probably the first widespread use of antimicrobial chemotherapy. The evolution of resistance to quinine was recorded by Sergent and Sergent (1921) in experiments on avian malaria. Their results were not accepted at the time, but in the following two decades supporting evidence came in from experiments with both *Plasmodium* species and *Trypanosoma* parasites. Resistance could be generated fairly easily by passaging these parasitic protozoa though treated and untreated animals and birds (Kritsehenwski and Rubinstein 1932). Today, resistance to all the commonly used antimalarial agents has evolved, and the last lines of defence for patients in hospital with with severe resistant malaria are derivatives of the oldest known antimalarial drug, ginghaosu.

Bacterial resistance to antibiotics was also anticipated and documented before the current era of wide-scale antimicrobial chemotherapy (Ryan 1993; Moberg 1996). Starting with the widespread use of sulpha drugs in the mid-1930s and the true antibiotics, natural products of fungi and bacteria like penicillin and streptomycin, in the 1940s, the number of resistant strains of previously susceptible pathogenic and commensal bacteria has risen steadily. Moreover, the frequency of strains with multiple resistance continues to rise. Modern antimicrobial chemotherapy has been likened to an 'arms race' between technology and bacterial evolution, a race in which the microbes are increasingly gaining the upper hand (Neu 1994). The rate at which antimicrobials with new modes of action are being put on the market has been declining over the past 30 years. Sometimes with frightening swiftness, pathogenic and commensal microbes resistant to these compounds arise, increase in frequency, and spread. The problem of resistance is perceived to be so widespread, and our ability to produce new antimicrobial agents so limited, that some predict an end to the era

of antimicrobial chemotherapy (Cohen 1992, 1994).

The seriousness of this problem is difficult to assess for two main reasons. The first concerns surveillance both at national and international levels. Few longitudinal data are available on trends in the frequency of organisms resistant to specific drugs (with careful definition of resistance). The excellent study by Nissenen *et al.* (1995), of β-lactamase-mediated resistance to penicillin in middle-ear isolates of *Moraxella catarrhalis* in Finnish children from 1978 to 1993, exemplifies what is needed (Fig. 11.1). A clear sampling strategy was defined, and the study lasted for many years during which records were kept of annual consumption of β-lactam antibiotics in the community under investigation.

To judge the magnitude of the resistance problem we also need more quantitative documentation of the clinical and epidemiological significance of drug resistance. Resistance is generally thought to enhance the morbidity and mortality induced by infections that had been treatable with antimicrobial agents. While studies of some infections, such as multiply-resistant *Mycobacterium tuberculosis* (Bloom and Murray 1992) and methicillin-resistant *Staphlococcus aureus* (Lewin Group 1997), do

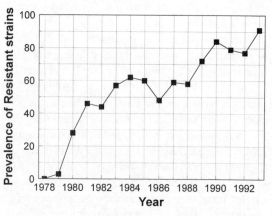

Fig. 11.1 Annual prevalence of β-lactam production among 1452 strains of *Moraxella catarrhalisis* isolated from middle-ear infections in children of less than 6 years of age at Tampere University Hospital between 1978 and 1993. (Data from Nissenen *et al.* 1995.)

demonstrate that resistance increases the morbidity, mortality, and economic burden of bacterial infections, in most cases we do not have quantitative information on how much resistance increases the human and financial cost of infectious diseases.

In this chapter, we explore anti-infective chemotherapy and drug resistance from the perspective of population and evolutionary biology. How does the theory developed in population biology and epidemiology help us understand the within-host dynamics of pathogens under drug pressure and the spread of, and competition between, susceptible and resistant organisms in populations of hosts? That the questions we raise are more numerous than the answers we can give shows the limitations of currently available data and theory. Our aim is to raise awareness of what population and evolutionary biology can contribute to:

(1) evaluating the nature and magnitude of the epidemiological and clinical consequences of resistance to anti-microbial agents;

(2) predicting the short- and long-term effects of different anti-infective chemotherapy regimes in communities of treated people; and

(3) optimizing the treatment of individual patients to control infections and minimize the rate of emergence of drug-resistant pathogens.

THE EPIDEMIOLOGY OF ANTI-INFECTIVE CHEMOTHERAPY AND RESISTANCE

The human cost of drug resistance

As mentioned above, the consequences of resistance for public health are poorly documented. Why, then, do we think that drug-resistant micro-organisms are bad news for society? The argument is that because drugs can reduce morbidity and mortality, resistance to drugs will result in failures in treatment and prophylaxis and will increase the burden of diseases caused by previously susceptible microbes (McGowan 1983; Gentry 1991; Baquero and Reig 1992; Bloom and Murray 1992; Cohen 1992; Berkowitz 1995; Tenover and Hughes 1996). This argument is compelling but insufficient, for it provides no information on how much excess morbidity and mortality is caused by drug resistance or the magnitude of the economic costs of infections and their treatment that are attributable to resistance.

Perhaps the major difficulty in getting the data needed to evaluate these human and economic costs is that most antimicrobial chemotherapy, particularly that directed against bacteria, is empirical, the agent causing the malady has not been identified, or treatment is gratuitous (Palmer and Baucher 1997). As a result, chemotherapy may be ineffective either because the microbe responsible for the symptoms has evolved resistance or is intrinsically resistant (like that of viruses to antibiotics) or because physiological factors, like the stage and location of the infection, preclude effective chemotherapy. Only rarely is the failure of a patient to respond to chemotherapy followed by the detailed microbiological procedures needed to establish whether drug resistance was responsible. Also problematic are the ethical and technical questions raised by conducting case–control studies in populations of patients infected with pathogens (e.g. Edmond et al. 1996). Untreated control groups are unthinkable, and it is not possible to control for all the factors contributing to the outcomes of infections among a heterogeneous group of patients. Indeed, because of the poor documentation of the human and economic costs, there is some debate about whether resistance to antibiotics constitutes a substantial public health problem in developed countries (e.g. Gaynes 1995; Lorian 1995)

Nevertheless, it is clear that strains of bacteria and parasitic protozoa are emerging that are resistant to all, or almost all, of the available drugs; these include *Mycobacterium tuberculosis*, several species of *Enterococcus* in hospitals, and *Plasmodium falciparum* in South-East

Asia. However, even for humans infected with microbes that, because of evolved resistance, are completely refractory to chemotherapy, there are surprisingly few quantitative data on how much excess mortality and morbidity is attributable to resistance.

We are not suggesting that the lack of quantitative data on the human and economic costs of resistant microbes renders meaningless the concerns about the public health and clinical consequences of resistance. Those concerns are well founded, even if their basis has yet to be described in the quantitative manner we would like to see. However, if we want to convince the public and policy makers that resistance to antimicrobial agents poses a serious threat, the human and economic costs of resistance must be well documented. That will require carefully designed longitudinal investigations that combine data on changes in the frequency and distribution of resistance with data on the volume of drug use and clinical and microbiological measures of treatment success and failure.

Volume of drug use and the frequency of resistant strains

Evolutionary theory is based on the tenet that the intensity and duration of selection determines the frequency of any given gene or combination of genes. The precise relationship between gene frequency and selective pressure may, however, be complex, and depend on the frequency of the gene, the absolute size and density of the population, the type of mating system, and patterns of linkage with other genes subject to the same and different selective pressures. For antibiotic resistance, anecdotal evidence suggests a close association between the frequency of a resistant pathogenic (and commensal) microbe and the volume of drug use, which therefore seems to measure the intensity of selection for resistance. An opportunity for a quantitative assessment is provided by the longitudinal changes in annual use of β-lactam antibiotics in the Finnish community mentioned earlier (Nissenen *et al.* 1995: see Fig. 11.2). There was a direct relationship between the frequency of isolation of resistant bacteria (*M. catarrhalis*) and annual

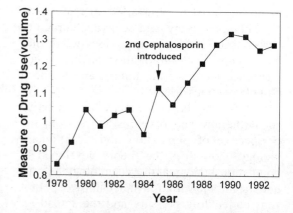

Fig. 11.2 Annual consumption of β-lactam antibiotics, calculated as defined daily doses in the community in Finland between 1978 and 1993. (Data from Nissenen *et al.* 1995.)

drug usage patterns (Austin *et al.* 1998*b*). This study, however, is exceptional for its quantitative detail and duration. Much more longitudinal research is needed, but this will be difficult in any country where records are not maintained on drug usage patterns. This includes most developing countries and many developed countries.

The interpretation of data on the relationship between the intensity of selection, measured by antibiotic use, and the frequency of resistance is greatly facilitated by mathematical models that meld epidemiological and population genetic processes (Gillespie 1975; Anderson and May 1991). Sensible frameworks, based on known epidemiological and biological details of drug resistance, can be used to examine how best to use different drugs to delay the onset of resistance, to minimize its frequency, or simply to quantify how drug volume relates to the frequency of resistance.

Simple epidemiological models have already been used to explore several aspects of the resistance problem. Some of these models consider organisms that persist in their hosts despite treatment, like those with commensal or symbiotic relationships, such as *E. coli* and other normal members of the intestinal flora, various species of *Streptococci* and other bacteria, protozoa, and fungi that maintain populations in the nasopharyngeal passages and vaginal tract (Austin *et al.* 1998*b*). These normally benign organisms are often the innocent bystanders of chemotherapy directed at pathogens or used gratuitously. Other approaches include a more classical epidemiological framework for microparasites that are transmitted within populations of hosts directly or indirectly—by vectors in the case of malaria. For these cases, models where hosts move from infected classes to either susceptible or immune compartments (Anderson and May 1991) are useful. Such models can accommodate the circulation of many distinct strains (i.e. resistant and susceptible types) and the genetic basis of the spread of resistance (i.e. by recombination, plasmid transfer, etc.). Next we briefly consider examples of each of these approaches to the epidemiology of drug resistance.

Resistance in commensal microbes

Levin *et al.* (1997) and Austin *et al.* (1997) modelled the population genetics of antibiotic use and resistance in commensal bacteria. The model employed by Levin and his collaborators assumes a population of hosts, each with a constant probability of being treated with a drug. Bacteria flow at constant rates from individual hosts to a common reservoir and from that reservoir to the hosts. In the absence of treatment, resistant organisms are assumed to be at a selective disadvantage in both the host population and the common environment and decline at a rate that depends on the magnitude of this disadvantage as measured by the traditional selection coefficient, s, $(0 < s < 1)$ of population genetics (Crow and Kimura 1971). When a host is treated, it is assumed that the frequency of resistance in that individual increases to 1.0, no matter what it was initially.

Both numerical (Levin *et al.* 1997) and analytical (Stewart *et al.* 1997) approaches suggest that moderate levels of resistance (i.e. of the order of 10 per cent) will be maintained even when the cost of resistance is high (e.g. $s = 0.04$ per generation) and patients are treated only

every second year. When the cost of resistance is low, high frequencies of the resistant type are anticipated to persist even when antibiotic use is infrequent. Such models also yield formulae that relate the change in the frequency of resistant organisms to changes in the rate of drug use and the fitness cost of resistance. Figure 3 plots some predictions of the Stewart *et al.* (1997) model concerning the rate at which the frequency of resistance changes under either a fourfold increase, or a fourfold decrease, in the rate of use of drug treatment. The model suggests that when the cost of resistance is low, the change in the frequency of resistance is more rapid with the fourfold increase in antibiotic use than with the fourfold decrease. In other words, it is easier to increase the frequency of resistance by promoting drug use than it is to reverse the process by restricting its use. Nevertheless, at least two recent epidemiological studies, one in Iceland (Kristensen *et al.* 1992) and one in

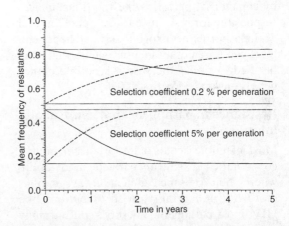

Fig. 11.3 The rate of approach to equilibrium when the frequency of treatment is raised and when it is lowered from the model of Stewart *et al.* (1997). The change is from twice a year or once every 2 years or vice versa. The solid lines show what happened when the rate is decreased from 2.0 to 0.5; the dashed lines show the behaviour when it is increased from 0.5 to 2.0. The thin lines show equilibrium values. Migration rate from the common reservoir is 0.1 per cent per day. The frequency of resistance in the wild in this model is assumed to be constant at 0.1 per cent. (In the similar model in Levin *et al.* (1997), the frequency of resistance in the environment varies.) The mean duration of treatment is 20 days. The maximum resistance frequency is 0.99.

Finland (Seppala *et al.* 1997), suggest that the tide of ever-increasing frequencies of resistance may be reversed by restricting drug use.

Drug resistance in commensal microbes can be problematic for at least three reasons. First, strains of many commensal bacteria, like *E. coli*, *Streptococcus pneumoniae*, *Neiserria meningitidis*, *Staphloccous aureus*, and *Enterococcus faecalis*, are also opportunistic pathogens. Resistance in these bacteria may result in treatment failure for cystitis, pyleonephritis, meningitis, and otitis media and the various hospital-acquired infections attributable to these organisms. Second, when resistance is encoded by transmissible plasmids (Falkow 1975), the likelihood of resistance genes being transferred to previously susceptible pathogens is proportional to the frequency and abundance of the resistant strain in the commensal population (Stewart and Levin 1977). Finally, some resistance-inducing enzymes that are produced by commensal bacteria, like the β-lactamases responsible for resistance to the pencillins and cephalosporins, act exogenously and hence can inactivate chemotherapeutic agents that may be used to treat otherwise susceptible pathogens (Lenski and Hattingh 1986).

Resistance in pathogens that cause acute infections

Most infectious agents of importance in human communities lead a truly parasitic mode of life where their presence almost always implies morbidity or mortality. With exceptions like HIV, most pathogenic microbes are normally either controlled or cleared from infected humans by the immune system. Malnutrition or other causes of immunodeficiency can induce persistent or lethal infections from microbes that are innocuous in immunocompetent hosts. In immunocompromised hosts, antimicrobial chemotherapy can be essential to prevent morbidity and mortality from normally benign micro-organisms. When these organisms are resistant to the drugs used in treatment, chemotherapy will be less effective and may fail. Efficacy is here defined in terms of reductions both in pathogen population size and in the typical duration of infection.

In the past five years several papers have described compartment models of the spread and persistence of drug-resistant strains of viruses, bacteria, and protozoa. Massad *et al.* (1993) used a three-equation model to describe changes in the densities of susceptible people infected with a drug- sensitive bacterial strain, and individuals infected with a drug-resistant strain, in hospitals. The resistant strain was assumed to arise naturally from mutation or plasmid transfer and to be at a competitive disadvantage in the absence of drug treatment. This simple model exhibited a broad range of behaviours including the possibility of co-existence of the two strains, the fixation of the resistant strain, and the dominance of the susceptible strain in the absence of treatment. Analytical studies permitted the definition of parameter regions leading to each of these three outcomes.

More recently Antia and the EcLF (in preparation) and Austin *et al.* (1997) found similar patterns. However, the possibility of coexistence depends on the complexity of the model. That of Antia and the EcLF (1997), for example, assumes similar properties for resistant and susceptible strains except that in the absence of treatment the resistant strain is more rapidly cleared from the host than the susceptible strain (Fig. 11.4). In this case coexistence is not possible. The model of Austin *et al.* (1997) is more complex and allows for dual infection and a degree of cross-immunity after recovery. In this case coexistence is possible for a wide range of parameter assignments.

These models are only a first step towards a general framework for understanding the epidemiology of resistance. There is particular need for infection-specific models constructed with parameters that can be independently estimated in communities of treated humans. They should yield predictions that can be tested by comparing the predicted epidemiology of resistance with that observed.

Evaluating different strategies for drug use in the community

Mathematical models can be used to evaluate the effects of antimicrobial chemotherapy on

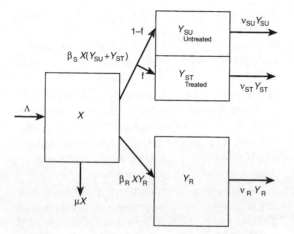

$\beta_S X(Y_{SU}+Y_{ST})$

$1-f$

Y_{SU} Untreated

$v_{SU}Y_{SU}$

f

Y_{ST} Treated

$v_{ST}Y_{ST}$

Λ

X

$\beta_R XY_R$

Y_R

$v_R Y_R$

μX

Fig. 11.4 The Antia and EcLF model for the epidemiology of antibiotic resistance for an acute infection. Variables: X, density of unifected hosts; Y_R, Y_{SU}, Y_{ST}, density of host infected with drug-resistant microparasites, and susceptible microparasites that are untreated and treated, respectively. Parameters: rate at which unifected hosts enter the population, S, R- transmission rate constants for susceptible and resistant parasites, respectively, f—fraction of hosts infected with susceptible bacteria that are treated (treatment is assumed to have no effect on the rate of clearance of hosts infected with resistant microbes), SU, ST, and R—rates of clearance for untreated and treated hosts infected susceptible microbes and hosts infected with resistant.

the epidemiology of an infectious disease (e.g. Blower *et al.* 1996). These models can also provide a template for evaluating, *a prori*, how best to use either a single drug or a combination of drugs within a community to minimize morbidity and mortality of microparasite infections while reducing the rate of increase of resistance. Is it better, for example, to cycle different chemotherapeutic agents or to use them simultaneously? Two recent studies explore these issues. The first (Austin and Anderson unpublished) considers a hospital setting and the use of two different antibiotics. Their model defines the number of uninfected patients, the number of patients infected with strains sensitive to both drugs, and patients infected with strains resistant to either drug 1, to drug 2, or to both. Resistance is assumed to be mediated by plasmid transfer, and each bac-

terial strain has a basic defined reproductive number (average number of secondary cases generated by one primary case in a susceptible population). In the absence of treatment, the fully sensitive strain has the greatest reproductive success.

The model is used to compare three strategies: (i) monotherapy, where a single drug is used at all times and only people infected with a resistant strain are given the alternative antibiotic; (ii) dual therapy, where both drugs are used with a certain ratio; and (iii) alternate therapy, where each drug is used exclusively for a defined period before switching to the other. Both analytical and numerical studies reveal that the best option, in terms of minimizing the frequency of resistant strains over a 4-year period (particularly resistance to both drugs), is most often to use monotherapy and only to use the alternate drug when treatment fails following the use of the drug of choice.

The second study, Bonhoeffer *et al.* (1997), considers two criteria for evaluating the efficacy of different antibiotic regimes. The first is the duration of a drug's effectiveness, defined as the time before the frequency of resistance exceeds a specified level. The second is the extent to which a particular regime of antibiotic use reduces the number of infected hosts over a specified period. For monotherapy, their model predicts that the long-term benefit, measured by these two criteria, is essentially independent of the pattern of drug use. However, it also predicts a slight benefit in heavy use in the early years, but although the initial benefit is increased in this way, the 'life expectancy' of the drug is reduced due to the more rapid ascent of resistance. For dual therapy, their model predicts that 50:50 treatment will always be better than cycling between the two drugs; the optimal way to use both drugs is to treat patients simultaneously with a combination of the two. As in most epidemiological models of antibiotic use, the rate of increase in the frequency of resistant strains is always faster than its decline after the cessation of selection by antibiotics.

The major conclusion from the models of Austin and Anderson (unpublished) and Bonhoeffer *et al.* (1997) is that cycling different antibiotics is not neccessarily the best solution and may actually be the worst (McGowan 1986; Swartz 1994).

Parameter assignments and the comparison of prediction with observation

How useful mathematical models of the transmission and persistence of drug-sensitive and drug-resistant pathogens will be in practice, and whether they will be widely employed for predicting outcomes and in designing treatment protocols, depends on their realism and on how well their predictions stand up to comparisons with retrospective and prospective studies of the epidemiology of drug resistance. Among the most significant stumbling blocks to the use and acceptance of mathematical models is the widespread distrust of simple mathematical models amongst clinicians and experimental and observational biomedical scientists, and the failure of model builders to work closely with empirical researchers. More studies are needed that combine theory with experimental and epidemiological study, and they are of two types.

First, we need to improve the quality and increase the number of epidemiological studies of resistance using molecular and other tools to monitor, in well-defined populations, the changes over time of the distribution and frequency of resistant lineages (assuming clonality) and the rates of use and mode of administration of specific anti-infective agents. Studies of this type are particularly important when a new drug is involved and when the pattern of drug use changes. Molecular, biochemical, and immunological procedure (polymerase chain reaction, restriction fragment length polymorphism, RAPD, pulse gel electrophoresis, multienzyme electrophoresis, and serotyping) provide exciting possibilities to quantify the origin and spread of resistant organisms. With rare exceptions (Coronado et al. 1993; Kreiswirth et al. 1993; Small and Moss 1993), few studies of the use of chemotherapeutic agents have been co-ordinated with investigations of the outcomes of treatment and the molecular genetic epidemiology of microbes.

Such studies need to be conducted concomitantly at a series of levels, from hospital or day care centres, though local communities to national and international geographical scales.

Second, we need experimental, clinical, and epidemiological studies to test the assumptions made in models of the spread of resistance and to estimate their parameters. The most important biological parameters are the fitnesses of drug-sensitive and drug-resistant strains of a given pathogen. In epidemiology, fitness is best defined as the basic reproductive number of a defined strain, R_0, which measures the average number of secondary cases generated by one primary case in a population of susceptible patients. R_0 is a composite parameter with components that measure transmission efficiency, the typical duration of infection, and average infectiousness over this period (Anderson and May 1991). Few have tried to measure R_0 for drug-sensitive and drug-resistant strains (Massad et al. 1993), but new immunological tools and the concepts and methods used for vaccine-preventable infections (typically viruses) may lighten the task.

For models of resistance in commensal bacteria, the fitness cost of resistance can be estimated by competition experiments (performed in experimental animals as well as in vitro) using resistant and sensitive variants of the same strain or species (Levin 1980; Bouma and Lenski 1988; Modi and Adams 1991; Borman et al. 1996). While it is unrealistic to perform such experiments for large numbers of microparasites in many different ecological settings, much can be achieved by simple studies that measure the fitness cost of specific resistance mechanisms (or genetic elements) that are common to many different microparasites. Are those costs large (in excess of 2 per cent) or small (undetectable experimentally)?

For commensal bacteria, human volunteers or rodent models and molecular methods could be used to monitor the frequency of resistance to a given antibiotic in hosts subjected to different treatment regimes. This approach is now widely used to study resistance to antiviral agents, particularly in patients with HIV treated with various drug combinations (Ho *et al.* 1995; Wei *et al.* 1995) and in patients with *falciparum* malaria (White 1997).

In the study of acute infections, strain fitness can be measured by clinical studies of pathogen growth and clearance rates and by R_0, the composite measure of transmission efficiency. In virology (e.g. HIV) and protozoology (e.g. malaria) much clinical work relies on measuring pathogen abundance in the host and how this changes in response to the host's immunological defence or to drug treatment. For example, for malaria one can compare the rate of population growth of the parasite in recently infected patients (percentage of red blood cells infected) for drug-sensitive and drug-resistant strains. This can be checked *in vitro* with red blood cell cultures (White 1997). Such studies will be strengthened by closer collaboration between the clinician or experimental scientist and the theoretician or evolutionary biologist.

An issue of great importance in the study of pathogen fitness concerns its evolution under selection by chemotherapy. Will the fitness cost associated with resistance decline under continued selection by drug use (Levin and Lenski 1983)? Experimental studies with bacteria and viruses suggest that an evolutionary adjustment of the fitness costs of resistance is likely and that the mechanism could either be modification of the resistance genes themselves, changes in the accessory elements that carry these genes, or second-site modifier mutations (Bouma and Lenski 1988; Modi and Adams 1991). The fitness costs of resistance may even decline in the absence of drug use due to the evolution of second-site mutations rather than reversion to sensitivity: examples include the streptomycin-resistant *rpsL* mutations in *E. coli* (Schrag and Perrot 1996) and the gene responsible for resistance to a protease inhibitor in HIV (Borman *et al.* 1996). Moreover, the mutations compensating for the costs of resistance can establish a genetic background in which the original genes responsible for susceptibility to that antimicrobial are at a selective disadvantage in the absence of the drug (Borman *et al.* 1996; Schrag *et al.* 1997).

THE WITHIN-HOST DYNAMICS OF PATHOGEN POPULATIONS

Growth and chemotherapy

For many pathogens drug resistance arises as a direct effect of treatment and is selected in or on treated patients. Resistance may rise by mutation, recombination, or exchange of genetic elements. In some cases resistance may lead to treatment failure and transmission of resistant organisms to new hosts (e.g. *Plasmodium falciparum*). To date, however, there is little formal theory of the within-host population dynamics of parasites under drug treatment. In contrast to the many quantitative studies of the pharmacokinetics of chemotherapeutic agents, less attention has been paid to pharmacodynamics, where pathogen population parameters are measured under different chemotherapeutic regimes. However, we now have several reliable studies of quantitative pharmacodynamics in viruses, bacteria, and protozoa (Garrett and Heman-Ackmah 1973; Garrett 1978; Gerber *et al.* 1983; Ho *et al.* 1995; Wei *et al.* 1995; Austin *et al.* 1998b; Berg *et al.* 1996; Hetzel and Anderson 1996; Bonhoeffer and Nowak 1997; Perelson *et al.* 1997).

The within-host dynamics of parasites has become a hot topic amongst mathematical biologists, with particular attention paid to the interaction of pathogen population growth and the host's immunological defences (Nowak *et al.* 1991; McLean and Nowak 1992; Antia *et al.* 1994; Essunger and Perelson 1994; Frost and McLean 1994; Hetzel and Anderson 1996; Mittler *et al.* 1996; Austin and Anderson 1998, Austin *et al.* 1998b, c). To study within-host pathogen dynamics and chemotherapy, models need to co-ordinate three things: the population biology and population genetics of the pathogen population, the host's cellular (and humoral) immune defences, and the pharmacokinetics and pharmacodynamics of chemotherapy. Few have attempted to meld the terminology and concepts of the different disciplines, although a start has been made (Austin *et al.* 1998c; Lipsitch and Levin 1997a, b, c).

It is not hard to define simple models of parasite population growth within the host that incorporate various regimes of chemotherapeutic treatment and hence describe changes in drug concentration over time. Such models allow us to relate the reproductive potential of the pathogen to the average drug concentration in the target tissue required to eliminate the parasite (Nowak and Bangham 1996; Austin and Anderson 1998). For protozoa and viruses that utilize specific host cells (e.g. erythrocytes for *Plasmodium* species and CD4+ cells for HIV), the within-host reproductive number is defined as the number of secondary cells infected by one primary infected cell in a susceptible cell population (Anderson and May 1991).

These models can also use concepts from pharmacokinetics and pharmacodynamics to define the minimum inhibitory concentration (MIC)of a drug required to eliminate the pathogen in terms of R_0 (Austin and Anderson 1998). In some cases values estimated in experiments can be assigned to the intrinsic growth rate of growth of the parasite (prior to immunological attack), to the kill rate induced by a specific drug, and to the treatment regimes used for specific infections. Such models often provide a reasonable description of the observed course of infection under chemotherapy (Ho *et al*. 1995; Wei *et al*. 1995; Berg *et al*. 1996; Austin *et al*. 1998*b*) and help to define optimal treatment patterns in the absence of the evolution of resistance.

They also underline how poorly we understand how the immune response of the host acts quantitatively to regulate—or not to regulate—pathogen population growth in the absence of chemotherapy. For most acute infections, whether viral, bacterial, or protozoan, the immune response succeeds in constraining population growth or eliminating the parasite. Even in persistent infections, immune responses typically keep pathogen populations successfully at levels that do not induce severe morbidity (HIV is an obvious exception). In the models persistence often arises via antigenic variation (e.g. Nowak *et al*. 1991; Mittler *et al*. 1996) and chemotherapy shortens the typical

duration of infection, in some cases, only eliminating the pathogen when acting together with the host's immunological defences. Present models represent the action of the immune defences on the pathogen rather crudely and call for more quantitative experimental studies of the population dynamics of the interaction between pathogens and both the constitutive and the inducible immune defences of mammals.

Preliminary models of such interactions reveal a rich array of possible dynamical behaviours while capturing the three main outcomes: failure to constrain pathogen population growth (leading to the death of the host), elimination of the parasite after an initial phase of exponential growth in the host, and regulation of pathogen abundance at a level that does not induce serious morbidity (Nowak and Bangham 1996; Austin *et al*. 1998*b*; Lipsitch and Levin 1997*a*). The addition of treatment to such frameworks would be straightforward; it would allow us to study the evolution of resistance under the selective pressure imposed by chemotherapy.

Temporal variation in drug concentration within the patient

Without continuous infusion of a drug to maintain concentrations above the MIC (Minimum Inhibitory Concentrations) level, there are periods during therapy when both sensitive and resistant microbes can proliferate. Under some conditions, temporal variation in the concentrations of antimicrobial agents can promote the increase of resistance and multiple resistance in treated hosts.

To better understand these conditions, recent studies have used simple mathematical models of within-host pathogen population growth and the dynamics of change in drug concentration. In the model of Lipsitch and Levin (1997*a*, *b*), antibiotics are introduced at periodic intervals between which their levels decline, possibly below the MIC. If resistant bacteria are not already present when treatment starts, and, if as a consequence of treatment the rate of decline in the density of infecting bacteria is comparable with the rate of cell division, resistance to single (much less multiple) antibiotics will rarely emerge during treatment. When these conditions are not met, and the concentration of the antibiotic is too low to keep the bacterial population declining (becomes 'subinhibitory'), resistant mutants will appear and increase. The resistant population increases most rapidly when the concentration of the antibiotic is holding the rate of bacterial growth to half its maximum. The model also predicts when large infrequent doses of single antibiotics will be more effective than more frequent smaller doses and the reverse.

When multiple drugs are used, the Lipsitch–Levin model of temporal variation in drug concentration predicts treatment failure due to the growth of strains resistant to single antibiotics or the appearance of multiply resistant mutants when subpopulations resistant to single antibiotics can grow for any reason, including dosing at inadequate concentrations or at too distant intervals. Non-compliance or non-adherence to a specified drug regime have the same effect as inadequate dosing. Lipsitch and Levin consider two forms of non-adherence: (i) random, or uncompensated failure to take the drugs at the prescribed interval; and (ii) thermostatic, or temporary cessation of treatment when the symptoms of the infection are in remission.

Austin and Anderson (unpublished) considered one specific case, the use of imipenem to control *Pseudomonas aeruginosa*, for which there are good experimental data on kill rates of the bacteria for a range of drug concentrations in a mouse model allowing estimates of MIC values for sensitive and resistant strains (Gerber *et al.* 1983; Berg *et al.* 1996). With a simple model representing the immune response of the mammalian host and different treatment regimes, they found optimum dosing levels and times between treatment that minimize the growth of resistant mutants and the likelihood of treatment failure. If treatment is delayed, the impact of the host's immunological response increases, and there was a subtle interplay between this delay and the frequency of resistant pathogens.

Their main conclusion was that early and frequent treatment minimized the overall severity of the infection but increased both the frequency of resistant organisms and the likelihood that they would be transmitted to new hosts. Late treatment provided little reduction in the severity of the infection (total bacterial abundance in the host integrated over the entire duration of infection) but minimized the frequency of resistant forms. More generally, the model provides precise guidelines for determining the frequency and level of drug use required to minimize the duration and severity of infection for the specific host–parasite interaction considered.

Heterogeneity in the distribution of drug concentrations within the host

Models can also be employed to examine more complex situations in which drug concentration varies in different tissues or sites within the patient. Such heterogeneity is thought to be of importance in the evolution of drug resistance, where concentrations above the MIC in some tissues permit resistant organisms to thrive in certain sites. Mitchison argued for the importance of this process in the development of drug resistance in *Mycobacterium tuberculosis*, where bacteria can move between sites or change to different physiological states when the efficacy of different drugs varies (Mitchison 1979, 1984).

Lipsitch and Levin (1997*a*, *c*) have recently produced a two-drug model of this idea using two sites between which bacteria can migrate. In one site both drugs are effective; in the other, only one drug is effective. They found that if a population of bacteria arises that is resistant to the single drug effective in the protected compartment, and if it grows fast enough in that compartment to achieve net growth in the system at large, mutants that are also resistant to a second drug will eventually be produced, rapidly leading to treatment failure.

Whether this happens depends on the positive and negative growth rates of the singly resistant bacteria in the two compartments, the rate of flow between these compartments, and the size of the protected compartment. If the rate of flow to the unprotected compartment is sufficiently small and the inhibitory effect of the second antibiotic sufficiently great, then even though the singly resistant mutants may proliferate in the unprotected compartment, their population will decline in the host at large. The smaller the protected compartment, the less likely it is that strains resistant to both antibiotics will emerge.

Random and thermostatic non-adherence in this model can promote treatment failure by allowing for growth of the bacteria in the unprotected compartment and the emergence of resistant mutants during the period of non-adherence. Paradoxically, the combination of non-adherence and a protected compartment can actually reduce the rate of ascent of multiply resistant mutants compared with non-adherence alone. This arises because the protected compartment serves as a reservoir for bacteria susceptible to the drug that is effective only in the non-protected compartment where the concentration is above MIC.

These studies are only first steps towards a theory of the within-host population dynamics of microparasites under antimicrobial chemotherapy. They highlight shortcomings in the available data on the within-host population dynamics of bacterial infections and their control by the host's immune defences and by chemotherapy. They illustrate how mathematical models help us understand how to treat infections with drugs of known pharmacokinetic and pharmacodynamic properties and how to reduce treatment failure due to the evolution of resistant organisms. They also provide a rough method of relating the abundance of sensitive and resistant strains within the host to their rate of transmission in populations of hosts. Future work should combine studies of the within-host population dynamics of infections and their control with the traditional models used to study the epidemiology of microparasite infections in host populations.

THEORY AND OBSERVATION

The discussions presented above make clear that using mathematical models to study the population biology of drug resistance, whether within host populations or within single hosts, is a relatively new enterprise. The simple models so far developed contain parameters for which no quantitative estimates are available for most host–parasite–drug interactions. The use of models for designing and evaluating drug treatment regimes will depend not only on the availability of such data, but also on how well these models predict observed outcomes. This is just as true for the within-host framework as it is for models of transmission and persistence of resistant organisms.

In both cases it is possible to rectify the mismatch between the data needed to test the models and the data currently available. We have already discussed this issue for the epidemiological models. It is easier to see how the within-host dynamic models can be tested experimentally, for progress has already been made using laboratory animals to study drug resistance in malarial parasites: defined mixtures of susceptible and resistant malarial parasites have been treated with different regimes of single and multiple drugs (Peters 1987). In a clinical context, HIV shows what can be achieved when there is great urgency to find better ways of using single drugs and combinations of drugs to slow the progression of a fatal disease. Quantitative virology with molecular methods for identifying sensitive and resistant viruses has recently provided many data on how different treatment regimes influence viral load in patients at different stages of progression to AIDS. The current excitement surrounding the obvious benefits of triple therapy, with drugs targeted at different biological functions of the virus, has evolved from trial and error in the use of different treatment regimes (Wein *et al.* 1997).

The urgency of finding a cure for a lethal disease has driven research on AIDS. In other cases much could be achieved in enhancing our understanding of how best to use antibiotics or antimalarial drugs by experimental studies of animal models that use classical pharmacological approaches to test the models of population and evolutionary biologists. We hope that recent theoretical studies that illustrate the benefits to be gained from an understanding of the population level processes that drive the evolution and spread of drug-resistant organisms will stimulate such interdisciplinary research in the future.

SUMMARY

The ascent and spread of pathogenic bacteria, viruses, protozoa, and fungi resistant to the chemotherapeutic agents used to control them represents the single greatest threat to the continued success of medical intervention. While the absolute magnitude of this resistance problem, as measured by excess mortality, morbidity, and financial costs, has only begun to be estimated, it is becoming clear that this human cost of resistance is potentially substantial; people are dying of previously treatable infections. In this chapter we have considered the questions that microbial resistance to chemotherapy poses to population and evolutionary biologists, current efforts to address these questions, and some of the answers that have been obtained. One line of questions we considered concerns the epidemiology (population genetics if you prefer) of resistance in communities of hosts treated for either chronic or acute infections. What is the relationship between the incidence of treatment and the frequency of resistance in microparasite populations? What is the rate and extent to which those frequencies will respond to changes in the incidence of use of these agents? Other questions of the epidemiology of resistance we considered deal with the consequence of different regimes for the use of anti-infective agents in communities of treated hosts. What regimes for the use of single and multiple drugs maximize the long-term efficacy of these agents in the face of resistance? The second line of questions we addressed involves the treatment of infections in individual patients. How does temporal variation in the concentrations of drugs and tissue heterogeneity in drug action affect the ascent of resistance to single and multiple chemotheraputic agents? At this time, these questions of the epidemiology and within-host population dynamics of resistance have been addressed primarily at a theoretical level by mathematical modelling. The assumptions behind the construction of these models and the predictions obtained from the analysis of their properties raise a number of empirical questions of the population biology and evolution of resistance which are just beginning to be considered.

12

DEVELOPMENT AND USE OF VACCINES AGAINST EVOLVING PATHOGENS: VACCINE DESIGN

Angela McLean

OVERVIEW

This chapter asks why vaccination has been so successful and why infectious agents have been unable to evolve resistance to vaccine-induced immunity. The answer to this question is in three parts. First, it is possible that vaccine-resistant strains will arise in the future: in the models investigated here it can take a very long time for enough vaccine recipients to accumulate for a vaccine-resistant strain to achieve competitive dominance. Second, and I believe most likely, current-generation vaccines are effective enough and cross-reactive enough that a vaccine-resistant strain will never become the competitive superior. A third, unlikely possibility is that so few individuals are vaccinated that the vaccine-resistant strain remains competitively inferior. This last point is, however, worth bearing in mind for new vaccines that one might wish to target at particular risk groups, whilst leaving the rest of the community unvaccinated. Subject to a limit on the aggregation and accumulation of vaccine recipients this might prove a method for avoiding the emergence of vaccine-resistant strains in the future.

Can we expect these successes to continue? The answer to this question has two parts. The first is in the situation where the target infectious agent is monotypic. To prevent the emergence of new vaccine-resistant strains, a vaccine should have high community-level impact. The community-level impact is a composite measure of different aspects of vaccine failure. It is (i) particularly sensitive to the duration of protection, but (ii) equally affected by vaccine-induced reductions in susceptibility

or infectiousness. These points have two practical implications. The duration of protection afforded by a vaccine and the planning of revaccination become very important components of the protection of communities using vaccination. Even when vaccine-resistant strains can evolve within vaccine failures, the community-level impact of a vaccine is equally affected by reductions in infectiousness or reductions in susceptibility. The only exception is when total protection is a possibility, and then it should be pursued.

For infectious agents where a vaccine gives differential protection across pre-existing competing strains, vaccination against one strain can allow another strain's prevalence to increase. An understanding of the prevaccination rules of competition between strains can help in the design of vaccines. In particular, the number of susceptibles can be maximized by using the cross-reactivity endowed by competing strains.

Infectious agents that can evolve to circumvent naturally acquired immunity may well do the same for vaccine-induced immunity. However, vaccine-resistant strains only gain competitive advantage when their natural competitors are suppressed by vaccination. The way to keep them out is to search for naturally occurring, minimally pathogenic, highly cross-reactive strains and cultivate their spread by attacking their existing competitors.

INTRODUCTION

Vaccination has pushed several childhood infectious diseases to the verge of extinction without their pathogens having adapted to circumvent this evolutionary pressure. How has this happened, and can we expect such successes to continue? Most vaccines in current use give highly effective, long-term protection against infections of childhood (Clements *et al.* 1992). Such infections do not exhibit wide antigenic variation. Many of the infectious agents now targeted for vaccine development do show wide antigenic variation, and many novel vaccines are of low efficacy and give protection of short duration (Sabin 1993; Bloom 1995; Targett 1997). This chapter addresses these two facets of modern vaccines using simple models of the transmission dynamics of infectious agents.

In the first section I survey several studies concerned with trade-offs between different types of vaccine imperfection. For example: if one had to choose between a poorly immunogenic vaccine giving long-term protection and a highly-immunogenic vaccine giving only short-term protection, what should one do? The same models can be used to consider the appearance and spread of strains of infectious agents that escape vaccine-induced immunity.

For infectious agents that have a rich strain structure even before vaccination a different family of models is required. The second section introduces two such models and investigates the predicted impact of vaccination on the prevalence of target and competing strains. These models are then used to ask: what are the trade-offs between depth (highly immunogenic against one strain) and breadth (cross-reactive immunity across several strains) in vaccine design for infectious agents with complex strain structure? Asking such questions highlights the pivotal question: what are the constraints on vaccine design?

In the third section I consider the within-host dynamics of infectious agents with the specific aim of developing sensible constraints on vaccine design. In an ideal world one would like to construct perfect vaccines against everything. Given that this is not possible, what is a rational constraint?

The fourth section compares the broad predictions of these mathematical models with some observations.

TRADE-OFFS IN IMPERFECT VACCINES

The most elegant way to model vaccination is by shifting a proportion of susceptible people into the immune class, permanently (Fig. 12.1a) (Anderson and May 1991). The fraction moved represents people successfully vaccinated, that is, the product of the proportion of susceptibles vaccinated and the efficacy of the vaccine. Thus one assumes that if a vaccine fails it fails completely: everybody is either fully protected or not protected at all. In what follows I consider several refinements to this simple assumption (Fig. 12.1b): imperfect or short-lived protection, different natural history of infection in vaccine failures, vaccine resistance arising in vaccine failures, and vaccine resistance that arises exogenously (Anderson and May 1991; McLean and Blower 1993, 1995; McLean 1995).

In considering the community-level impact of vaccines it is useful to introduce the vaccinated reproductive rate R_p. This represents the number of secondary cases caused by one infectious individual introduced into a community where a fraction p have been vaccinated and everybody else is susceptible. R_p is therefore a generalization of the basic reproductive rate R_0. For vaccines that protect completely R_p is related to R_0 by:

$$R_p = (1 - p)R_0 \qquad (12.1)$$

p being the fraction vaccinated. The eradication criterion for an infectious agent is derived by setting $R_p = 1$. Thus for a perfect vaccine, infection is eradicated when p, the fraction vaccinated, exceeds the critical vaccination proportion p_c, where

$$p_c = 1 - \frac{1}{R_0} \qquad (12.2)$$

This simple equation. has had widespread impact in thinking about the eradication of infectious diseases through vaccination (MacDonald 1952; Anderson and May 1991). Equation (12.1) can also be used to think about control of infections that does not attain eradication. In particular, calculation of R_p for two different candidate vaccines allows direct

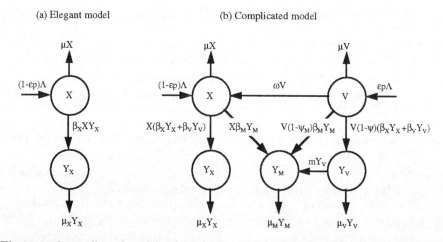

Fig. 12.1 Elegant and complicated models of vaccination. (a) In the past it was standard practice to assume that all vaccine recipients are permanently protected from infection. Such models were used to investigate questions such as the post-vaccination age-distributions of cases of mumps and rubella. (b) For vaccines that have recently become available and are in production it is unlikely that the degree of protection (ψ) will be perfect or that immunity will be permanent. More complex models are needed to investigate the probable impact of such vaccines. This 'complicated' model is a generalized structure allowing four different aspects of vaccine failure to be investigated. Its equations and biological assumptions are given in detail in the text. Briefly, five populations are modelled: susceptibles, X; vaccinated, V; infected in the absence of vaccine-induced protection, Y_X; infected despite vaccine-induced protection, Y_V; and infected with a vaccine-resistant mutant, Y_M. For each of the four questions asked different parts of the model are 'disabled'.

comparison of their predicted community-level impact.

For vaccines that are less than perfect, a more generalized model than Fig. 12.1a is required. Figure 12.1b provides a framework within which several different aspects of vaccine imperfection can be explored. The model's biological assumptions and equations are as follows. The model tracks the dynamics of five populations: susceptible, X; vaccinated, V; infected having been susceptible, Y_X; infected after vaccination Y_V; and infected with vaccine-resistant mutant, Y_M. Recruitment to the community is at a constant rate Λ and deaths are at constant per capita rate μ. A fraction p of new recruits are vaccinated, and vaccine immunizes a fraction ε of these, but vaccine-induced immunity wanes at per capita rate ω.

Two types of infectious agent are assumed to exist, wild type and vaccine resistant, and before vaccination the vaccine-resistant strain is outcompeted by the wild-type strain. The per-susceptible risk of becoming infected is

therefore the sum of the risk of catching wild-type infection and the risk of catching a vaccine-resistant infection. For susceptibles the rate of catching wild-type infection is the sum of the number of cases of wild type amongst people who were susceptible before being infected (Y_X) multiplied by their infectiousness β_X plus the number of cases of wild type amongst people who were vaccinated before being infected (Y_V) multiplied by their infectiousness β_V. The rate of being infected with the vaccine-resistant mutant is the number of cases of vaccine-resistant mutant (Y_M) multiplied by their infectiousness β_M. Taken together these assumptions yield the following equation describing the rate of change in the number of susceptibles:

$$\frac{dX}{dt} = (1 - \varepsilon p)\Lambda - \mu X - X(\beta_X Y_X + \beta_V Y_V) - X\beta_M Y_M + \omega V \qquad (12.3)$$

Those in whom the vaccine takes enter the vaccinated class. In vaccinated individuals, who still retain their vaccine-induced protection, the rates of acquiring wild-type and mutant infection are reduced by the amounts ψ and ψ_M, respectively. ψ and ψ_M are the degree of protection afforded by the vaccine against wild-type and vaccine-resistant mutants, respectively (if the mutant completely avoids vaccine-induced immunity, ψ_M is zero). The number of vaccinated individuals, V, therefore obeys the equation:

$$\frac{dV}{dt} = \begin{array}{l} \varepsilon p \Lambda - \mu V - V(1-\psi)(\beta_X Y_X \\ + \beta_V Y_V) - V(1-\psi_M) \\ \beta_M Y_M + \omega V \end{array} \quad (12.4)$$

Upon infection, susceptible individuals enter the class Y_X where they stay until they die or recover at per capita rate μ_X. Vaccinated individuals enter the class Y_V after becoming infected and leave it at per capita rate μ_V. In some cases I consider the possibility that vaccine-resistant mutants can only arise within vaccine failures, so vaccinated infected individuals (Y_V) spontaneously switch to become infected with vaccine-resistant mutant at per capita rate m. The equations for people infected with wild type are therefore:

$$\frac{dY_X}{dt} = X(\beta_X Y_X + \beta_V Y_V) - \mu_X Y_X \quad (12.5)$$

and

$$\frac{dY_V}{dt} = \begin{array}{l} V(1-\psi)(\beta_X Y_X + \beta_V Y_V) - \\ m Y_V - \mu_V Y_V \end{array} \quad (12.6)$$

Individuals infected with the vaccine-resistant mutant die or recover at per capita rate μ_M so their dynamics are governed by the equation:

$$\frac{dY_M}{dt} = \begin{array}{l} X\beta_M Y_M + V(1-\psi_M)\beta_M Y_M \\ + m Y_V - \mu_M Y_M \end{array} \quad (12.7)$$

In what follows this model is used to investigate questions about the community-level impact of vaccines:

(1) the importance of duration of protection;

(2) the impact of vaccines that change the natural history of infection in vaccine failures; and

(3) the emergence of vaccine-resistant mutants either within vaccine failures or exogenously.

Take, degree, and duration

Let us consider three ways in which a vaccine could fail even before the target pathogen evolves to escape vaccine-induced immunity. A vaccine may have no effect upon some people, only taking in a fraction ε; may only reduce (without eliminating) the probability of infection upon exposure, only giving a degree of protection ψ; or may confer protection that wanes over time allowing vaccinated individuals to return to the fully susceptible state, only conferring protection that wanes at per capita rate ω. These possibilities are modelled using Eqns (12.3–12.7) with the simplifying assumptions that $Y_M = 0$ and $m = 0$ (no vaccine-resistant mutants arise) and that $\beta_X = \beta_V$, $\mu_X = \mu_V$ (natural history of infection is independent of vaccination status). Under these assumptions it is straightforward to calculate the vaccinated reproductive rate R_p:

$$R_p = (1 - p\varepsilon\psi \frac{\mu}{\mu + \omega})R_0 \quad (12.8)$$

What does Eqn (12.8) tell us? First it tells us that take ε and degree ψ are interchangeable in terms of community-level impact. A vaccine that takes in 80 per cent of people and gives complete protection has exactly the same vaccinated reproductive rate as a vaccine that takes in everybody but gives only 80 per cent protection. The counterintuitive part of Eqn (12.8) is the relationship between the community-level impact of a vaccine and the duration of protection it affords (McLean and Blower 1993; Anderson et al. 1995). The relevant measure of duration is not the rate of decay but the quantity $\mu/\mu+\omega$. Biologically this represents (in the absence of infection) the proportion of successfully vaccinated individuals who retain their vaccine-induced protection until they die. If an average lifespan is 50 years and the average duration of vaccine-induced protection is 10 years, then community-level impact is reduced to less than one-fifth of that of a similar vaccine giving life-long protection.

Equation (12.8) also makes explicit that different kinds of vaccine failure compound in a multiplicative manner. Thus a rather promising

sounding vaccine can have a quite small community-level impact. For example, a vaccine
that protects to a 90 per cent degree, takes in
80 per cent of people, and wanes with a half-
life of 20 years cannot (with a single round of
immunization) eradicate infection with any
infectious agent for which R_0 is greater than
1.5. It is the combination of sensitivity to duration of protection and multiplicative com-
pounding of different sources of vaccine failure
that might allow relatively modest increases in
risk behaviour to overwhelm the beneficial
effects of a prophylactic vaccine against HIV
(Blower and McLean 1994).

Definition of vaccine failures

In all of what follows I define vaccine failures
as people in whom the vaccine takes (they
initially seroconverted) but who nevertheless
become infected. For this modelling framework to be consistent with this definition I
henceforth set $\omega = 0$ (ω is the rate at which
protection wanes) so that different natural
history or excess selection pressure for the
evolution of a vaccine escape mutant apply to
all individuals ever successfully vaccinated.

Different natural history in vaccine failures

It is quite plausible that the natural history of
infection in vaccine failures will be different;
they might be (over the lifetime of their infectiousness) either more (Anderson et al. 1991)
or less (Longini and Halloran 1995) infectious.
Pessimism about the prospects for a vaccine
against HIV able to confer a high degree of
protection have led to the suggestion that by
reducing the levels of circulating virus during
primary infection a failed prophylactic vaccine
could indirectly affect the epidemic by reducing infectiousness rather than susceptibility
(Longini and Halloran 1995). Conversely, if
vaccine failures survive (and are infectious)
for longer than people not vaccinated, then a
vaccine could actually worsen the spread of
infection (Anderson et al. 1991).

In general it is worth considering the possibility that the natural history of infection may
be different in vaccine failures. This is repre-

sented in Fig. 12.1b by splitting those infected
with wild-type infectious agent into Y_X and Y_V
depending on which class they originate from,
and then allowing failures to have different
death/recovery rate (μ_V) and infectiousness
(β_V). As explained, $\omega = 0$. Let us define a basic
reproductive rate for vaccine failures, R_V (put
one vaccine failure into a wholly susceptible
population and count the number of secondary
cases). R_V summarizes all changes in natural
history in vaccine failures (or at least all that
pertain to their lifetime infectiousness). In this
model

$$R_V = \frac{\beta_V}{\beta_X}\frac{\mu_X}{\mu_V}R_0 \qquad (12.9)$$

If we assume that, overall, vaccination reduces
the lifetime infectiousness of a vaccine failure
by amount ξ we can write,

$$R_V = (1 - \xi)R_0 \qquad (12.10)$$

We can now investigate the relative role of
protection from susceptibility and protection
from infectiousness upon the community-level
impact of a vaccine. It follows that,

$$R_p = (1 - p\varepsilon(\psi + \xi(1 - \psi)))\,R_0 \qquad (12.11)$$

This has the reassuringly commonsensical
biological corollary that reductions in susceptibility and reductions in infectiousness are
interchangeable in their community-level impact (McLean and Blower 1995; Anderson and
Garnett 1996). Would this still be the case if
vaccine resistance were a problem?

Vaccine-resistant strains that arise in vaccine failures

What if replication of the infectious agent
within a vaccine failure were to lead to the
emergence of a vaccine-resistant strain that
could then be transmitted? To address this
question I use the general model with $\beta_V =
(1-\xi)\beta_X$, $\mu_X = \mu_V$, and $\omega = 0$ (as before,
natural history is different in everyone ever
successfully vaccinated). Intuition might suggest that if vaccine escape mutants evolved
within vaccine failures, it would be much better
to search for vaccines that give strong protection from infection rather than reducing infec-

tiousness. It turns out that this is only partially true. For the special case where $\psi = 1$ is attainable (i.e. successfully vaccinated individuals cannot be infected), then this is always the best option, for a vaccine with a perfect degree of protection can permanently prevent the emergence of a vaccine-resistant mutant. In any other case, in terms of R_p, ψ and ξ are still interchangeable,

$$R_p = \left(1 - p\varepsilon\left(1 - \frac{\mu}{\mu + m}(1 - \psi)(1 - \xi)\right)\right) R_0$$

(12.12)

Even if the wild-type infectious agent can be eradicated ($R_p < 1$), transmission of infection can only be halted if the vaccine-resistant mutants that arose in the vaccine failures are themselves unable to cause an epidemic in the vaccinated community. In the general case $\psi < 1$ the intuitive reasoning—that it is better to avoid vaccine failures—is therefore incorrect

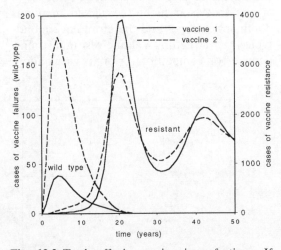

Fig. 12.2 Trade-offs in vaccine imperfections. If vaccine-resistant strains arise in vaccine failures, then: in the short term, vaccines with a higher degree of protection are preferable; but in the longer term, reductions in susceptibility or infectiousness are interchangeable. A numerical simulation of the complicated model using a step-length of 0.1 years and a fourth order Runge Kutta algorithm. Parameter values were $\Lambda = 2000$, $\mu = 0.02$, $m = 0.2$, $\mu_R = \mu_X = \mu_V = 1$, $\omega = 0$, $p = 0.8$, $\varepsilon = 1$, $\psi_M = 0.5$, $\beta_X = 0.0001$, $\beta_V = (1-\xi)\beta_X$ and $\beta_M = 0.9\,\beta_X$. For vaccine 1 $\psi = 0.9$ and $\xi = 0.5$, whilst for vaccine 2 $\psi = 0.5$ and $\xi = 0.9$.

in terms of R_p. It is, however, correct in the short term as illustrated in Fig. 12.2, which shows numbers of vaccine failures and of cases of infection with the vaccine-resistant strain for two vaccines. One gives a high degree of protection and a moderate reduction in infectiousness of vaccine failures ($\psi = 0.9$, $\xi = 0.5$, solid lines), the other protects only moderately but gives greatly reduced infectiousness in vaccine failures ($\psi = 0.5$, $\xi = 0.9$, dashed lines). The latter vaccine rapidly allows a large number of vaccine failures infected with wild-type vaccine, and these in turn quickly give rise to cases infected with the vaccine-resistant strain. However, *de novo* infection with the vaccine-resistant strain very quickly dominates the generation of new cases infected with vaccine-resistant strain so that the two vaccines are soon comparable in the number of cases of infection with vaccine-resistant strain. Ultimately, unless the within-host selection of such resistance is completely suppressed by a completely protective vaccine, it does not matter whether the failed vaccine gave strong protection from infection or greatly reduced infectiousness.

Vaccine-resistant strains that arise exogenously

The complicated model can be used to ask another question: what would happen if a vaccine-resistant strain arose from some outside source (McLean 1995)? In the absence of vaccination such a vaccine-resistant strain would be outcompeted if it had a lower R_0 than the wild type. Vaccination acts to shift the competitive balance between wild-type and resistant strains. If vaccine-induced immunity is less cross-reactive ($\psi_M < \psi$) than naturally acquired immunity, there may be a level of vaccine coverage above which the second strain will emerge as a result of the vaccination campaign. This situation is illustrated in Fig. 12.3a. Vaccination begins at time 3 years. There follows a period of very low incidence (the honeymoon period) before epidemics of the wild-type strain restart (McLean and Anderson 1988). Notice (Fig. 12.3b) that vaccine efficacy remains at 80 per cent during these post-honeymoon epidemics. The post-honeymoon epidemic that starts at time 15 years is a result of the slow accumulation of unvaccinated susceptibles. A small number of those who have been vaccinated are also infected because of the incomplete protection conferred by the vaccine. Several decades later a much larger epidemic occurs and vaccine efficacy plummets. The vaccine-resistant strain has achieved competitive dominance as a result of the growing number of vaccinated individuals. These vaccinated people are well protected against the wild-type strain, but have only minimal protection against the vaccine-resistant strain. The vaccinated reproductive rate for the vaccine-resistant strain is larger than that for the wild-type strain.

It takes several decades of accumulation of vaccinated people before this shift in competitive advantage manifests itself in epidemics of vaccine-resistant strain, but for this combination of parameters the effect is inevitable. It is not, however, an unavoidable consequence of vaccination. Highly cross-reactive and immunogenic vaccines can lead to the eradication of both strains at coverage levels below those at

which the vaccine-resistant strain gains the competitive advantage. Alternatively, low levels of vaccination leave the wild-type strain the competitive superior (McLean 1995).

Thus there are three possible explanations why we have not seen outbreaks of vaccine resistance in response to the major vaccination campaigns against childhood infectious diseases. The first (Fig. 12.3a) is that we haven't—yet. The second is that coverage is too low to give the competitive advantage to resistant strains. The third is that current vaccines give enough cross-immunity so that resistant strains will never emerge.

INFECTIOUS AGENTS WITH PRE-EXISTING STRAIN STRUCTURE

In the previous section I asked questions about the impact of vaccine resistance in infectious agents that bear highly conserved immunodominant epitopes and therefore have monotypic strain structure. Vaccine-resistant strains, being poor competitors prior to vaccination,

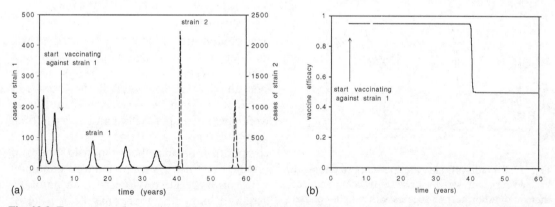

Fig. 12.3 Emergence of a vaccine-resistant strain. (a) At time 3 years, a vaccination campaign is introduced that reaches 80 per cent of newborn infants. The vaccine is 95 per cent efficacious against the circulating strain (strain 1) but only 50 per cent effective against the vaccine-resistant strain (strain 2). A 10-year honeymoon period ensues followed by post-honeymoon outbreaks of strain 1. Almost 40 years after introduction of vaccination there is an outbreak of the vaccine-resistant strain. ($\Lambda = 2000$, $\mu = 0.02$, $\mu_X = \mu_V = \mu_M = 25$ $\beta_X = \beta_V = 0.0025$ $\beta_M = 0.00125$, $m = 0$, $\omega = 0$.)

(b) During the post-honeymoon outbreaks, vaccine efficacy is unchanged. These outbreaks are a natural consequence of the non-linear nature of interactions between susceptible and infectious individuals. At time 40 years, when strain 2 emerges, vaccine efficacy falls from 95 to 50 per cent signalling the arrival of the new strain.

only emerge after vaccination is in place. But for many infectious agents complex strain structure is apparent before vaccination is introduced, and even before vaccination has acted to shift the competitive balance between strains, it is clear that the vaccine will give better protection against some strains than others (Nitta *et al.* 1995; Obaro *et al.* 1996).

In such a situation one would like to ask about the changes in incidence of infection with different strains subsequent to vaccination (McLean 1995; Gupta and Anderson 1997; Lipsitch 1997*a*). Since the strain structure is apparent before vaccination, it is also interesting to ask what would be the best way to design vaccines against such infectious agents. Is it best to target just one strain or is cross-reactivity the top priority in vaccine design?

To address such questions requires models which allow the coexistence of multiple strains. Two examples are illustrated in Fig. 12.4; they are simplifications of models introduced by Nowak and May (1994) and Gupta *et al.*

(1994*b*). In the first model, an individual infected with strain 1 can be superinfected by strain 2 leading to clearance of their strain 1 infection. Thus a strain with a lower basic reproductive rate can compensate for its lower transmission rate between hosts by an ability to outcompete the other strain within hosts. In the second model, infection with a second strain allows a double infection with doubly infected individuals being doubly infectious. In this model, coexistence is possible because cross-protection is incomplete: existing infection with one strain reduces the probability of infection (for a given exposure) with the other, without making it impossible.

For the first model it is possible for vaccination to cause the emergent dominance of a strain with higher R_0 (McLean 1995). What is more, there are circumstances where vaccinating against the most prevalent strain causes the overall prevalence of infection to increase. How can this be so? Before vaccination people infected with strain 1 could be superinfected

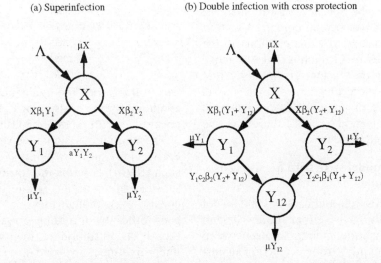

(a) Superinfection (b) Double infection with cross protection

Fig. 12.4 Different models of infectious agents with complex strain structure. (a) The superinfection model of Nowak and May (1994) allows competing strains to coexist through superinfection. (b) The cross-protection model of Gupta *et al.* (1994*b*) links strains through cross-protection, but allows coexistence if cross-protection is not perfect. The equations for the double-infection model (Lipsitch 1997*a*) are:

$$\dot{X} = \Lambda - \mu X - X\beta_1(Y_1 + Y_{12}) - X\beta_2(Y_2 + Y_{12})$$
$$\dot{Y}_1 = X\beta_1(Y_1 + Y_{12}) - Y_1 c_2 \beta_2(Y_2 + Y_{12}) - \mu Y_1$$
$$\dot{Y}_2 = X\beta_2(Y_2 + Y_{12}) - Y_2 c_1 \beta_1(Y_1 + Y_{12}) - \mu Y_2$$
$$\dot{Y}_{12} = Y_1 c_2 \beta_2(Y_2 + Y_{12}) + Y_2 c_1 \beta_1(Y_1 + Y_{12}) - \mu Y_{12}$$

with strain 2. Strain 2 has a shorter infectious period (hence its lower R_0 despite greater infectiousness). One consequence of superinfection is a lower prevalence of people infected with strain 1. Vaccination against strain 2 makes this 'rescue by superinfection' much less likely. By both direct and indirect effects, vaccination against strain 2 leaves a large pool of vaccine recipients who have been vaccinated against strain 2, then infected with strain 1, and will not be superinfected with strain 2. These individuals can cause an excess of cases in the years following vaccination.

Vaccination with one target strain in the cross-protection model

Lipsitch (1997a) has published an elegant study of the impact of vaccination using the cross-protection model. He investigates the impact of a vaccine that gives perfect protection against one strain, but little or no protection against the other. He shows that when only two strains are present, vaccination will always act to reduce the overall prevalence of infection, but that when three or more strains are present, strain-specific vaccination can increase the total number of people colonized with at least one strain. A key assumption in the model as applied by Lipsitch is that it is active carriage rather than acquired immunity that mediated cross-protection. However, the model as applied here assumes persistent infection and cross-protection could therefore be through acquired immunity.

A simplified model structure for the cross-protection model

The next step is to consider vaccines that do not protect perfectly against either strain. The aim is to be able to ask about the trade-offs between a vaccine with depth (strong protection against one strain, weak against the other) versus one with breadth (moderate protection against both strains). At this point it is useful to perform some algebraic jiggery pokery in order to see woods not trees. Lipsitch's study was based on the model in Fig. 12.4b, that is, the equations as in the legend. (He also allowed recovery which I have excluded for simplicity).

Since the total population is fixed, this four-dimensional model can be reduced to just three-dimensions, and with a suitable choice of new variables, coexistence and eradication criteria are easy to derive. Define x to be the fraction of individuals susceptible and λ_1 and λ_2 to be the forces of infection for strain 1 and strain 2, respectively: $\lambda_i = \beta_i(Y_i + Y_{ij})$. Furthermore, c_1 and c_2 are cross-reactivities. Under these definitions it is trivial to show that:

$$\frac{dx}{dt} = 1 - x(\lambda_1 + \lambda_2 + 1) \tag{12.13}$$

$$\frac{d\lambda_1}{dt} = \lambda_1[R_{01}(x + c_1 + c_1 x) - c_1\lambda_1 - 1] \tag{12.14}$$

$$\frac{d\lambda_2}{dt} = \lambda_2[R_{02}(x + c_2 + c_2 x) - c_2\lambda_2 - 1] \tag{12.15}$$

Coexistence is assured if and only if Y_1 can invade a community where Y_2 is already present. If Y_1 is absent then $x = 1/R_{02}$ and $\lambda_1 = 0$. Under these circumstances

$$\frac{d\lambda_1}{dt} = \lambda_1\left[R_{01}(\frac{1}{R_{02}} + c_1 + \frac{c_1}{R_{02}}) - 1\right] \tag{12.16}$$

which is positive precisely when:

$$\frac{R_{01}}{R_{02}} > \frac{1}{(1 + c_1 R_{02} + c_1)} \tag{12.17}$$

which along with the analogous, symmetrical condition for the invasion of Y_1 by Y_2 gives the coexistence criteria as previously defined.

These equations, simplified versions of those derived previously by Gupta (1994b) and Lipsitch (1997a), describe the constraints which, if satisfied, allow a pair of infectious agents to coexist, even though they may compete with each other by affording cross-protection to their hosts. Furthermore, their generalizations under treatment give guidance about the effect required to eradicate one or both agents from a host population. Thus they represent the equivalent of Eqns (12.1 and 12.2) for infectious agents with strain structure.

Eradication criteria

The eradication criteria for this model are precisely the coexistence criteria in the presence of vaccination. With imperfect vaccines and incomplete coverage a seven-dimensional system needs investigating with little hope of algebraic insight. So I have traded in the assumption of perfect vaccines for an assumption of perfect coverage. Assuming that everybody is successfully vaccinated (although the vaccine does not give full protection) is a major supposition, and represents a move away from an era when the limit on the community-level impact of a vaccine was how many people received it, to a situation where community-level impact is limited by properties of the vaccine itself. If everybody is vaccinated I can use Eqns (12.13–12.15) simply replacing R_{01} with $(1-\psi_1)R_{01}$ and R_{02} with $(1-\psi_2)R_{02}$. The eradication criterion for strain 1 is then:

$$(1-\psi_1) < \frac{(1-\psi_2)R_{02}}{R_{01}(1+c_1R_{02}(1-\psi_2)-c_1)} \quad (12.18)$$

Of course, if strain 2 were absent, the criterion for eradicating strain 1 would simply be $(1-\psi_1)R_{01} < 1$. As Lipsitch points out, when strain 2 is present, its competitive effect makes the eradication of strain 1 easier. This is because once an individual has been infected with strain 2 it becomes harder for that individual to be infected with strain 1. Exactly symmetrical results exist for the eradication of strain 2. Figure 12.5 plots these eradication criteria as a function of ψ_1 and ψ_2.

Vaccine design

Cross-protection can be exploited in targeting vaccine design for breadth or depth. In the simplest case, one might simply want to reduce the incidence of one strain to a minimum (if for example one strain were particularly pathogenic). It has been suggested that in this case the vaccine should be designed to target that strain and that strain alone in order to exploit the useful competitive effect of the other strain (Ewald 1996), although it is possible that cross-reacting antigens might still be useful. In the more general case, a reasonable goal would be to maximize the number of susceptibles

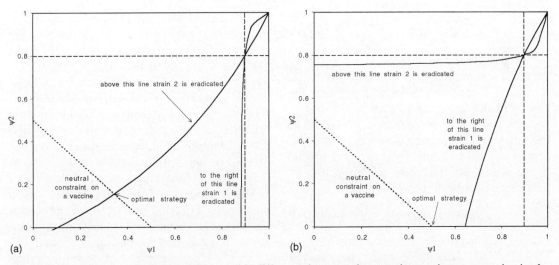

Fig. 12.5 Eradication criteria for vaccines with different degrees of protection against two strains in the double-infection model. The figures are reminiscent of the plots of coexistence criteria of Gupta *et al.* (1994*b*), except that the axes are transformed. In the context of vaccination, the converse of coexistence is eradication. Dashed lines show the eradication criterion for each strain in the absence of the other: $\psi_i = 1-1/R_{0i}$. As Lipsitch (1997*a*) points out, the presence of competing strains makes the eradication threshold lower (solid lines are lower and to the left of dashed lines). Dotted lines show an example of the neutral constraint, $\psi_1 + \psi_2 = 0.5$. The vaccine design strategy that maximizes the fraction susceptible is always to concentrate first on one strain (until it is eradicated or protection is complete) and then to use excess capacity on the other. The figure is from Eqn (12.18) with $R_{01} = 10$, $R_{02} = 5$. In (a) $c_1 = 0.8$ and $c_2 = 0.1$; in (b) $c_1 = 0.1$ and $c_2 = 0.8$.

(synonymous with vaccine recipients when coverage is 100 per cent, but degree of protection is less than 100 per cent). Solving Eqns (12.13–12.15) under vaccination for x gives:

$$x = \frac{-b + \sqrt{b^2 + 4ac}}{2a} \qquad (12.19)$$

where:

$$a = c_2(1-\psi_1)R_{01} + c_1(1-\psi_2)R_{02} - c_1c_2((1-\psi_1)R_{01}$$
$$+ (1-\psi_2)R_{02}$$
$$b = c_1c_2((1-\psi_1)R_{01} + (1-\psi_2)R_{02} + 1) - (c_1 + c_2)$$
$$c = c_1c_2 \qquad (12.20)$$

x is maximized when both infectious agents are eradicated giving $x = 1$. The question is only interesting when there are limits on the degree of protection that vaccines can impart, so that this ideal is not achievable. In that case the question becomes: what choice of allocating 'vaccine resources' between ψ_1 and ψ_2 will maximize x? The answer to this question depends on what limits vaccine construction. If building a given degree of protection is equal for all target strains then the constraint is simply

$$\psi_1 - \psi_2 = K \qquad (12.21)$$

I will call this the neutral constraint. It represents the assumption that it is equally difficult to endow vaccine recipients with a given degree of protection against either strain. In the next section I shall present an argument that the difficulty of achieving a defined degree of protection for a strain is proportional to the between-host infectiousness of that strain, in which case the relevant constraint is:

$$R_{01}\psi_1 + R_{02}\psi_2 = K \qquad (12.22)$$

I will call this the R_0-weighted constraint. It represents the assumption that (other things being equal) it is more difficult to achieve a given degree of protection against an infectious agent which has a greater basic reproductive rate. If the constraint in Eqn (12.22) is applied then x is maximized by concentrating on the strain with less cross-reactivity until its eradication criterion is met and then using any spare capacity to target the other strain. This is formally demonstrated by substituting Eqn (12.22) into Eqn (12.20), noticing that b and c are then constants and that under such conditions x is maximized by minimizing a. Does it mean anything in terms of biological intuition? The R_0-weighted constraint is a strong biological assumption because it says that any benefit accrued by constructing a vaccine against a strain with higher R_0 is offset by the difficulty in achieving a given degree of protection when R_0 is higher. Therefore (given that total eradication has been ruled out) it is best to benefit from the competitive effects of the more cross-reactive strain and target the less cross-reactive strain.

If both strains are equally cross-reactive, then under the neutral constraint the one with higher R_0 should be targeted until its eradication threshold is reached and any spare capacity used on the strain with lower R_0.

WITHIN-HOST DYNAMICS OF INFECTIOUS AGENTS

A discussion of optimal vaccine design is ultimately empty without an accompanying discussion of constraints on vaccine design. It is hardly worth saying that the optimal vaccine against any infectious agent is one that permanently renders all recipients totally immune after a single shot. Because of herd immunity, infection can be eradicated with vaccines that fall short of this ideal. Where eradication is not attainable, an optimal vaccine can only be defined in the context of a target (e.g. maximizing the number of people never infected) and a constraint. For the multistrained infectious agents discussed in the last section, the constraint is a rule that specifies how vaccine efficacy against one strain is traded-off for vaccine efficacy against the other. The neutral constraint is simply $\psi_1 + \psi_2 = K$. This represents the assumption that it is equally difficult to induce a given reduction in susceptibility to either strain: increasing the degree of protection against strain 1 from 10 to 20 per cent is exactly as difficult as increasing the degree of protection against strain 2 by the same amount.

This assumption certainly has the benefit of simplicity, but might different strains not vary in the effort required to endow a given degree of protection?

Relating infectiousness to constraints on vaccines

To address this question requires relating within-host dynamics of infectious agents to their epidemiology. First I ask: what is vaccine-induced immunity in terms of the within-host dynamics of a replicating antigen? Suppose a replicating antigen grows exponentially, but for clearance by effector cells and molecules of the inducible immune response,

$$\frac{dA}{dt} = rA - eAE \qquad (12.23)$$

A is the replicating antigen, r its exponential growth rate in the absence of specific immune responses, eE its clearance rate when a quantity E of effectors are present, and e the efficiency with which effectors act to clear antigen (McLean and Kirkwood 1990; Anderson and May 1991; Antia *et al.* 1994; Austin and Anderson 1996; Nowak and Bangham 1996). In this model, vaccine-induced immunity consists of forcing the effector population E to remain above a threshold size E^* defined by:

$$E^* = \frac{r}{e} \qquad (12.24)$$

So long as $E > E^*$, a dose A of the infectious agent trying to infect the host will have negative net growth rate and will rapidly become extinct: the host is immune. I take the size of the threshold required for immunity against strain i, E_i^*, as a measure of the effort required to induce protection against strain i.

To my knowledge we do not currently have direct measures of r and e for any infectious agent. Fortunately, for the purpose of constructing constraints, only relative measures are required. Turning to parameters we may be able to measure, consider the R_{0i}'s for different strains. By definition R_0 is the number of secondary cases caused by one case introduced into a wholly susceptible population. In the models used here, R_0 is the product of three quantities:

$$R_{0i} = N_i \times p_i \times d_i \qquad (12.25)$$

N_i is the number of potentially infectious contacts per unit time, p_i is the probability a potentially infectious contact between a case and a susceptible leads to infection of the susceptible, and d_i is the duration of infectiousness. For the same route of transmission in the same community, N_i is fixed across all strains of an infectious agent. p_i is largely determined by the within-host growth rate of the infectious agent in the absence of the induced immune response, r_i. In particular, r_i determines the probability that a given dose can establish an infection (the susceptibility of a susceptible), but it also plays a role in determining the infectiousness of the primary case. The third component inversely reflects the speed with which inducible immune responses clear infection— directly related to the efficiency with which immune effectors remove the infectious agent, e_i. If p_i is proportional to r_i and d_i is proportional to the reciprocal of e_i, then the thresholds E_1^* and E_2^* will fall in the same ratio as R_{01} and R_{02}. If, furthermore, the 'cost' of producing a given degree of protection is proportional to the threshold effector population required for vaccine-induced immunity, then the relevant constraint on vaccine design is:

$$R_{01}\psi_1 + R_{02}\psi_2 = K \qquad (12.26)$$

Antia *et al.* (1994) have studied the theoretical relationship between within-host pathogen dynamics and transmission. They find that infectiousness increases much more than linearly as a function of within-host growth rate and would therefore debate the assumption that p_i is proportional to r_i. As more data on within-host pathogen growth rates accumulate it should become possible to make an empirical study of this relationship.

OBSERVED AND PREDICTED EPIDEMIOLOGY

This chapter has investigated two different scenarios in which vaccine escape mutants might arise. The first is for monotypic pathogens in which the vaccine escape mutant would be an entirely new variant, not observed before the introduction of vaccination. The conclusion from the models is that, so long as vaccines continue to exploit the opportunities for broad cross-reactivity offered by the biology of such pathogens, the emergence of vaccine escape mutants is unlikely. The second scenario is when rich strain structure exists before vaccination is introduced. In this case, any vaccine that targets only a subset of the strains already observed is predicted to lead to increased circulation of the untargeted strains. This is because the removal of competing strains by the vaccine increases the opportunity for the untargeted strains to circulate. A third possibility, not discussed in the context of the models, is that a pathogen may start to exploit susceptible hosts after eradication of a competing infection and subsequent cessation of vaccination. Whilst dealing with large numbers of eradicated pathogens is not yet our problem, it seems worthwhile to consider the possibility that vaccination will have to continue post-eradication to prevent infection with antigenically related zoonoses.

Is measles evolving towards vaccine resistance?

Comparison of the sequences of currently circulating measles virus with historical samples shows that there has been an increase in the rate of nucleotide change in the measles haemagglutinin gene since vaccination became widespread (Rota *et al.* 1992). Furthermore, this sequence variation translates into antigenic differences between currently circulating strains and the strains that make up the vaccine (Tamin *et al.* 1994). Serum from individuals infected with current wild-type strains reacts four- to fivefold more effectively with wild-type strains than it does with the vaccine strain. Fortunately, the reverse is not true, serum from people who have recently been vaccinated has an equally strong antibody response either to the strain with which they were vaccinated or to current wild-type strains.

Thus, for the moment, there is no evidence that measles vaccine escape mutants are about to emerge.

Hepatitis B vaccine escape mutants

In contrast, vaccine-induced escape mutants of hepatitis B virus have already been observed (Carman *et al.* 1990). Since antigenic subtypes of hepatitis B occur naturally, it belongs in the section on models for infectious agents with complex strain structure. Since the vaccine is relatively new, there is no large pool of vaccine recipients to act as fuel for an epidemic of vaccine-resistant hepatitis B. However, as the number of people vaccinated against hepatitis B grows, the transmission of the variant hepatitis B virus must be considered. It is already being suggested that the variant sequence should be included in future vaccines.

The emergence of previously outcompeted zoonoses

The eradication of smallpox and consequent cessation of vaccinia vaccination is often held up as the holy grail of goals for vaccine strategies. As pointed out in this volume and elsewhere by Aaby (Aaby *et al.* 1995), vaccination has poorly understood but quantifiable benefits over and above the prevention of infection. There is a further reason why one might consider continuing with vaccination even after eradication of an infectious agent. The patterns of competition amongst strains discussed in this section apply to any group of infectious agents that share cross-reactive antigens—not just different strains of the same pathogen.

Monkeypox, smallpox, and vaccinia give an intriguing example. Before the eradication of smallpox, infection of humans with monkeypox virus was rare, and human-to-human transmission rarer still. Vaccinia immunization protects against monkeypox virus infection, and so, presumably, did immunity to smallpox. A recent outbreak of monkeypox virus in Zaire was characterized by large numbers of human cases (mostly amongst smallpox-naïve individuals) and long chains of human-to-human transmission. Thus it may be that first smallpox and then vaccinia immunization were protecting exposed individuals from infection with monkeypox virus. Now that smallpox has been eradicated and vaccination has ceased, a pool of individuals susceptible to monkeypox virus infection has accumulated and appears to have fueled an epidemic. Reintroduction of vaccinia immunization is being considered (Anon 1997).

13

EVOLUTION OF INFECTIOUS DISEASES: THE IMPACT OF VACCINES, DRUGS, AND SOCIAL FACTORS

Charles Bangham (rapporteur), Roy Anderson, Fernando Baquero, Richard Bax, Ian Hastings, Jacob Koella, Marc Lipsitch, Angela McLean, Tom Smith, François Taddei and Bruce Levin (chair)

Twenty-five years ago it was widely believed that infectious diseases had, at least in the developed world, been overcome. While most of this achievement was due to better hygiene, nutrition, and living conditions, the remaining infectious diseases could, it was thought, be prevented with vaccines or cured with antibiotics. But the emergence of diseases such as HIV, and the increasing problem of antibiotic resistance in bacteria such as those responsible for tuberculosis, pneumonia, or meningitis, have made it clear that this belief was exaggerated. The rapid replication and huge population sizes of microparasites (viruses, bacteria, protozoa, and fungi) give them enormous potential for the rapid evolution of variants that escape the immune response, become refractory to drugs, and adapt quickly to new ecological niches.

We must therefore design new and subtler strategies to control infectious diseases. In designing these strategies, evolutionary theory and mathematical analysis play an important role, for intuition is often inadequate to predict or quantify the effects of specific factors on the extent of spread and the nature of an infectious disease.

Recent improvements in both microbiological and mathematical techniques have made it possible to ask questions on the epidemiology of infectious diseases with unprecedented precision. In this chapter we summarize the main questions that we believe must be answered for a full understanding of the impact of vaccines, drugs, and social factors on the evolution of microparasites. The evolution of the host in response to selection pressure exerted by parasites is considered in Chapter 5.

VACCINATION

Introduction

Vaccines against infectious diseases have been developed almost exclusively by empirical means: a full understanding of the immune response to the parasite has been reached—if at all—only later. Several vaccines, such as those against smallpox and polio, have been remarkably successful, although it is not fully understood how these vaccines induce a protective immune response while other vaccines do not. However, there have also been important failures, such as the respiratory syncytial virus vaccine that was tested in the 1960s; and many existing preparations should be replaced with safer or more effective vaccines. There is also now an increasing need to develop vaccines against parasites that are highly antigenically variable, such as malaria, HIV, and hepatitis C virus. Until now, most successful vaccines have been developed against pathogens in which there is little antigenic variation between isolates.

Can we expect the past success of certain vaccines to be repeated with these parasites? There are two broad issues to consider. First, are there any adverse effects of vaccines on the distribution or the genetic composition of pathogens in the population? Second, how can we measure the impact of a vaccine, not only on the disease but also on the ecology of the parasite? The following specific questions should be considered:

1. What is the natural distribution of the parasite and its different strains in the population (Smith *et al.* 1993)? To understand the impact of a vaccine on a pathogen or the disease it causes, we need longitudinal data on the pre-existing prevalence of both the parasite and the disease in the population, carefully stratified by host age. For example, certain potentially pathogenic bacteria, such as *Escherichia coli* and *Streptococcus pneumoniae*, are widespread as commensal organisms and only occasionally cause disease (Finch and Phillips 1986).

2. What factors determine the 'strain structure' of the parasite population, that is, the genotypes and the relative frequencies of the parasite strains? Many parasites have genetic variants or strains that differ in their antigenicity, distribution, and virulence. For example, dengue fever virus has four major strains or serotypes, defined by antibody reactivity. These serotypes coexist in endemic areas, and there is evidence that type 2 is more pathogenic than the other types. However, it is not known how infection with one serotype affects the outcome of a subsequent infection with another serotype. Certain parasites have many more variants—there are over 100 distinct serotypes of rhinovirus, the common-cold agent.

3. Parasite strains are often characterized by certain combinations of alleles at several genetic loci. How do these multilocus genotypes persist in the face of the very frequent recombination that is often observed in parasite populations (see Chapter 16)? In particular, is the degree of antigenic similarity or cross-reactivity between parasite strains important in determining strain structure? Gupta *et al.* (1996) suggested that the pattern of antigenic cross-reactivity between alleles of a parasite gene determines both the gene composition of different strains and their distribution in the host population.

Even where distinct strains of a pathogen exist, is recombination between strains frequent enough to allow alleles that affect vaccine/drug resistance to segregate (and evolve) freely across the strain structure (Hastings and Wedgwood-Oppenheim 1997)? The frequency of recombination between strains depends on the extent of coinfection or superinfection of the host, which in turn may be affected by control strategies, such as vaccination, and social changes.

4. How does vaccination change the distribution of parasites or parasite strains in the host population? More specifically, vaccination against one strain of a parasite may, under certain conditions, change the strain structure or distribution in such a way that more (or different) disease results (Gupta *et al.* 1997; Lipsitch 1997*a*; and Chapter 12). This consideration may be important both in community-wide vaccination, such as the *Haemophilus influenzae* type b vaccine, and in the design of vaccines targeted at high-risk groups, such as vaccines for HIV and malaria.

5. If we remove a common parasite from the host population, are we exposing an ecological niche that could be occupied by another organism—perhaps a new pathogen? For example, in 1996 there was an outbreak of monkeypox infection in humans in Zaire: could the eradication of smallpox have left an ecological niche that could be filled by a new pox virus (Anon 1997)?

6. How might the impact of a vaccine on parasite ecology be affected by the precise regimen of vaccination? Selective vaccination of certain hosts, or changing from one-stage to two-stage vaccination programmes, might profoundly alter the impact of the vaccine on the epidemiology and transmission dynamics of the parasite. For example, children might be vaccinated against certain common infections either once when 2 years old, or at 2 years with a booster when 6 years old.

7. The widespread application of a strong and uniform selection pressure from a monotypic vaccine could favour the establishment of a vaccine escape mutant in the population.

Certain vaccines which have been highly successful, such as the live attenuated measles virus vaccine, could lose their efficacy if vaccine escape occurs and spreads (see Chapter 12).

8. Immunization against a single antigen or epitope of a parasite may be particularly liable to lead to the emergence of vaccine escape mutants. There is evidence in mice that immunization against a single virus epitope can lead to rapid vaccine escape, but simultaneous immunization against two epitopes prevents such escape, presumably because the frequency of simultaneous escape mutations in the two epitopes is too low (Weidt *et al.* 1995).

There is a clear analogy here between multi-drug treatment and multi-epitope immunization (see Anti-infective drugs and parasite evolution, below). Recent theoretical studies on the evolution of resistance in malaria (Dye and Williams 1997; Hastings 1997) have shown that increasing the number of components (i.e. drugs or epitopes) greatly decreases the rate of evolution of resistance, and indeed may prevent the development of resistance altogether. Many of the potential malaria vaccines currently under development contain three or four antigens.

9. Vaccination may alter the parasite distribution in unvaccinated as well as vaccinated hosts by changing the intensity of transmission in the population (Anderson and May 1991). An example of this is the effect of mass vaccination on the age-specific incidence of measles virus infection and its associated diseases. Again, careful prospective studies are necessary to ascertain this potential consequence of vaccination, particularly in the long term.

Potential confounding factors

The distribution of the parasite may change over time because of factors unrelated to vaccination, such as genetic drift. Such changes may, for example, be caused by changes in the host population size or movements (see below). Careful choice of controls is always important (Barbour *et al.* 1995).

In some cases one infection can directly alter

the host's susceptibility to a quite different infection. Thus respiratory virus infections can alter the bacterial flora in the nasopharynx, predisposing the host to a bacterial infection, and the immunosuppression that results from HIV-1 infection increases the likelihood of many other infectious diseases. Therefore, intervention in one infection may have unforeseen consequences for other infections.

Conclusion

To answer the questions raised here, two types of data are required. First, baseline data are needed on the abundance and distribution of parasite strains and the factors that determine this distribution. Public health surveillance should be expanded to incorporate the new scientific techniques such as polymerase chain reaction (PCR), to monitor not only the prevalence and incidence of infection but also the genetic changes and diversity in the parasite population (Felger *et al.* 1994; Holmes *et al.* 1995*a*). Second, surveillance data are needed to measure the vaccine coverage, and the immunological, microbiological, and clinical impact (see below) of vaccination. To be useful, this surveillance must be long-term, stratified by variables such as host age, include a detailed identification of different types of vaccine failure (see Chapter 12), and designed to answer specific questions. Dengue fever virus may be a particularly suitable test case for such a study, for it has a small number of well-defined, cocirculating serotypes. The improved experimental and mathematical techniques now available will make it possible to evaluate the evolutionary consequences of vaccination in a way that has hitherto been impossible.

ANTI-INFECTIVE DRUGS AND PARASITE EVOLUTION

Introduction

Tuberculosis, HIV-1, and malaria are three examples of serious infectious diseases where drug resistance is clearly detrimental both to the individual, because of failure of treatment, and to the community, because of the transmission of resistant microbes to uninfected hosts. There is widespread recognition that drug resistance is common (Bloom and Murray 1992; Cohen 1992, 1994; Levy 1992, 1994; Neu 1992, 1994), and can appear quickly in parasites in response to drug treatment. Less certain is the importance of the problem in infections that are not immediately life-threatening. The issues are analogous in many respects to those in vaccination (see above), and the top priority is also a full assessment of the extent of the problem.

An accurate assessment of the human costs of drug resistance in infectious diseases requires a careful, specifically designed study. A consideration of the methods needed to assess the problem, and of the difficulties involved in such studies, is given in Box 13.1 and in Chapter 11. We summarize here the questions that we believe should be considered first.

Specific questions

1. How quickly will a given change in the quantity or pattern of drug use reduce the present problem of drug resistance in parasites (Levin *et al.* 1997; Stewart *et al.* 1997; and Chapter 11)? Naïve recommendations to reduce antibiotic prescribing are unacceptable without a reasoned alternative and in the absence of secure evidence that a given reduction in antibiotic use will materially lower the prevalence of drug resistance.

2. What protocols for the use of single and multiple anti-infective drugs will: (i) maximize the efficacy of treatment of individual patients, (ii) minimize the rate of appearance and spread of resistance, and (iii) minimize the number of infected hosts before a drug loses its efficacy because of resistance? Two specific models that address these problems are discussed in Chapter 11.

Box 13.1: How to measure the impact of antibiotic resistance on human morbidity and mortality

Despite the interest that resistance to antibiotics has created in the scientific and lay press, both data and criteria are lacking to measure the effects of antibiotics on the clinical evolution of symptoms in individual infected patients and to measure the transmissibility of the pathogens. Most standard clinical trials do not provide evidence for effectiveness, differences between antibiotics, or appropriate treatment, at least in part because we do not have quantitative measures of outcome, for two reasons: the complex nature of the host–pathogen interaction during infection, and the variation inherent in clinical and laboratory results. In particular, short-term outcomes are difficult to measure, particularly in mild, self-limiting, acute infections. If it is hard to measure the effects of antibiotics on the outcome of common infections caused by susceptible bacteria, then evaluation of the impact of antibiotic resistance on human morbidity or mortality may appear unattainable. Here we analyse how to measure the impact of antibiotics on human infectious diseases at three levels: clinically, microbiologically, and socially.

Clinical evaluation strategies

1. Quantitative assessments should be independently formulated for each type of infection, taking into account the causative organism(s), the patient's condition, and the antibiotic that is being evaluated. Quantitative assessments should be conducted first in infections where the clinical outcome is clearly dependent on the microbiological outcome: for example, bacteraemia in neutropenic patients, meningitis, endocarditis, sexually transmitted diseases, and urinary tract infections.
2. The multifactorial outcomes of infections require composite markers for assessment of the progression of the infection, including objective measures of signs and symptoms. For instance, in respiratory tract infections, a rating scale can be developed for: body temperature; cough sputum volume, colour, and cytology; pleuritic pain; respiratory function parameters; white blood cell count; acute-phase proteins (including cytokines); data from X-rays; other non-invasive, lung morphological analysis procedures; and newly emerging symptoms. These markers should be combined, for example through a discriminant function analysis, in accordance with their ability to discriminate and their association with the pathogenesis of the infection. The resulting composite score would define the overall status of the patient at a given time and would be expressed in numbers reflecting the distance of the current status of

Box 13.1: Continued

the patient from complete health (score 1) or death (score 0). Thus the clinical score (CS) should be 1 $\geq CS \geq 0$.

3. The 'basal health' should be first defined for each type of patient, considering their age and other factors. When a quantitative assessment of the effect of antibiotics on the clinical outcome is planned, the basal health status of each patient should correspond to that expressed by the patient before the onset of symptoms of that infection. For instance, in a patient with a chest infection who has a ventricular dysfunction, the basal health status would be less than 1.

4. The assessment of the antibiotic effect should be measured by the change of the score over time in periodic observations, producing a curve tending towards cure or deterioration and representing the progression of the disease. Analysis of the trends may provide useful information, for instance in the form: time to recovery to the basal score, time to recovery to 50 per cent of the basal score, time to 75 per cent deterioration, or time to death.

5. Only when these analyses are complete can a measurement of the antibiotic effect on susceptible or resistant organisms be made. Antibiotics that are active against the infecting organisms should yield scoring curves quantitatively different from those obtained from patients treated with antibiotics inactive against the causative bacteria. These differences estimate the impact of antibiotic resistance on the dynamics of the patient's status and indirectly estimate the dynamics of the infection. This method quantifies the effects; it does not directly measure the causes.

Microbiological evaluation strategies

The study of the within-host population dynamics of infective organisms should provide a direct quantitative estimation of the activity of the antibiotic. This task is often limited by the difficulty of identifying the causative organism and the uncertainty of its sequential recovery from clinical samples. Nevertheless, in some cases (as in urine specimens, blood cultures, cerebrospinal fluid samples) conventional quantitative cultures may provide the required data. In multicontaminated samples, molecular typing procedures must be used to identify the bacterial population. Eventually, new antigen- or DNA-based quantitative technologies may be useful. Sequential data obtained throughout the infective process may reveal the appearance of strains with different *in vitro* antibiotic resistance.

The transmission rate from the infected patient to the family, community, or hospital may correlate with the antibiotic susceptibility of the pathogen. Evaluating such transmission rates is of particular interest in infections where the primary goal of the treatment is to prevent the dispersal of the infective agent (e.g. *Streptococcus pyogenes* infections).

Quantitative scoring should be used to express the results of studies of population dynamics and transmission rates so that they can be compared and integrated.

Social evaluation strategies

Social factors, often neglected in the analysis of antibiotic efficacy, frequently influence the appreciation of the importance of antibiotics—and consequently the importance of antibiotic resistance—by the patient, the patient's family, the doctor, and the community. Some social 'symptoms' have particular impact: number of days of school or work loss, number of days of hospitalization, number of hours per day of sleep loss in families, and number of doctor consultations. If the patient is spreading bacteria, secondary cases will multiply these problems. Interestingly, little has been done to evaluate the impact of antibiotic resistance on social 'symptoms'. Cases of death or permanent incapacitation caused by the pathogen should be included in the scores.

Another important social factor which can be quantitatively evaluated is the economic cost of resistance—the cost of failure. This includes both the extra cost of interventions—new consultations, new diagnostic procedures, new prescriptions that are frequently of more expensive drugs, progression of the disease, and new hospitalizations—and the direct social costs, such as time off work.

Conclusion

We do not know how much impact antibiotic-resistant pathogens are having. Data from these three strategies will help us assess the clinical, microbiological, and social effects of antibiotic use and thereby the impact of antibiotic resistance both on individual patients and on our society.

3. Most anti-infective drug treatment is empirical: it is based on the clinical judgment of the doctor, without a direct identification either of the causative organism or of its pattern of sensitivity and resistance to drugs. Would rapid identification of resistant pathogens influence the choice of drugs? If every doctor could rapidly identify resistant organisms in the clinic (Jacobs *et al.* 1993; Kapur *et al.* 1995), could this reduce the problem of drug resistance, and if so, how quickly? More rationally based drug therapy has obvious potential to reduce the frequency of treatment failure due to resistance and to allow the use of drugs with a narrower spectrum of activity. Such drugs should in turn reduce the intensity of selection for resistance in commensal organisms.

4. When should combinations of anti-infective drugs be used? The benefits of multidrug therapy of tuberculosis have been known for decades (Rist 1964; American Thoracic Society 1992), but there is still some debate as to the best strategy in HIV-1 infection. Elementary population dynamic theory suggests that the best strategy is to use combination drug therapy from the outset, rather than serial single-drug therapy. In practice, this clear recommendation must be tempered by considerations of toxicity and cost and (therefore) duration of treatment.

5. What, if any, is the contribution of parasites with intermediate levels of drug resistance to the evolution of clinically important resistance (Kaatz *et al.* 1991; Negri *et al.* 1994; Lipsitch and Levin 1998)?

6. The large-scale use of antibiotics in agriculture and fish farming carries particular possible dangers (Witte 1997). The use of several antibiotics may lead to the emergence of multidrug-resistant organisms in foodstuffs, and widely prevalent drug-resistance genes could subsequently be transferred to pathogenic organisms. These antibiotics may also alter the species and the genetic composition of environmental and commensal flora, with unknown consequences. The size of the contribution of this drug use to the clinical problem of antimicrobial drug resistance urgently needs to be measured.

What data are needed?

Comprehensive data are needed on: (i) the intensity of drug-induced selection, that is, the volume of drug use; (ii) the frequency and dynamics of drug resistance in the parasite; and (iii) the impact of drug resistance, measured in morbidity, mortality, and economic cost. Typically only one of these (i, ii, or iii, above) is studied at a time; to establish the qualitative and quantitative relationships between them, all three should be studied at once. How quickly does a given amount of drug use produce a given level of drug resistance in the population, and what is the impact of the drug-resistant parasite on health? A more detailed analysis of how to measure this impact is given in Box 13.1.

The evidence gathered in such studies could be used as the basis of clear recommendations on the policy of drug use at all levels. Because the policy may be effective only if adopted internationally, a broad consensus will be necessary on such recommendations. At present, there are large differences between and within countries in the frequency of use of anti-infective drugs, particularly the common antibiotics.

Potential confounding factors

A frequent problem in any study of drug use is the variable compliance of the patients: do they actually take the drugs prescribed? For example, non-compliance has been cited as the most important single factor in the emergence of resistance and failure of treatment in tuberculosis (Addington 1979).

The rate of emergence of drug resistance in a parasite may be affected by incidental exposure to the drug. About 50 per cent of the total volume of antibiotic use occurs in farming and fisheries. It will therefore be important to control for environmental exposure when measuring the impact of antibiotic use on the prevalence of resistance.

The biological mechanism of resistance, which determines the selection pressure exerted by drugs, is a major determinant of the rate of appearance and the rate of spread of resistance

Table 13.1 Common bacterial resistance to antibiotics

Antibiotic class	Mechanism of resistance	Origin	Inheritance	Micro-organisms	Diversity of resistance genes	Frequency
β-Lactams (penicillins, cephalosporins)	Detoxification β-Lactamase	Housekeeping genes Penicillin-binding proteins?	Chromosomal Plasmid (transposon)	Gram positive (Streptococci no β-lactamase) Gram negative	Great (all Gram negative have their own β-lactamase 300+ plasmid enzymes)	High e.g. 20–60% of E. coli have plasmid-mediated resistance
β-Lactams (penicillins, cephalosporins)	Modification of target cell-wall biosynthetic enzymes	Mutations of housekeeping genes	Chromosomal	Gram positive (Streptococci, Staphylococci) Infrequent in Gram negative	Dispersal of a limited number of modified genes	High in S. pneumoniae (10–50%) Epidemic in S. aureus (methicillin resistant)
Aminoglycosides (streptomcyin, kanamycin, gentamicin)	Detoxification Modification of target ribosome	Antibiotic-producing organisms? Mutations of housekeeping genes	Plasmids (transposons, integrons) Chromosomal	Gram positive Gram negative Mycobacterium	3 classes of genes for detoxification (50+ different genes) Chromosomal ('6)	High (10–25% in E. coli)
Macrolides Lincosamides (erythromycin)	Modification of ribosomal target, active efflux, detoxification	Antibiotic-producing organisms? Mutations of housekeeping genes	Plasmids (transposons, chromosomal)	Gram positive Few Gram negative	About 20 methylase genes modifying ribosomes and about 5 ribosomal mutants	High in S. aureus and S. pneumoniae about 20%
Tetracyclines	Active efflux mechanisms Protection of ribosomal target	Modification of housekeeping genes? Antibiotic-producing organisms?	Plasmid (transposons)	Gram negative Gram positive	About 15 Tet-r genes	High (10–30% in E. coli)
Chloramphenicol	Detoxification Active efflux	Modification of housekeeping genes? Antibiotic-producing organisms?	Plasmid (transposons)	Gram negative Gram positive	About 7 chloramphenicol acetyl transferases	High (10–30% in E. coli)
Glycopeptides (Vancomycin)	Modification of target in cell wall	Unknown	Plasmid (transposons)	Gram positive (Enterococcus and S. aureus)	Four Van genes	Variable incidence in enterococci (1–20%)
Quinolones (nalidixic acid, Ciprofloxacin)	Modificaton of targets: DNA gyrase or topoisomerases IV Active efflux	Housekeeping genes	Chromosomal	Gram negative Gram positive	About 6 different mutations	Rapidly increasing (1–10% in E. coli)
Sulphonamides Trimethoprim	Overproduction of target Metabolic bypass	Housekeeping genes	Plasmids (transposons, integrons)	Gram negative Gram positive	About 10 genes	Frequent, E. coli (20–30%)
Rifamycin (rifampicin)	Mutation in target (RNA Polymerase)	Housekeeping genes	Chromosomal	Gram negative Gram positive	About 3 mutations	Rare (less than 1% in E. coli)

in the population, and of its rate of disappearance when the selection force—the drug—is removed. To emphasize this point, the variety of resistance mechanisms in Gram-positive and Gram-negative bacteria is shown in Table 13.1.

The number of parasite (or plasmid) genes involved in drug resistance is a particularly important determinant of the population dynamics that result. In meiotic organisms, the number of genes involved in drug resistance is likely to matter because of recombination between resistance loci. In asexually reproducing organisms, the number of genes may be important in determining the likelihood of multiple mutations or simultaneous transfer of resistance genes by plasmids or other accessory elements.

Drug resistance may spread very quickly once it is established, but the initial acquisition of resistance may be slow because several genes frequently have to be assembled in one organism (Arthur and Courvalin 1993).

A parasite's resistance to drugs is often a quantitative trait. A bacterium may be conventionally defined as resistant if the minimal inhibitory concentration of the antibiotic exceeds a certain agreed threshold. However, the antibiotic may still cure the patient.

The population dynamics of drug resistance are also strongly determined by the pharmacokinetics and the mode of action of the drug (see Chapter 11). Here again, because of the non-linear nature of the dynamics, simple intuitive arguments may fail and mathematical analysis is essential.

Firms that manufacture anti-infective drugs should be required to make a broad, environmental impact assessment for the drug. This is already recommended as best practice and is observed by some firms.

Many community-acquired infections are cured with no treatment or with a simple course of drugs. Treatment may be more complex and more difficult in hospital patients, who are often older, more severely infected, and ill for unrelated reasons. Failure to take account of this difference may give rise to spurious differences in the efficacy of treatment between hospital patients and outpatients.

A stress applied to a parasite can rapidly change its rate of mutation or recombination or its life-history strategy. For example, a strong selection force such as an antibiotic can select for a higher mutation rate or recombination rate in bacteria (Mao et al. 1997; Taddei et al. 1997).

Conclusion

As with vaccination, a full and accurate assessment of the scale of drug resistance must be followed by studies that are designed to quantify the impact of specific measures on the prevalence of resistance and its impact on health and the economy.

SOCIAL FACTORS IN PARASITE EVOLUTION

There is widespread belief that an increase in population size, density, and mobility favoured the emergence and spread of certain diseases such as HIV-1. It is also possible that these demographic changes and other social factors influence the evolution of infectious agents, changing their incidence or prevalence, their virulence, or their resistance to drugs or vaccines. Such social factors are difficult to control for, and very large sample sizes are required. These factors should also be borne in mind as potential confounding factors in studies of the impact of vaccines or drugs (see above).

Specific factors

1. Travel gives a parasite the opportunity to enter a new host species or population and to recombine with other strains of the parasite. Patterns of travel within and between countries have changed greatly over the past few decades and will continue to change. The introduction of a new strain may result in competition with a pre-existing strain and alter the strain structure of the parasite population (see above).

2. Rapid population growth is causing the emergence of 'megacities' with populations over 15 million. This may lead to an increased number of endemic infections, since a certain

population size is needed to maintain the endemicity of a given parasite (Bartlett 1957). The increased host population size may also cause an acceleration in the net reproductive rate and (therefore) the potential for evolutionary change in the parasite.

3. Measures to increase hygiene may reduce the frequency of parasite transmission and select for reduced virulence in the parasite population (see Chapter 18).

4. Social factors at the level of the individual may have a material effect on the expression and dynamics of an infectious disease. Aaby (see Chapter 18) has obtained evidence that measles virus infection is more severe when acquired from family members, possibly because close contact increases the infectious dose. Sexual behaviour, especially the number and frequency of change of partners, strongly influences the rate of spread of sexually transmitted diseases, and by implication the potential of the pathogen to evolve.

5. Changes in practice in the food industry may also have direct effects on the spread of infectious diseases: prion diseases are a notable example.

Assessment of the impact of vaccines, drugs, and social factors on the evolution of infectious diseases

There is no generally applicable list of criteria by which this impact can be measured. The assessment will depend on an agreed definition of health and an agreed trade-off between longevity and health. These factors may be perceived very differently by patients and their doctors, and by society.

Illness may be considered as the decrease in an individual's normal functions and capabilities. The result of treatment of that individual for an infectious disease is likely to be measured by a combination of subjective criteria—such as health, illness, and social impact—and objective criteria—such as parasite load, fever, body weight, lung function, and economic consequences. These objective and subjective measures of impact have quite different respective applications.

SUMMARY

Because of the non-linear dynamics that result from the complex interactions of pharmacokinetics, immunity, strain structure, and drug resistance, evolutionary theory and mathematical analysis are essential for an understanding of the evolution of infectious diseases. The top priority is to get good data on the extent of drug resistance, the present distribution of types or strains of variable pathogens, and the factors that maintain this distribution. Comprehensive longitudinal studies will then be needed to measure the impact of vaccines, drugs, and social factors on the evolution and emergence of infectious diseases.

14

THE EVOLUTION AND EXPRESSION OF PARASITE VIRULENCE

Dieter Ebert

Selection has nothing to do with what is necessary or unnecessary, or what is adequate, for continued survival. It deals only with an immediate better-vs.-worse within a system of alternative, and therefore competing, entities. It will act to maximize the mean reproductive performance regardless of the effect on long-term population survival. It is not a mechanism that can anticipate possible extinction and take steps to avoid it.

George C. Williams (1966)

Studies of the evolution of virulence attempt to understand the morbidity and mortality of hosts (i.e. the virulence) that is caused by parasites and pathogens as the result of an evolutionary process. The evolutionary perspective on virulence seeks to determine the costs and benefits of virulence from the points of view of both parasite and host, with the aim of identifying the selective processes at work. The degree of harm inflicted on the host is the trait of interest, ranging from avirulent (asymptomatic) to highly virulent (rapidly killing). An understanding both of the fitness costs and benefits of virulence, and of the host and pathogen physiology involved in virulence-determining interactions, allows the evolutionary biologist to predict the expression of and the evolutionary changes in virulence.

The first general hypothesis about the evolution of virulence stated that it is an initial side-effect of novel host–parasite associations. It was widely accepted that evolution would lead to a gradual reduction in virulence, because parasites (here defined to include pathogens) that harm their host also harm themselves. This hypothesis has been rejected on theoretical and empirical grounds and has been replaced by a whole set of new hypotheses. Depending on the costs and benefits of virulence, historical constraints, and variance in

biological, ecological, and epidemiological aspects of host–parasite interactions, everything is possible (but not always predictable), ranging from the expression of highly virulent 'andromeda strains' to mild, avirulent infections.

Understanding the evolution of virulence is not just academic; it offers a conceptual framework to professionals in many fields and can contribute substantially to decision making. In discussing different hypotheses I have tried to consider the needs and expectations of different fields. Public health workers and epidemiologists are concerned with the prevention of disease and with the reduction of its average virulence. Their interest has a broad overlap with evolutionary biologists, for both are interested in populations as a whole. Agricultural biologists concerned with pests of livestock and crops fall roughly into the same category. In contrast, most of human and veterinary medicine deals with case-to-case considerations, with the aim of reducing the harmful effects of parasites in individual patients. At first, evolutionary biology seems to have little to offer them, but as I point out below, this is not always the case.

In the first part of this chapter I outline three frameworks which have been proposed to understand the evolution and expression of virulence. More details on the different hypotheses can be found in recently published reviews (Levin and Svanborg-Edén 1990; Bull 1994; Read 1994; Ebert and Hamilton 1996; Ebert and Herre 1996; Frank 1996; Levin 1996; Lipsitch and Moxon 1997). In the second part I summarize experimental studies relevant to the evolution of virulence and in the final part I compare different approaches to the study of virulence.

THREE FRAMEWORKS FOR AN UNDERSTANDING OF VIRULENCE

1: Virulence is a coincidental by-product

This hypothesis suggests that disease is a coincidental by-product of unusual (often novel) host–parasite interactions (Table 14.1). Parasites sometimes accidentally colonize a 'wrong' host or a 'wrong' tissue. The genes responsible for the expression of virulence in such cases did not evolve under the conditions in which they are observed by the investigator, and their effect on the host is hardly predictable (Levin and Svanborg-Edén 1990; Read 1994). The effects of the botulinum toxin produced by *Clostridium botulinum* certainly did not evolve to kill people. *Toxocara canis* can cause blindness, but this symptom is not the result of evolution within human hosts. The often deadly Ebola virus most likely did not evolve in association with man.

Little effort has been made to draw predictions from the coincidental virulence hypothesis. However, based on reciprocal transplant experiments a rough rule of thumb has been suggested: the average virulence in novel hosts decreases with the phylogenetic distance between the normal host and the new host, that is, the parasite is *on average* most virulent in the host population where it evolved and is *on average* least virulent in hosts which are very different from its normal host (Ebert 1994; Morand *et al.* 1996). Individual cases of high virulence in novel host–parasite associations have been reported, for example Ebola virus, Dutch elm disease, and rinderpest. However,

most avirulent 'novel' infections pass unnoticed, leading to a sampling bias of highly virulent interactions. Furthermore, the picture is often unclear because it is difficult to distinguish between genetic and acquired resistance. For example, the reasons for the high susceptibility of American Indians to diseases introduced by Europeans (e.g. measles) are still under debate (Black 1992). Only controlled experiments can avoid these biases. The important question to ask is: what is the explanatory power of the rule of thumb? My guess is it is not very high, and a given relationship found for one parasite might not be true for a different parasite. More research is needed.

It has been suggested that an increase in the contact rate of humans and their livestock with novel pathogens will increase the number of accidental infections and coincidental virulence (Garrett 1994). This could be relevant to our exploration and exploitation of the tropical rain forest, which undoubtedly harbours many as yet unknown disease agents, the Ebola virus being one example.

Interestingly, high virulence in novel hosts is not necessarily associated with high transmissibility or high parasite reproduction. In fact, some parasites with virulent effects in wrong hosts cannot transmit at all (e.g. the fox tapeworm *Echinococcus multilocularis* and *E. granulosis* when found in humans).

On the other hand, most established diseases were novel at one time. A formerly novel parasite will adapt to its new host and evolve to an appropriate level of virulence. Whether this evolved level of virulence will be higher or lower than it was during the first interaction of host and parasite depends on various factors. A good example is the myxoma virus which was introduced to control the European rabbit in Australia. The virus was highly virulent when introduced to Australia and evolved to intermediate levels within a few years. After several decades the most common virus genotype is still of intermediate virulence, indicating stabilizing selection for virulence (Fenner and Ratcliffe 1965; Anderson and May 1982). An established disease evolved from a novel host–parasite interaction.

The relationship between virulence and host genetic similarity has been considered in applied fields. Two hundred years ago Edward Jenner used the cowpox virus to immunize humans against smallpox. The cowpox virus is avirulent in humans, but it is similar enough to the smallpox virus that it induces immunity. One of the aims of biological control is to find virulent parasites which are able to reduce the prevalence (i.e. proportion of infected hosts) of the pest. The search for such (exceptionally) virulent parasites typically concentrates in the immediate taxonomic neighbourhood of the pest organism (the host). Of central importance is that virulence is specific to the pest, ensuring that other 'innocent' organisms do not suffer coincidentally.

2: Virulence is adaptive for the parasite

The most commonly discussed hypotheses about the evolution of virulence are based on the idea that virulence evolves because it benefits the parasite (Table 14.1). The benefits can be either direct—symptoms increase parasite fitness—or indirect—virulence is an unavoidable side-effect correlated with some other fitness component of the parasite. The evolution of the disease phenotype is assumed to be parasite driven and host evolution is too slow to play a major role. Three major hypotheses have been proposed.

The direct-benefit hypothesis

This hypothesis is used to explain single aspects of virulence as a parasite adaptation, rather than fitness reductions in hosts as a whole. The idea is to associate certain symptoms (e.g. diarrhoea, coughing, skin lesions) as being beneficial for parasite survival, reproduction, or transmission. For example, parasitic castration might increase the resources available for parasite growth and reproduction (Baudoin 1975; Obrebski 1975). Likewise, aberrant behaviour of infected hosts is thought to increase parasite transmission (Holmes and Bethel 1972). In many cases it is not clear why a particular symptom is expressed; it can be, at least partially, a result of the host's response to the parasite (Ewald 1980). For example,

coughing may be an induced behaviour that benefits parasites of the respiratory tract by enhancing the chance of transmission (Bull 1994) or coughing might be a host adaptation designed to expel pathogens from the respiratory tract (Nesse and Williams 1994). However, two further possibilities exist. A disease symptom could benefit parasite and host or neither. Possibly, coughing benefits both host and parasite. This example illustrates the difficulties in categorizing a symptom as a host or parasite adaptation.

To distinguish these four alternative hypotheses is of both fundamental and applied interest. Medical treatment should not inhibit symptoms expressed by the host to fight the parasite (Williams and Nesse 1991; Nesse and Williams 1994). If fever is part of the host's immune response to fight the parasite, drugs which reduce fever might help the parasite and prolong the disease. Using evolutionary insight into the costs and benefits of disease symptoms for hosts and parasites has considerable potential as a tool in everyday medical treatment.

The cost-of-virulence hypothesis

Anderson and May (1982) introduced a mathematical concept to study the evolution of virulence. Their model differs in two important aspects from the direct-benefit hypothesis. First, it is based on an epidemiological framework. Parasite fitness is determined by different components of survival, reproduction, and transmission. Second, virulence is a fitness component of the parasite: everything else being equal, the more virulent the parasite, the less fit it is (virulence is beneficial in the direct-benefit hypothesis!). Functional constraints between fitness components presume that a fitness enhancing change in one trait leads to a fitness reducing change in another trait. Given such trade-offs, evolution would select the parasite strain with the optimal combination of costs (virulence) and benefits (reproduction and survival). Depending on the details of the relationship between virulence and other parasite fitness components, any outcome is possible. Empirical work has confirmed that virulence is indeed correlated with other fit-

Table 14.1 Classification of hypotheses proposed to explain the evolution of virulence (see Bull (1994) for a different classification scheme)

Concept of model	Example
1. Virulence is coincidental Genes responsible for virulence evolved under conditions different from those in which they are presently considered. Virulence is a chance product, determined by the interaction of the novel host and the parasite genotype	Ebola, *Clostridium botulinum* toxin, *Toxocara canis* in humans
2. Virulence is a parasite adaptation The evolution of disease phenotypes is parasite driven. Host evolution is considered to be so slow that it can be neglected and hosts are considered to be invariant	
(a) Benefit-of-virulence An expressed symptom is directly beneficial for the parasite because it is correlated with higher parasite fecundity, survival, or transmission	Diarrhoea?, coughing?, skin lesions, castration
(b) Cost-of-virulence Virulence itself is detrimental for the parasite, but correlated with a fitness-enhancing trait. A negative correlation between virulence and other parasite fitness components is usually assumed. Virulence evolves as a balance between within- and between-host evolution. For horizontally transmitted parasites only host death is regarded as virulence	Myxoma virus, bacteriophages, vector-borne parasites, sexually transmitted diseases
(c) Short-sighted evolution High virulence is a consequence of within-host evolution, but it is not associated with a higher transmission rate. A high mutation rate in the parasite is adaptive in the long term, but detrimental in the short term	HIV?, bacterial meningitis?, polio?
3. The coevolution hypothesis Virulence is determined by specific interactions of host and parasite genotypes and is an ever changing trait in an ongoing host–parasite arms race. Since attenuation occurs after each transmission, virulence is usually below the optimum of the parasite	?

ness components of the parasite, such as fecundity, survival, and transmission (Anderson and May 1982; May and Anderson 1983; Bull and Molineux 1992; Ebert 1994; Ebert and Mangin 1997; Lipsitch and Moxon 1997). While this alone does not validate the hypothesis, the cost-of-virulence hypothesis has proved a powerful tool in understanding the evolution of virulence, in particular that of microparasites (viruses, bacteria, protozoa). Numerous modifications of the model have been proposed to cover a wide range of conditions (Nowak *et al.* 1991; Kakehashi and Yoshinaga 1992; Nowak and May 1993; Antia *et al.* 1994; Ewald 1994; Lenski and May 1994; Lipsitch and Nowak 1994; Lipsitch *et al.* 1995; Mangin *et al.* 1995; Bonhoeffer *et al.* 1996; Ebert and Weisser 1997).

A disadvantage of most versions of the cost-of-virulence hypothesis for horizontally transmitted parasites is that virulence is defined as parasite-induced host death rate. Morbidity and reduction in host reproductive success are not considered, unless their strengths correlate with the parasite-induced death rate, an assumption which is questionable (Ebert and Mangin, submitted). Morbidity is often directly related to parasite transmission success, and some models take morbidity into account.

Vertically transmitted parasites should not reduce the lifetime reproductive success of their hosts, because their fitness is directly correlated with the fitness of their host (Fine 1975; Bull *et al.* 1991; Herre 1993, 1995; Agnew and Koella 1997). In this case virulence must be defined as any fitness-reducing effect of the parasites. Other cases where morbidity is taken into account are those in which transmission is associated with host behaviour. Ewald (1983) proposed for vector-borne parasites that transmission to the blood-sucking vector would be increased if the host shows reduced levels of physical activity (lethargy). For sexually transmitted diseases, the argument works in the other direction. The less lethargic the infected host, the more likely it will continue its normal sexual activities and thus transmit the parasite (Lockhart *et al.* 1996). While vector-borne diseases have been claimed to be associated with high levels of host morbidity and mortality (Ewald 1983), sexually transmitted diseases are associated with comparatively lower levels of parasite-induced mortality (Lockhart *et al.* 1996).

An important extension of the cost-of-virulence concept incorporates within-host evolution (Axelrod and Hamilton 1981; Levin and Pimentel 1981; Bremermann and Pickering 1983; Nowak *et al.* 1991; Bonhoeffer and Nowak 1994; Nowak and May 1994; van Baalen and Sabelis 1995*a, b*; Frank 1996). High frequencies of multiple infections or mutations in parasites generate within-host competition, which is predicted to lead to an increase of within-host growth rates and higher virulence. The advantage these more virulent strains have during within-host competition does not necessarily translate into improved between-host transmission (Ebert and Mangin 1997). Transmission rate (per day) could increase in association with a higher growth rate, but the parasite's overall transmission success (per infection) might be reduced, for higher within-host growth rates usually shortens the lifespan of the infection. As a consequence, virulence is expected to evolve to balance within-host and between-host competition. All published models on this combined effect suggest that

virulence will increase when multiple infections become more common. I return to within-host evolution in the second part of this chapter.

A problem of the cost-of-virulence hypothesis which has received little attention is that confounding factors may result in contradictory predictions for the evolution of virulence (Frank 1996; van Baalen and Sabelis 1995*b*). For example, increased levels of parasite-independent host mortality lead to shorter periods of infection and reduce the frequency of multiple infections. While the former favours higher virulence, the latter favours lower virulence. The number of confounding variables can be large, and in natural populations the situation becomes very complex. Confounding factors are even a challenge in experimental set-ups, because it is often impossible to disentangle them (Ebert and Mangin 1997).

Although the cost-of-virulence hypothesis has little to offer for professionals dealing with individual patients, its mathematical rigour is appealing for those concerned with whole populations: evolutionary biologists, epidemiologists, health officials, and agricultural biologists. Practices based on its predictions can help to reduce the prevalence and/or virulence of infectious diseases. Recommendations have been addressed to health officials, although the feedback is so far rather poor (Bull 1994; Nesse and Williams 1994).

The short-sighted evolution hypothesis

An extreme case of the cost-of-virulence hypothesis arises when within-host competition leads to high virulence without transmission. Levin and Bull (1994) hypothesized that within-host evolution can result in drastically increased virulence when the microparasite mutant which is the proximate cause for the observed damage is favoured during competition among (clonal) parasite lines. Mutants with a high growth rate are able to replace other mutants and by this gain dominance within the host or within parts of the host, such as in the cerebrospinal fluid. This hypothesis states that competitive ability is not transmitted to the next host. Thus, virulence is a local phenomenon, similar to a cancer, which can spread locally but cannot propagate beyond the individual patient.

The short-sighted hypothesis applies, first, to parasites which regularly produce mutants that migrate to host tissues from which they cannot be transmitted to new hosts, but where they produce considerable damage. Typically, only a few patients suffer from the harmful consequences of this type of within-host evolution, while for the majority of hosts the infection is harmless. Examples are bacterial meningitis caused by *Haemophilus influenzae*, *Neisseria meningitidis*, and *Streptococcus pneumoniae* and poliomyelitis caused by an RNA virus of the picornavirus group (Levin and Bull 1994). The second scenario differs in that every infected host suffers from the harmful effects of within-host evolution. In long lasting infections, such as infections with the human immunodeficiency viruses, HIV, an arms race between the immune system and the evolving pathogen population can be observed, during which selection favours new mutants with a higher net growth rate. For HIV, it is currently believed that transmission from infected hosts occurs mainly during the early phase of an infection. Mutants with a high within-host fitness, but which reduce the fitness of the host late during an infection are not selected against, because they do not reduce the virus's transmission success (Levin and Bull 1994). Evolution within the host is short-sighted, because genetic changes are not passed on to the next generation of infections.

At first, virulence produced by short-sighted evolution appears to be non-adaptive. However, virulence arises as a consequence of high mutation rates, and these might be adaptive during the within-host arms race between the immune system and the parasite (Nowak *et al.* 1991; LeClerc *et al.* 1996). Thus, virulence due to short-sighted evolution can be seen as a cost associated with high mutation rates. In contrast to other adaptive hypotheses, the short-sighted hypothesis does not predict an optimal level of virulence as is the case for the cost-of-virulence and the direct-benefit hypotheses.

The short-sighted hypothesis is more appealing to a medical person dealing with individual patients than are the previous hypotheses. It might be possible to monitor the rise of new parasite mutants within the patient and to predict the evolutionary course of an infection. If severe disease occurs only in a fraction of all infections, it would be helpful to distinguish early on whether an infection will be harmful or asymptomatic. The short-sighted hypothesis is less important for epidemiologists and health officials, since changes in the host population (social practices, demography, ecology) would hardly affect the evolution of the virulent phase of the disease. For example, if the terminal phase of AIDS is entirely due to short-sighted evolution, changes in transmission patterns would have little influence on morbidity or mortality (Lipsitch 1997*b*).

3: The coevolution hypothesis

The above hypotheses, which assume that the level of expressed virulence reflects an adaptation of the parasite, ignore the fact that virulence is usually the result of an interaction between host and parasite genotypes. Variation among hosts within populations is common, if not ubiquitous, and virulence across different combinations of host and parasite genotypes has consistently been shown to vary markedly (e.g. Hill *et al.* 1991; Ruwende *et al.* 1995; Meyer and Kremsner 1996; Singh *et al.* 1997). In the absence of parasite evolution, virulence would decrease as a result of host selection for reduced costs of parasitism. In the absence of host evolution, parasites would evolve to exploit their hosts optimally. With reciprocal selection, virulence is expected to be an ever changing trait, balanced by the antagonistic evolution of hosts and parasites (Ebert and Hamilton 1996). The expressed level of virulence observed in a natural population would tend to be between the hosts' and the parasites' optimum and would vary within the population.

The host-genotype specificity of virulence is enhanced when microparasites spend many generations on an individual host and adapt to deal specifically with the defence of their host's genotype. Interestingly, such adaptation to one host genotype reduces the parasites' ability to exploit other host genotypes. Thus, in a genetically diverse host population, a parasite has to adapt anew whenever it infects a different host genotype (Ebert and Hamilton 1996).

This hypothesis makes a series of testable predictions. When parasites adapt to host monocultures one would expect their virulence to rise. The notorious sensitivity of agricultural monocultures to parasites is consistent with this prediction (Holguin and Bashan 1992). The current investment in the cloning of livestock needs particular attention here, for harmless diseases could evolve into deadly epidemics if they infect herds of clonal sheep or cows. Other examples of increased virulence in hosts with little genetic diversity come from serial passage experiments and are discussed in the second part of this chapter.

The coevolution hypothesis is of little practical value for medicine in hospitals. It can explain variation in morbidity and mortality among patients, but this is of more concern to epidemiologists and health officials. Its main application is likely to be found in its predictions for monocultures in animal husbandry and crop sciences.

Virulence and intensity of infection in macroparasites

The virulence of many parasites depends on the intensity of the infection. The larger the number of transmission stages entering the host, the higher the virulence of the parasite. This phenomenon is probably most important in macroparasites, but is also known from microparasites (Garenne and Aaby 1990; Hochberg 1991; Ebert 1995), where this effect is usually referred to as dose dependence. The common finding is that virulence increases while parasite fitness decreases with dose. Although these effects are known, no formal model has been proposed to understand their evolution.

Any formal model of the evolution of virulence with respect to dose effects has to include within-host evolution. At one extreme, a host could be infected with several transmission stages of the same genotype (parasite evolution would be exclusively between hosts), at the other extreme, all individuals within a host would come from different parasite genotypes (evolution would mainly happen within hosts). While group selection would play an important role in the former, individual selection would dominate in the latter case. Models of dose effects would be of great value in understanding macroparasite virulence, a subject which has not yet attracted much attention.

Tissue specificity

Infections are almost always specific to certain tissues or organs, and the overall well-being of the host might depend on which tissue is infected. For example, gut parasites are usually less harmful than blood parasites, and these are less harmful than infections of the central nervous system. Can this observation be explained with natural selection? To some degree, tissue specificity might reflect phylogenetic constraints. For instance, tapeworms are usually confined to the host gut. However, much variation for site of infection exists even within species. For example, different subtypes of papilloviruses infect different host tissues and cause different diseases. The bacteria *H. influenzae*, *N. meningitidis*, and *S. pneumoniae* are nearly avirulent when infecting the nasopharyngeal tract but are highly virulent when infecting the cerebrospinal fluid.

These relations suggest that in a comparative analysis of virulence, site of infection might explain more variation in virulence than any other factor. Interesting questions emerge: Is the expression of tissue-specific virulence constrained, coincidental, or an adaptation? Is virulence a criterion a parasite should consider during selection of host tissue? It has been suggested that parasites should prefer to infect organs with the least importance for host survival (Obrebski 1975). Are the relative costs of virulence for hosts and parasites tissue specific? For example, it has been argued that infecting the reproductive tissue does not impose a cost for the parasite, but it does for the host (Baudoin 1975). Tissue specificity of parasites is certainly an important covariable to consider when making cross-species comparisons of virulence.

EXPERIMENTAL EVIDENCE FOR THE EVOLUTION OF VIRULENCE

The most powerful approach to understanding the proximate and ultimate causes of evolutionary change are selection experiments. Experimenters can alter the factors which are thought to be responsible for natural selection so that selection is intensified, weakened, or reversed. Few selection experiments have been published on the evolution of virulence (Bull *et al.* 1991; Ebert and Mangin 1997). However, many studies report an evolutionary change in parasite virulence as a consequence of altered conditions, most of them being 'serial passage experiments'. Typically, these experiments involve the artificial infection of a novel, but related host, and subsequent transmission from one host individual to the next (e.g. by syringe transfer of blood). The host strains used are usually well defined and of low genetic diversity (inbred lines, full sib families, clonally propagated cell lines). Host-to-host transmission is controlled by the experimenter, and selection for the natural mode of transmission is relaxed. Serial passage experiments have often been cited in discussions of the evolution of virulence (e.g. Bull 1994; Ewald 1994; Ebert and Hamilton 1996; Lipsitch and Moxon 1997), but no thorough account of their impact has been completed. In the following I summarize the results of a literature study on serial passage experiments (Ebert, in preparation) and outline the main conclusions. Despite huge variation in the purpose and methodology of serial passage experiments, several results are consistent.

1. *Virulence increases during serial passage experiments.* The most remarkable and consistent result of nearly all studies is that virulence increases with passage number through the new host.

2. *Increase in virulence during serial passage experiments is host-genotype dependent.* Parasites passed through one host-type become 'attenuated', that is, their pathogenic effects in hosts different from those in which they were passed are reduced. Attenuated parasites are useful vaccines, for they can elicit an immune response without causing harmful effects (e.g. Sabin's polio vaccine, smallpox, rubella, measles, mumps).

3. *Within-host evolution drives an increase in within-host growth rate and virulence.* Several studies have shown that within-host growth rate and virulence increased because of mutations and recombination (Muskett *et al.* 1985; Bull *et al.* 1991; Bull and Molineux 1992; Ni and Kemp 1992; Novella *et al.* 1995). Competition trials between strains with different within-host growth rates have shown that the most rapidly growing strains outcompete slower growing strains (Ni and Kemp 1992; Novella *et al.* 1995). Thus, it seems that within-host competition between parasite strains drives within-host growth rate and that the growth rate is positively correlated with virulence.

Given these results, the question arises: why does virulence not increase under normal, 'non-passage' conditions? A general solution to this problem seems to relate to the evolution of between-host transmission. As mentioned above, the experimenter usually ensures host-to-host transmission regardless of the level of virulence, so that selection for host-to-host transmission is relaxed. Thus, serial passage experiments mimic endless within-host growth. Under such conditions, within-host competition drives selection for increased within-host growth rates, a process which might have costs in terms of reduced host-to-host transmission. In the following I discuss three hypotheses which might explain these costs and why under natural conditions virulence does not escalate. These hypotheses are not necessarily mutually exclusive. I do not discuss the possibility that the experimenter selected directly for virulence, although this might be the case in some studies.

Hypothesis 1: virulence increases because there is no 'cost of virulence' in serial passage experiments

A central tenet of the cost-of-virulence hypothesis is that a parasite which kills its host kills itself. This simple statement describes the idea that parasite-induced host mortality adds to parasite mortality rate and shortens the period during which the parasite is able to be transmitted. This cost of virulence can be balanced by benefits the parasite gains from reproduction, transmission, or the reduction of another source of mortality, such as the immune defence of the host (Anderson and May 1982; May and Anderson 1983; Bull 1994; Ewald 1994; Read 1994; Ebert and Herre 1996; Levin 1996). In serial passage experiments the parasite does not pay a cost of virulence, for the experimenter ensures transmission regardless of the level of virulence. Thus, in the absence of a cost of virulence, as is the case in serial passage experiments, the cost-of-virulence hypothesis predicts that virulence should increase.

Hypothesis 2: virulence increases because selection favours the production of within-host transmission stages at the expense of between-host transmission stages

Many parasites have different life stages, each being adapted to particular conditions. The production rates of alternative stages may be traded-off against each other, such that an increase in the production of one stage dictates a decrease in the production rate of another stage. For example, the malaria agent *Plasmodium* propagates asexually within its vertebrate host but has to produce sexual stages for transmission to its insect vector. During serial passage of *Plasmodium berghei* directly from mouse to mouse (without a vector), asexual growth forms became dominant. After 14 passages, a purely asexual strain had evolved: sexual reproduction had ceased (Dearsly *et al.* 1990). Other examples of parasites with alternate forms for transmission within and between hosts include bacteria (Ebert *et al.* 1996), nuclear polyhedrosis viruses (Miller *et al.* 1983; Kumar and Miller 1987), and *Trypanosoma* (Contreras *et al.* 1994). It is reasonable to assume that natural selection has optimized the production of different transmission stages (either by switching at the right moment or at the right proportion) under normal conditions to intermediate levels.

The trade-off between within-host growth and the production of host-to-host transmission stages explains the increase in virulence in some serial passage experiments, but this may not be a comprehensive explanation. The prediction which begs testing is whether transmission rates decrease in hosts with higher within-host growth rates. This prediction is the opposite to that derived from the cost-of-virulence hypothesis. Both hypotheses predict that lifetime transmission success is maximal at intermediate transmission rates.

Hypothesis 3: virulence increases because parasites adapt to specific host genotypes and there is no attenuation after being passaged to the next host

During serial passage experiments, parasites are exposed to a narrow range of host genotypes. Infection of a novel host, usually a different host species, or a different cell line, results in parasite attenuation, indicating that growth and virulence are adaptations to the host genotype it evolved in. Ebert and Hamilton (1996) proposed that virulence does not usually escalate in natural populations because genetic diversity among hosts prevents the parasite from evolving host genotype-specific virulence. In other words, because the parasite suffers from attenuation whenever host-to-host transmission occurs, it rarely has sufficient time to evolve high virulence on a single genotype. Genetic diversity among hosts hinders the escalation of virulence. Under such conditions rare host genotypes have a selective advantage.

Conclusions from the serial passage experiments

Serial passage experiments show that virulence can evolve rapidly and on ecological time scales. Observations of parasite virulence under natural conditions suggest that virulence usually does not change as drastically as it does in the laboratory. This difference is probably due to the unconstrained within-host evolution that occurs during serial passage, while in natural systems, between-host evolution would limit the escalation of virulence. The three hypotheses suggested rely on different mechanisms. At present, only the coevolution hypothesis can account for attenuation. The cost-of-virulence and the alternative-stage hypothesis do not consider host genetic diversity and host evolution. As the three hypotheses can be tested experimentally, we should soon be able to answer the question: what prevents virulence from escalating in natural systems?

HOW TO STUDY THE EVOLUTION OF VIRULENCE

The evolution of virulence is a young field. Consequently, only a few research projects have been devoted to hypothesis testing, and few examples have been published. However, the three main methods used in evolutionary biology to test hypotheses—observations, experiments, and the comparative method—have all been applied successfully. It is often difficult, if not impossible, to recommend a particular method, as many decisions will be based on practical considerations: Which hypothesis is to be tested? What experimental system is available? Are ethical considerations to be taken into account? What background information is available?

Observational methods

Observational approaches, which compare whether predictions match observed cases, are widely applicable but usually not conclusive. Cause–effect relationships are difficult to elucidate, for correlations do not allow us to exclude alternative hypotheses. Despite its shortcomings, the observational method is still the most commonly used method to infer evolutionary processes. Classic examples for the use of the observational method for virulence are myxomatosis (Fenner and Ratcliffe 1965; Anderson and May 1982) and the study of virulence in novel hosts (Holmes 1982). The strength of the observational method lies in its ability to provoke discussion and stimulate new hypotheses.

Experiments

The experimental method is certainly the most powerful approach to study evolution. Cause–effect relationships can be elucidated and alternative hypotheses tested. Unfortunately, the feasibility of experimental methods to study the evolution of virulence is inversely related to the body size of the host and often to its phylogenetic proximity to humans. Experimental studies have been carried out with small organisms, such as bacteria–phage systems (e.g. Bull *et al.* 1991) and a protozoan–arthropod system (Ebert and Mangin 1997). But can we conclude that what is true for a planktonic crustacean such as *Daphnia* is true for humans? The answer might be 'Yes, but ... ' for the evolutionary biologist, or 'No, but ... ' for most other professionals interested in virulence. The difference between *Daphnia* and humans is certainly large when hosts are considered, but both can suffer from viral, bacterial, or protozoan infections. What answer we accept depends largely on our expectations and on what methods are available. The speed with which microparasites and small hosts can evolve in nature and in the laboratory suggests a great potential for microparasite–host systems for use in selection experiments. The experimental method is the only method which can distinguish between the effects of confounding factors, and detailed studies of specific host–parasite system are certainly most convincing for decision makers in agriculture and medicine.

Comparative methods

The comparative approach is best suited for understanding large-scale patterns of virulence and is a powerful tool in testing and generating hypotheses. However, in contrast to experimental studies, general statements derived from comparative efforts are often vague and lack explanatory power, although they are of interest in basic research. Comparative evidence is not very convincing for professionals dealing with individual diseases.

An ongoing debate concerns whether comparative approaches must correct for lineage effects (Harvey and Pagel 1991). Most comparative studies on the evolution of virulence have not corrected for clade-specific effects (Ewald 1983, 1991, 1994; Herre 1993; Lockhart *et al.* 1996), and it is not clear whether a correction would alter the results obtained. In a cross-species comparison of sexually and non-sexually transmitted diseases, Lockhart *et al.* (1996) tested for possible effects of host and parasite clades. They concluded that disease characteristics are largely a function of transmission mode and disease ecology rather than the taxonomic status of either the host or the pathogen. Herre (1993, 1995) studied a group of closely related fig wasp–nematode systems and found a good fit to the predicted positive relationship between virulence and the degree of horizontal transmission. Phylogenetic correction would not alter his results since expressed virulence showed no clade-specific effects (Herre 1995). Correcting for clade-specific effects might still reveal more information and give more explanatory power, and I recommend the use of proper comparative methods where possible. However, patterns of the evolution of parasite virulence can be convincing without it. In contrast, phylogenetic components seem to play an important role in deciphering patterns of host evolution with respect to parasite virulence (Møller and Erritzøe 1996).

Molecular phylogenies will be increasingly relevant to understanding the evolution of virulence, and new uses for phylogenies are appearing at a rapid rate (e.g. Harvey *et al.* 1996; and Chapter 15). Powerful applications include the analysis of the origin of new infectious diseases (Sharp *et al.* 1995), within-host evolution of rapidly evolving pathogens (Holmes *et al.* 1992), and the understanding of host–parasite coevolution (Hafner *et al.* 1994).

CONCLUSIONS

While there is no lack of hypotheses there is a lack of data, including data on the virulence of human parasites and pathogens, even though we know more about human infectious diseases than about any other host–parasite system.

The evolution and expression of virulence can be analysed in three different frameworks: virulence is a chance product, virulence is an adaptation of the parasite, and virulence is the result of an interaction between hosts and parasites. Aside from preventive measures, there is little we can do to fight the effects of coincidental virulence. The two latter hypotheses, however, can help us to use evolution as a tool to reduce the harmful effects of infectious diseases.

The benefit-of-virulence and the short-sighted evolution hypotheses are both appealing for everyday medical practice in that individual patients can potentially experience short-term benefits. In contrast, the cost-of-virulence and the coevolution hypotheses are more appealing for long-term strategists such as population biologists, epidemiologists, and agricultural biologists. Decision makers can influence both within- and between-host evolution by modulating transmission, host population structure, hygienic standards, and demography. Agricultural biology is the first applied field which has profited from the emerging knowledge. Biological pest control and mixed crop approaches are just two examples. The success of attenuated viruses as live vaccines is a strong case for the application of evolutionary biology to fighting human diseases. New technologies and the incorporation of evolutionary thinking into traditional medicine will certainly produce more examples soon.

SUMMARY

Hypotheses on the evolution and expression of parasite virulence and their relevance to applied research are reviewed. The hypotheses fall into three frameworks. In the first, virulence is a coincidental expression of genes which evolved under conditions different from those in which they are currently considered. In the second, virulence is regarded as a parasite adaptation, while host evolution is considered to be too slow to be important. In the third, virulence is the product of the interaction between the host and the parasite genotypes and thus is a constantly evolving trait in an ongoing host–parasite arms race. Hypotheses can be further categorized with respect to the role epidemiology plays. Those hypotheses which include epidemiology are useful for public health officials and agricultural planners, while hypotheses which consider virulence in individuals are of greater interest for applied professions dealing with patients.

I further review evidence for the evolution of virulence from serial passage experiments. Typically, virulence increases during serial passage experiments because these experiments mimic prolonged within-host growth. Three hypothesis shed light on the factors which might limit the escalation of virulence under natural conditions. All three are based on the assumption that virulence evolves as a balance between within- and between-host evolution. First, increased virulence produces disproportionately large costs for the parasite and is therefore selected against. Second, there exists a trade-off between the production of within- and between-host transmission stages. Third, specific adaptations to one host genotype lead to attenuation of virulence in the next host genotype. Predictions are proposed which can be tested experimentally to distinguish between the three hypotheses.

ACKNOWLEDGEMENTS

I thank Natasha Sokolova, David Haig, Allen Edward Herre, Marc Lipsitch, Tom Little, Sven Krackow, Jacob Koella, Rustom Antia, Eddie Holmes, Bernhard Haubold, and Steve Stearns for valuable comments on earlier versions of the manuscript. Jürgen Hottinger provided help during various stages of preparing the manuscript. This work was supported by Swiss National Fond grant Nr. 3100–043093.95.

15

MOLECULAR PHYLOGENIES AND THE GENETIC STRUCTURE OF VIRAL POPULATIONS

Edward C. Holmes

It is difficult to exaggerate the impact of the HIV pandemic on the health sciences. As well as rekindling a general interest in infectious disease, it ushered in a new way to think about pathogen variation where the nucleotide sequence, rather than the serological profile, became the primary source of information. In HIV the amount of genetic variation observed was astounding: even single infected individuals harbour a diverse population of variants, often called a 'quasispecies'. However, HIV is not unique in this respect: sequence data have begun to accumulate from other viral populations, making it clear that genetic diversity, rather than being the exception, is perhaps the rule. This extensive genetic diversity, most apparent in viruses with RNA genomes, allows viruses to adapt quickly to changing environments, including immunological surveillance and drug treatments, and hinders our attempts to develop effective vaccines.

Given the practical importance of genetic variation in viral populations, it is crucial that this diversity be analysed in such a way that its full implications are understood. Traditionally, studies of viral population structure were only concerned with an important minority of variation, that recognized by the host immune system and hence detected in serological assays. However, with the development of the polymerase chain reaction (PCR) and nucleotide sequencing technology, analyses based on reconstructing the phylogenetic, or genealogical, relationships between sequences have come to the forefront. The phylogenetic tree, now an important analytical tool, is perhaps the most obvious way in which evolutionary ideas

have entered the virological community. Even a cursory glance at virology journals will indicate that phylogenies, once conspicuous by their absence, are now a fundamental part of any sequence analysis.

The development of phylogenetic analysis in virology mirrors that in population genetics, where a growing appreciation of the value of reconstructing evolutionary relationships between sequences (alleles), in the form of gene genealogies, has led to a quiet revolution. As in virology, the proliferation of phylogenies in population genetics was initiated by the realization that simple summary statistics, such as allele frequencies and average heterozygosities, could not adequately describe the full extent of genetic variation revealed by new, high-resolution molecular techniques. The advantage of the phylogenetic approach is that it takes into account *all* of the evolutionary information contained within a sample of sequences —both the frequency of any particular mutation and its relationship to the others in the sample —and therefore represents a more complete way to depict population structure. This does *not* mean that other analyses, such as those based on serology, should be replaced, as they obviously provide a critical functional perspective, but rather that a phylogeny depicts the contribution of every nucleotide and in doing so gives a better reflection of what has happened in the evolutionary past.

Coupled to the more widespread use of molecular phylogenies is a new awareness that they are more than just visual classification schemes. The multitude of new uses to which trees can be put is one of the most notable developments in recent evolutionary biology (Harvey *et al.* 1996; Huelsenback and Rannala 1997). We now know that different evolutionary processes, such as natural selection, genetic drift, and past history of population growth, have different effects on the pattern of evolutionary relationships between sequences depicted in a phylogeny. Consequently, a deeper analysis of phylogenetic tree structure can sometimes decipher which of these processes have happened in the past, and even when they occurred. Indeed, the success of phylogenetic analysis in evolutionary biology is that it comes complete with a well-formulated theoretical basis, called coalescent theory, which shows how different evolutionary processes affect the shape of phylogenetic trees. As theory develops it is likely that we will be able to recover even more detailed evolutionary signals contained in the historical sequence of mutations.

In this chapter I aim to illustrate how phylogenetic trees can provide important information about the genetic structure of viral populations. I focus on three specific areas where molecular phylogenies have already proven their worth and which may eventually assist in the control and eradication of viral diseases.

1. When and how have viral infections emerged?
2. What evolutionary forces structure viral populations?
3. What are the consequences of viral genetic variation for the clinical outcomes of infection?

EMERGING VIRAL INFECTIONS

Evolutionary biology and virology have already fused in the study of 'emerging' viral infections. In this broad area, much emphasis has been given to understanding the factors responsible for the seemingly increased burden of infectious disease faced by our species, particularly the impact of changes in human ecology, including greater urbanization, improved networks of global transportation, and de-forestation (Morse 1994). Where have these 'new' infections come from and when did they first appear? Are these diseases newly emerging or merely newly recognized?

The source of emerging viruses

A fruitful way to answer the where and when questions of viral emergence has been the phylogenetic analysis of nucleotide sequence data. Sometimes these studies have simply sought to reveal the taxonomic position of an emergent virus, while others have looked for the specific reservoir(s) occupied by a new infection. A dramatic example of the power of this latter form of molecular detective work was the discovery that a respiratory illness of high mortality affecting people in the 'Four Corners' region of the south-western United States was caused by a hantavirus circulating in the local deer mouse population (Nichol *et al.* 1993).

Much attention has focused on HIV-1 and HIV-2, potentially the most dangerous of the emerging viral infections of humans. Phylogenetic trees of sequence data from these and the related SIV viruses isolated from a number of other primate species have indicated that HIV-1 and HIV-2 have entered human populations independently, on more than one occasion, and probably from different simian sources (Fig. 15.1). For example, in the case of the more virulent HIV-1, it is clear that some people in parts of West Africa are infected with divergent viruses, referred to as 'subtype O', which are separated from the other HIV-1 sequences by the viruses found in chimpanzees (Sharp *et al.* 1995). Although a simian past is without doubt (movement from humans to monkeys is much less likely given the high numbers of infected animals and genetic diversity observed in the latter), it is less certain exactly which primate species have acted as reservoirs, particularly as many other African monkey species evidently harbour related viruses which have not yet been sequenced in large quantities (Hirsch *et al.* 1995).

Also unclear is when HIV-1 and HIV-2 first jumped into human populations. One way to answer this question is to compare nucleotide

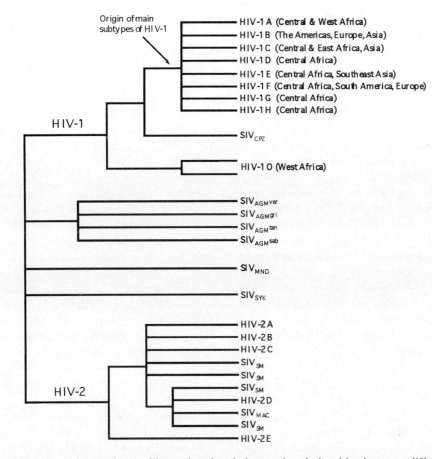

Fig. 15.1 Schematic phylogenetic tree illustrating the phylogenetic relationships between different primate immunodeficiency viruses. Branch lengths are not drawn to scale. Abbreviations for simian immunodeficiency viruses are as follows: SIV$_{AGM}$, African green monkeys (ver, vervet; gri, grivet; tan, tantalus; sab, sabaeus); SIV$_{MAC}$, macaque; SIV$_{MND}$, mandrill; SIV$_{SM}$, sooty mangabey; SIV$_{SYK}$, Sykes' monkey. For more detail refer to Hirsch *et al.* (1995) and Sharp *et al.* (1995).

sequences under the assumption that they diverge at a constant rate. Unfortunately, these 'molecular clocks' can be fickle time keepers, and appear especially erratic if the viruses in question differ in virulence, as less immunogenic viruses seem to evolve more slowly (Shpaer and Mullins 1993). These difficulties have so far prevented a consensus being achieved for estimates of the divergence times of the deepest nodes on the tree, such as the split between HIV-1 and HIV-2, although most workers place the origin of HIV-1 (with the exception of the subtype 'O' viruses) within the last 50 years.

Viral emergence and human ecology

As well as providing valuable information about the emergence of specific viruses, molecular phylogenies also help to answer questions about the factors facilitating the spread of infections. For example, did the transition from preagricultural foraging in the palaeolithic to farming in the neolithic greatly increase the disease burden on human populations (Chapters 8 and 22)? It has been suggested that the agricultural revolution at around 10 000 years ago gave rise to larger population sizes, increased sedentism, and long-lasting associations with animals, all of which would have aided the emergence and spread of infectious diseases. In contrast, palaeolithic hunter–gatherers may have suffered a more restricted range of infections than their agricultural descendants because their smaller population sizes set a limit on the availability of susceptible hosts.

Despite the appeal of this hypothesis, and theory which suggests that host population size is often an important factor in pathogen transmission, there is little direct supporting evidence. Although studies of palaeopathology have revealed an apparent increase in malnutrition, anaemia, and possibly in viral infections such as measles in farmers compared with their hunter–gatherer ancestors, only a small number of infections affect skeletal pathology, and the lesions observed are often not diagnostic. However, the revolution in molecular biology has provided us with another retrospective tool by which to uncover the history of human infectious disease—the analysis of the genomes of pathogens themselves. Through the techniques of phylogenetic analysis it should be possible to determine when in time a particular pathogen emerged or started to spread epidemically, even given the neurotic nature of molecular clocks. If the development of agriculture did increase the load of infectious disease on humans, we should expect our molecular estimates of emergence times for different pathogens to converge to this point. Similarly, the archaeological record preserved in the pathogen genome will make it possible to determine whether the spread of infectious diseases can be correlated with mass movements of human populations. Molecular data have already suggested that the increased spread of dengue virus and HTLV-I are associated with the exploration and colonization of the Americas which began in the fifteenth century (Dekaban *et al.* 1995; Zanotto *et al.* 1996*b*).

WHAT FORCES STRUCTURE VIRAL POPULATIONS?

Genetics and geography

One of the most important tasks facing the emergent science of molecular epidemiology is to determine the evolutionary forces which have built the structure of viral populations. Studies of this kind are now becoming commonplace, particularly those asking how and why viral populations are structured on a geographical level. HIV-1 again provides a leading example. Trees built from sequences of the structural *gag* and *env* genes have shown that the branches are arranged in a series of clusters, generally referred to as 'subtypes', sometimes as clades (Louwagie *et al.* 1993). At least 10 subtypes (denoted A to J) have now been identified world-wide, but it is unclear how far below the surface we have scratched. These subtypes are not phylogenetically equivalent: subtype A is more diverse than the others, and subtypes B and E are more homogeneous, probably because they represent the outcome of founder effects of viruses that only recently entered new populations. In the case of subtype B, the subtype associated with the western AIDS epidemic which began in the early 1980s, the host populations were in Europe, Japan, and North America, whilst subtype E has entered and spread quickly within South-East Asia, especially Thailand, within the last 10 years. That the majority of subtypes are found in Central Africa, along with most of the related monkey infections, makes this region the likely place of origin of the virus. Studies of HIV-2 have also uncovered the existence of five distinct genetic subtypes (A to E; Gao *et al.* 1994), but the data are not yet sufficient to determine their geographical distribution.

Mapping the global patterns of genetic variability in viral populations, as has been done for HIV, enables us to reconstruct the pathways of geographical spread and may aid in the design of population-specific vaccines and antiviral agents. These uses are also illustrated by studies on hepatitis C virus (HCV), another blood-borne and emerging human pathogen which, on a global scale, shows even more genetic variation than HIV-1. Although the origins of HCV remain a mystery, phylogenetic analysis has been used to classify this virus into six major genotypes, each containing a series of more closely related subtypes (Simmonds *et al.* 1993). These genotypes and subtypes have an interesting pattern of world-wide distribution. In the developed world most HCV infections are due to parenteral transmissions, such as the transfusion of blood products and needle sharing by injecting drug-users. Such restricted transmission routes have resulted in a restricted number of genotypes, which appear to be recently introduced, causing epidemics in individuals infected within the last 50 years or so (Holmes *et al.* 1995*a*). The situation is very different in developing countries where, in a specific locality, infections are predominantly due to a single major genotype, although this genotype is often very diverse and may contain a large number of subtypes. Such a phylogenetic pattern suggests that the virus has been evolving in these areas for an extended period of time, long enough for diversification at the subtype level to have taken place (Simmonds *et al.* 1996).

Retracing transmission networks

The examples presented so far have dealt with broad-scale patterns of genetic variation. Phylogenies can also be used to retrace, more precisely, the paths of transmission among different groups of people in a population, even between single individuals. Such analyses allow us to address questions of direct epidemiological importance. Do different behavioural groups possess characteristic viral variants, do viruses move from one behavioural group to another, and are new variants entering the population?

Once again, HIV has proved to be a valuable testing ground for studies of this sort. Intensive analysis of the HIV epidemic in Amsterdam indicated that viral sequences from injecting drug-users cluster tightly on a phylogenetic tree, probably because of needle sharing. Viral sequences from people infected through sexual contact form another distinct cluster (Kuiken *et al.* 1993). Injecting drug-users also form a distinct group in Edinburgh, where people thought to be infected through heterosexual contact also cluster with the drug-users. This suggests that injecting drug use is a bridge by which the virus can move from a 'high-risk' to a 'low-risk' behavioural group (Holmes *et al.* 1995*b*).

Similar studies are accumulating for other viruses. Molecular investigation of an HCV epidemic in Ireland revealed that up to 900 women may have been infected with the virus following exposure to contaminated batches of anti-rhesus D immunoglobulin administered in 1977 (Power *et al.* 1995). A phylogenetic analysis showed that sequences from the recipients formed a tight cluster on the tree, which included sequences from one of the implicated batches and the original HCV-infected donor, strongly supporting infection by this route. Given their power of resolution, molecular analyses are likely to form the backbone of future forensic studies in the health sciences.

Viral population dynamics

As outlined above, we can recover more evolutionary information from nucleotide sequences than simply the patterns of geographical variation in viral populations. Much progress has been made in understanding the dynamics of viral spread by examining the branching structure of phylogenetic trees (Harvey *et al.* 1994; Holmes *et al.* 1995*a*). The key insight, with roots in coalescence theory from population genetics, is that different evolutionary processes leave different signatures in the branching structure of phylogenies. For example, alleles maintained by balancing selection exist as polymorphisms for longer periods than neutral alleles whose fate is determined by genetic drift and hence produce longer branch lengths. Conversely, advantageous alleles fixed by a 'selective sweep' exist as polymorphisms for shorter periods than neutral mutations and generate a 'star-like' phylogeny with many short branches that represent variation that has arisen since the sweep (Takahata and Nei 1990). Different rates of population growth can also be distinguished by how they affect the shape of phylogenies: compared with a population that has maintained a constant size through time, a growing population will exhibit more splitting (coalescent) events near the root of the tree because population sizes will have been smaller at this time (Fig. 15.2).

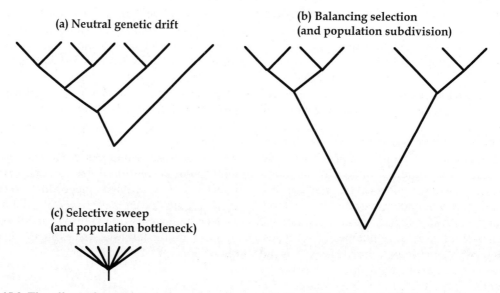

(a) Neutral genetic drift

(b) Balancing selection (and population subdivision)

(c) Selective sweep (and population bottleneck)

Fig. 15.2 The effects of several evolutionary processes on the branching structure of phylogenetic trees and the coalescent times of lineages.

The evolution of dengue virus, a mosquito-borne RNA virus which annually infects about 100 million people in tropical and subtropical countries (Monath 1994), illustrates how phylogenetic tree shape can be a rich source of information about the dynamics of disease transmission. In an analysis of the envelope (E) genes from many dengue isolates, Zanotto *et al.* (1996*b*) observed a rapid increase in the number of viral lineages in the recent past, suggesting a rapid acceleration in the rate of population growth (Fig. 15.3a). By plotting the number of dengue lineages against the place at which they appear in the phylogeny, and using molecular clock estimates to date nodes on the tree, it was shown that this dramatic expansion in viral population growth began about 200 years ago, perhaps because of a sharp increase in the size and mobility of the human host population at about the same time (Fig. 15.3b).

Analyses of this sort are being applied to other viral infections. For example, the branching structure of trees of HIV-1 shows that this virus has grown at an approximately constant exponential rate throughout its entire history (Holmes *et al.* 1995*a*). Our ability to infer underlying evolutionary processes is one of the

most profitable new uses for molecular phylogenies.

GENETIC VARIATION AND DISEASE

Their high mutation rates give viruses the potential to adapt rapidly to new and changing environments. The resistance speedily developed by HIV-1 to the antiretroviral drug AZT (zidovudine) serves as a sad illustration (Kellam *et al.* 1994). Because this adaptability is the result of variation, it is not surprising that viruses from the same population differ in their capacity to cause disease and in the symptoms they induce. Associating viral mutations with specific clinical outcomes is an important aspect of molecular medicine and another area where molecular phylogenies play a supporting role.

An example is presented by recent work on hepatitis B virus (HBV), one of the most common and serious viral diseases of humans. Infection leads to a variety of disease outcomes, ranging from subclinical infections to severe liver disorders, such as cirrhosis and

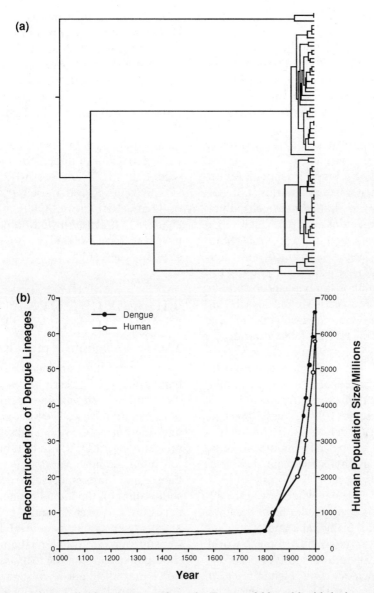

Fig. 15.3 (a) Phylogenetic tree linking sequences from the E gene of 66 world-wide isolates of dengue virus. To infer evolutionary processes one must analyse the distribution of coalescence times, which means that the tree must be reconstructed using a molecular clock. (b) Association between the logarithmic number of lineages of dengue virus against the time at which they appear, estimated from the tree shown in (a) (and drawn to the same time scale), and the corresponding growth of the world's human population. (From Zanotto *et al.* 1996*b*.)

hepatocellular carcinoma, and acute manifestations like fulminant hepatitis. What causes this variability in clinical outcome is under intense investigation. Some cases of serious HBV disease may be caused by a mutation in the precore region of the HBV genome (denoted A_{1896}), which produces a premature stop codon and hence inhibits the production of e antigen. A phylogenetic analysis revealed that A_{1896} is associated with several viral genotypes (six have been identified world-wide), particularly those found in South-East Asia, suggesting that these might represent more virulent viruses (P.L. Bollyky *et al.* unpublished). Furthermore, and contrary to previous expectations, close clustering of some of the A_{1896} mutations on the phylogenetic tree indicates that these mutations are directly transmitted between individuals, rather than arising *de novo* in each new case, and that superinfection with divergent viral lineages is a common occurrence, sometimes leading to recombination events.

Do the different subtypes of HIV-1 differ in transmissibility and virulence? For instance, the rapid spread of subtype E through South-East Asia has been attributed to this subtype's increased tropism for Langerhans cells, abundant in oral and genital mucosa, which might aid sexual transmission of the virus (Soto-Ramirez *et al.* 1996). If verified, this may greatly increase the transmission potential of the virus in populations previously considered as 'low risk'. However, it should also be remembered that an earlier and well-documented difference between HIV isolates, their ability to produce syncytia in cell culture, and therefore a marker of viral virulence (Schuitemaker *et al.* 1992), does not recognize subtype boundaries. Given the remarkable turnover rate of this virus, with some 10^{10} virions being produced each day (Perelson *et al.* 1996), it is likely that other biologically important mutations in HIV are as phylogenetically flexible as those leading to syncytia.

A key case for assessing whether viruses differ in virulence is dengue, the most common vector-borne virus of humans and a growing threat in many parts of the tropical world. Dengue virus exists in four serotypes (DEN-1 to DEN-4), which show a high degree of genetic variation—up to 40 per cent in the amino acid sequences of the E gene. What causes the most serious clinical manifestations of dengue infection—dengue haemorrhagic fever (DHF) and dengue shock syndrome (DSS)—is currently debated. The most popular hypothesis involves the coinfection of two different serotypes and the enhancement of the second infection with antibodies at subneutralizing levels developed from the first (Halstead 1988). If this hypothesis proves correct, then future antibody-based dengue vaccines must be highly cross-protective. If not, they may cause more virulent infections. Alternatively, certain serotypes may be inherently more virulent than others, with epidemics of DHF/DSS in Thailand, for instance, most frequently associated with DEN-2 (Burke *et al.* 1988). Furthermore, *within* each serotype viruses may differ in virulence. For example, two 'genotypes' of DEN-2 are present in the Caribbean, one introduced in the late 1960s and associated with mild forms of disease, and a second, more recent invader from South-East Asia, which may have triggered more serious epidemics of DHF (Rico-Hesse 1990). The nucleotide changes that cause this difference in virulence have yet to be isolated; phylogenies will help structure the search.

THE FUTURE OF EVOLUTIONARY VIROLOGY

The use and abuse of phylogenies

Phylogenetic trees are one of the most informative ways to dissect viral population structure, but, as with any new technique, there is potential for misuse. The biological meaning, if any, of groups observed in a phylogeny, such as the subtypes of HIV, has at times been misinterpreted. Some have assumed that, because they exist, clusters of sequences must have some common functional property. This is not necessarily the case: such groupings could equally well depict a neutral history of the transmission events linking them. Investigators should take care that by focusing on the mutations that distinguish clusters, they do not ignore important variation which has arisen within them. Rather than an endpoint in itself, the phylogenetic tree should be treated as a tool with which to understand other aspects of viral biology.

There is also a tendency to use phylogenetic trees to impose a structure on viral populations that does not exist in reality. At one extreme, some have tried to use molecular phylogenies of polymerase sequences to construct global classifications of RNA and DNA viruses (Koonin 1991; Ward 1993). These sequences are often so divergent that reconstructing phylogenetic relationships is more often an act of faith than of scientific endeavour (Holmes *et al.* 1996; Zanotto *et al.* 1996*a*). To complicate matters further, recombination, even between distantly related viruses, has clearly been so rampant (Gibbs 1995) that the concept of a continually diverging phylogenetic tree is undermined. Phylogenies at this scale have little practical use.

At the other extreme, genotyping viruses through phylogenetic analysis can produce a false sense of security. While such classification schemes are useful labels, there is a danger that the labels will acquire a status that distorts natural variation. The evolutionary process is unlikely to throw up evenly spaced clusters of sequences that can be neatly termed 'genotypes'. Those interested in viral classification need to adopt a policy of flexible response and accept that their system will doubtless include an arbitrary component.

Another way in which the genetic variation observed in viral populations can be shoehorned into ill-fitting concepts concerns the quasispecies. There is an understandable tendency to refer to a variable viral population as a quasispecies without having considered what this actually means. Whilst this may sound like a semantic issue, the term quasispecies comes laden with heavy theoretical baggage that is usually left in the hallway. The cornerstone of quasispecies theory is that viral genomes evolve at such elevated rates that even the sequence of highest fitness (the 'master sequence') cannot preserve itself intact for long against the mutation pressure (Nowak 1992). This process eventually structures the population into a 'cloud' of variants just a few mutational steps away from the master sequence, but which are able to mutate back to it with some probability. Therefore, as an outcome of the rapid mutation rate, selection acts on the whole distribution of mutants—the master sequence and its near neighbours—rather than on a single variant.

While the quasispecies is an appropriate and valuable concept for viruses as mutagenic as HIV, it is less clear that it is applicable to viruses in which mutation rates and levels of genetic variation are much lower and selection is absent. Such populations are simply old-fashioned 'polymorphic'.

Phylogenies of the future

The misuse of terminology is less important than developing methods to improve phylogenetic analyses. Whilst methods of tree-building are continually being refined to reflect more accurately what actually happened, and methods that assess the robustness of trees, such as the bootstrap, are more often applied, there is still much to learn about how processes like natural selection affect our ability to reliably recover the true phylogeny.

Although the methods are not yet perfect, a new sense of optimism in the field has been engendered by the recognition that they are already quite good. One particularly enlightening study concerned an HIV-1 transmission network established by a man, infected in Haiti, who then passed the virus on to several women in Sweden in a known sequence at known times. Trees reconstructed from sequences taken from members of this contact network were able to recover the correct topology, even though some of the transmission events were only a few months apart, although branch lengths were less accurate (Leitner *et al.* 1996). Furthermore, simulation studies using model tree topologies suggest that even complex phylogenies with large numbers of taxa can be reconstructed with surprising efficiency provided that a large and unbiased sample of lineages is available (Hillis 1996).

Another factor complicating the analysis of genetic variation through molecular phylogenies is recombination. Like bacteria (see Chapter 16), viral populations show considerable variation in their sexuality, from the promiscuous recombination that characterizes HIV to the celibacy of flaviviruses. Recombination complicates phylogenetic analyses, producing sequences with evolutionary relationships better depicted as an interconnected 'network' than as a continually bifurcating tree (Fitch 1997). Indeed, genes from the same viral isolate that produce very different phylogenetic trees are evidence for recombination, as is the case for some variants of HIV-1 and HIV-2 (Robertson *et al.* 1995). Recombination also makes it hard to draw any correlation between phenotypic character and phylogenetic position. There are two partial solutions to this problem. One is to detect recombinant sequences and remove them from the sample, the other is to devise evolutionary (coalescent) models which have the probability of recombination built in (these will be mathematically complex).

However, the quality of the sample of sequences used is the single most important issue in molecular epidemiology. Too many studies of viral population structure, including some done by me, have been happy fishing expeditions with little consideration of how sampling bias may have affected the quality of the inferences drawn. Here those working at the molecular level can learn from classical epidemiology. We must base our analyses on a properly structured sample in which variables such as geographical location, age of infection, and clinical condition are controlled.

The perils of biased sampling are illustrated by early studies of genetic variation in HIV-1 that did not capture the global diversity of this virus. The first HIV-1 isolate to be cultured and sequenced, the French virus HIV_{LAI}, almost became a 'type specimen'—the reference point for many later studies of viral biology, including work in vaccine development. Unfortunately, not only is HIV_{LAI} greatly dissimilar to the viruses circulating in those parts of the world where HIV-1 is most prevalent, it is also unlike most of its relatives in the western hemisphere. Vaccines based on HIV_{LAI} may therefore provide little benefit for the vast majority of HIV infections. Just as the notion of a 'wild-type' allele in population genetics no longer reflects the true extent of molecular variation, 'type specimens' in virology are no longer a valuable concept. Depicting the genetic variation with a phylogeny is much more informative and less misleading.

Despite the deficiencies mentioned, and the future improvements required, phylogenetic trees represent the most effective way to determine the genetic structure of viral populations. They can depict the relationships between all the mutations that have arisen in the history of a viral infection and be used to uncover the footprints of past evolutionary events. Phylogenetic trees are a tool emerging from the revolution in molecular biology that has transformed the medical and evolutionary sciences alike. More than that, they define an area where the two disciplines come together, for both are concerned with the origins and outcomes of the genetic variation observed at the molecular level. Already a central metaphor for the evolutionary process, in time, phylogenetic trees may acquire equal status in the health sciences.

SUMMARY

Genome sequence data have transformed biological science. Medical microbiologists can now pinpoint the mutations that give pathogens their distinctive properties whilst evolutionary biologists can use the pattern of mutations to reconstruct the taxonomic relationships between species. An analytical tool which links both these research goals is the phylogenetic tree. This chapter illustrates how molecular phylogenies can be used to reveal the population structure of viruses, organisms characterized by their genetic diversity, and why the phylogenetic tree is one of the main ways in which evolutionary ideas can assist those working in the health sciences.

ACKNOWLEDGEMENTS

My thanks to Denis Shields and Peter Simmonds for comments on the manuscript and to The Royal Society and The Wellcome Trust for support.

16

THE GENETIC POPULATION STRUCTURE OF PATHOGENIC BACTERIA

John Maynard Smith and Noel Smith

Bacteria are frequently cited as a good example of asexual reproduction. In fact, parasexual reproduction can be common. There are three parasexual processes: transduction, conjugation, and, perhaps most important, transformation. In transduction, a fragment of chromosomal DNA is accidentally incorporated in a virus particle. In conjugation, a chromosomal DNA fragment is transferred by a plasmid. In transformation, which appears to be an evolved bacterial adaptation, chromosomal DNA in the medium is actively taken up by related bacteria. These mechanisms of recombination in bacteria are fundamentally different from sex in higher eukaryotes in the following ways: sex and reproduction are not linked in bacteria; genetic material is unidirectionally transferred from a donor to a recipient strain; only a small part of the chromosome is transferred at a time. These differences in the mechanism of sex make profound differences in the effects of recombination on the genetic population structure of bacteria.

THE CLONAL PARADIGM

Until recently, it was widely believed that bacterial populations consist of a set of independently evolving clones (Selander and Levin 1980; Orskov and Orskov 1983). Some genetic properties could be horizontally transferred, but, with rare exceptions, chromosomal genes were vertically transmitted. The opinion was reasonable. Protein electrophoresis showed that some multiple-locus genotypes (ETs) cropped up repeatedly. For example, a survey (Caugant *et al.* 1987) of 15 loci in 650 isolates of *Neisseria meningitidis* found all loci polymorphic, with an average of 7 alleles per locus, yet 61 ETs were found more than once, and 19 were found in more than one continent and up to 15 years apart. If alleles associate randomly (that is, are in linkage equilibrium), as is expected if recombination is frequent, this would be impossible.

If true, the clonal hypothesis had two important implications. First, any trait, or set of traits, if sufficiently rare, could be used to characterize a strain. This provides a justification for the use of biotyping, serotyping, or electrophoretic typing to identify bacterial strains. Second, there could be no real hope of identifying 'species' of bacteria: we will return to this point later.

DOUBTS ABOUT THE CLONAL PARADIGM

Mosaic sequence structure

For one of the most widely studied genera, *Salmonella*, the clonal hypothesis has turned out to be near enough true and is supported by extensive sequencing studies of housekeeping genes from many strains (Selander *et al.* 1996). However, as DNA sequence data became available for other bacteria, it was noticed that two strains might resemble one another over much of the sequence but differ substantially in particular regions (Dykhuizen and Green 1986; DuBose *et al.* 1988; Stoltzfus *et al.* 1988).

The evolution of antibiotic resistance in *Neisseria* and *Streptococcus* illustrates processes responsible for this pattern (Dowson *et al.* 1989; Spratt *et al.* 1989). It appears that 'recipient' strains have become resistant to penicillin by the transfer of small blocks of DNA from a resistant 'donor' strain. In *Neisseria*, the donors have been identified as closely related non-pathogenic strains. The effect is particularly striking in *Neisseria* and *Streptococcus* because the sequence divergence between the donor and recipient is so large (up to 20 per cent nucleotide divergence) that the transferred blocks of DNA can readily be identified. Furthermore, because the transfer has presumably occurred since the widespread use of penicillin, the origin of the transferred blocks has not been obscured by subsequent mutation in the recipient strain.

In some cases it is possible to trace not the strain but a particular gene responsible for successive outbreaks world-wide. For example, genes coding for a modified penicillin-binding protein (PBP) in *Streptococcus* first appeared in New Guinea and Australia immediately after the Second World War and have since been responsible for outbreaks in Spain, Britain, and the United States. The gene has not, however, remained in the same genetic background, and has even transferred to previously sensitive non-pathogenic streptococci.

These facts suggested the 'clonal frame' model of bacterial populations (Milkman and Bridges 1990): a set of strains related by cellular descent will have relatively uniform 'frame' or 'background' DNA, peppered by insertions from various sources. Several statistical methods have been developed to identify the occurrence of such events (Stephens 1985; Sawyer 1989; Maynard Smith 1992). These methods are effective in recognizing rare recombinational events between rather distant strains (more than 5 per cent nucleotide divergence) rather

than frequent recombination between more similar strains.

Reanalysis of electrophoretic data

The first grounds for reanalysing multilocus electrophoretic enzyme data from a number of bacterial groups came from a survey of variation in *Neisseria gonorrhoea* (O'Rourke and Stevens 1993). These data turned out (Maynard Smith *et al.* 1993) to fit closely the hypothesis of random assortment between alleles at different loci (linkage equilibrium). For *N. gonorrhoea* the population structure suggests that the transfer of DNA sequences between strains is frequent in nature; it appears to be panmictic. Following Brown *et al.* (1980), Maynard Smith *et al.* proposed a measure of generalized linkage disequilibrium, appropriate for multiple loci and alleles per locus, which they called the 'index of association', I_A. For random association, the expected value of I_A is zero. (Its standard error is perhaps best estimated by generating random matrices with the same numbers of alleles at each locus.) The index of association primarily detects departures from linkage equilibrium, but the absolute value of I_A is a poor measure of how clonal a bacterial population is, for it depends on the number of loci studied and on the number of alleles found.

As an example of the use of I_A, Table 16.1 compares *N. gonorrhoea* with *Salmonella*. Although genetic variability is similar in the two groups, the gonococcus does not depart from linkage equilibrium, but *Salmonella* does, implying that the rate of recombination is higher in gonococci than in strains of *Salmonella*. One reason might be that '*Salmonella*' includes several 'species'. After all, if a similar test were made of a group of birds including blue, great, marsh, and coal tits, I_A would be significantly positive. In a further analysis *Salmonella* is

Table 16.1 The index of association

	Isolates	Loci	Number of different ETs	Mean genetic distance	I_A
N. gonorrhoea	227	9	89	0.31	0.04 ± 0.09
Salmonella	1495	24	106	0.35	3.11 ± 0.04

Table 16.2 Further analysis using the index of association

	Mean genetic distance	I_A
N. gonorrhoea	0.31	0.04
Salmonella (five 'species')	0.355	3.11
S. panama	0.04	1.34
S. paratyphi B	0.07	2.68
S. typhimurium	0.035	1.03
S. paratyphi C	0.05	4.66
S. choleraesuis	0.04	2.57
Bacillus subtilis	0.53	1.47
'B'	0.37	0.2 (ns)
'D'	0.38	0.46

ns, not significant.

divided into species inhabiting different hosts (Table 16.2). The reduction of genetic variance indicates that these are real genetic entities, and the I_A values indicate the there is little or no recombination within them.

In contrast, the *Bacillus* 'species', B and D, were erected on the basis of the electro-phoretic data (subsequently confirmed by sequence data). Recombination between the two species is rare (although possible in experiments), but the species themselves are close to linkage equilibrium. Lenski *et al.* (1993) have pointed to statistical problems that arise if the data are first used to identify 'species' and are then calculated for the species so identified. In *Bacillus*, the procedure is justified because species have since been confirmed by sequence data. In *Salmonella*, the species were identified prior to collecting the data. Further subdivision of the data would not lead to apparently pan-mictic groups because, after such subdivision, there would be insufficient variability left for statistical analysis.

What of *N. meningitidis*? The data quoted above show that the species is not in linkage equilibrium. Further analysis (Maynard Smith *et al.* 1993) suggests an 'epidemic' structure: that is, the species approaches linkage equilibrium, but occasionally a particular genotype increases dramatically in numbers and will ultimately be merged by recombination into the population as a whole. This interpretation

is plausible but needs to be confirmed by repeated studies. There is also room for improved methods of statistical analysis.

Gene trees

If sequences of two or more genes are available for the same set of strains, a phylogenetic tree can be constructed for each gene. Dykhuizen and Green (1991) suggested that, if different genes give the same tree, then the strains should be regarded as belonging to different species; if the trees are different, the strains belong to the same species. We doubt whether this can be a guide to the naming of species: in a clonal bacterium it would lead to ever finer splitting, whereas in *Neisseria* it would lead us to lump together very different bacteria. But the method is a valuable way of detecting recombination, or its absence. For example, Dykhuizen *et al.* (1993) showed that different genes in *Borrelia* give the same phylogenetic tree, suggesting that the species is truly clonal.

Applying the method to *Neisseria* shows just how complex bacterial population genetics can be. Sequences of *recA*, *ArgF*, and *Adk* have been obtained for seven named 'species'. The first two genes agree in placing four of these species (including the two pathogens: *N. gonorrhoea* and *N. meningitidis*) together in a closely related group (5 per cent or less nucleotide divergence): resolution within this group is uncertain, and occasional mosaic structure is observed. This group and the other three species differ from one another by 15–20 per cent of nucleotides: further resolution is impossible. Thus there is enough similarity between the two trees to suggest that recombination between the four main groups is rare. *Adk* makes nonsense of this conclusion. All seven species are about equally different (5–7 per cent), and no branch points are significant. The 'close' species are more distant for *Adk* than for the other genes, and the distant species are more similar. This only makes sense if recombination between all seven species has continued at the *Adk* locus, but not at the other two loci. We have no idea why this should be so.

The homoplasy method

This method (Maynard Smith and Smith 1998) is a sensitive way of detecting recombination between a set of closely related strains when sequence data are available. It is based on the idea that if (i) there has been *no* recombination and (ii) if the number of sites that could change is effectively infinite, then the number of steps in a maximum parsimony tree should equal the number of polymorphic sites: that is, there will be no 'double hits' or homoplasies. Thus homoplasies would be evidence of recombination. The snag, of course, is that there is not an infinite number of sites. Therefore, we have developed a method of estimating the 'effective site number' by using an outgroup, and hence estimating an expected number of homoplasies, given no recombination: this can be compared with the number expected if the population is clonal and with the number expected with random assortment between sites.

Some results given in Table 16.3 confirm the finding from comparing gene trees that recombination is absent in *Borrelia* and common in *Neisseria* (the data are for four closely related species). As expected, the rate of recombination in *Escherichia coli* is intermediate.

These methods are complementary rather than alternative. In particular, analysis of mosaic structure is particularly useful when seeking relatively rare transfers between strains, whereas the homoplasy method is effective in detecting frequent recombination between close relatives. The application of these methods to sequence data from bacteria has clearly shown that differences occur not only in

the population structure of bacterial groups but also in the frequency of recombination of genes from the same chromosome.

SPECIES, ECOTYPES, AND STRAINS

Are there species?

Two apparently contradictory points need to be made.

1. In practice, we cannot avoid using the term 'species', and we need to give names to populations. Despite the difficulty of defining bacterial species, we have not been able to avoid either practice in writing this chapter.

2. There are no entities in the bacterial world corresponding to the species of sexual organisms. In higher organisms, we can place organisms into distinct categories, called species, in such a way that interbreeding is common within a species but rare or absent between different species (or between their descendants or recent ancestors). The distinction is sharp (so long as one does not move about in space). At first sight one might hope that, in bacterial groups with extensive horizontal transfer, sexual species might be identified, but there are reasons for thinking that this is a false hope. There is no sharp distinction between cases in which two strains can exchange genes and cases in which they cannot. Thus Roberts and Cohan (1993) found that, in *Bacillus*, the frequency of exchange decreases continuously with genetic distance: there is no sharp cut-off. Vulic *et al.* (unpublished manuscript) found a similar continuous decline in enterobacteria.

Table 16.3 The homoplasy test

	Number of homoplasies:			Homoplasy ratio $(h-h_c)/(h_r-h_c)$
	expected if no recombination h_c	observed h	in shuffled matrix h_r	
Borrelia flagellin	0.43	1	8.4	0.07
E. coli mdh	2.74	8	26.7	0.22
Neisseria recA	5.99	17	25.0	0.58

We should not get excited about the 'species problem' in bacteria. Let us describe patterns of variation, and use names as we find them convenient. But we have to recognize that the term 'species' has no precise meaning, and that Latin binomials will have different implications in different groups of bacteria.

Ecotypes

Even in an entirely asexual taxon, we would not expect phenotypic or genetic variation to be continuous. Groups of bacteria will be adapted to different ecological niches: for example, in endosymbionts to particular host species or to particular sites within the host. The appropriate term for such a group is an 'ecotype'. Possible examples include:

1. 'Species' of *Salmonella* are often confined to different host taxa, although the situation is inconsistent: *S. typhi* is confined to humans, but *S. typhimurium* has a wider host range.

2. Two genetically distinct populations of *Pseudomonas*, described by Haubold and Rainey (1996), living on the leaves of sugar beet in the same field: each subpopulation was in linkage equilibrium, and there were differences between the leaf types on which they were found.

3. *N. gonorrhoea* is an ecotype adapted to the human urogenital tract. The extreme uniformity of its 'frame' DNA suggests that it is of recent origin, although it has acquired considerable variation by occasional horizontal transfer with other *Neisseria*.

4. Two genetically distinct populations of *Bacillus subtilis* were found living together in the same square metre of soil, although both have a world-wide distribution (Istock *et al.* 1992). At present, we do not know what ecological difference permits them to coexist.

As this list illustrates, there is a tendency to give Latin names to ecotypes if they are pathogens but not if they are free-living. The advantage of the concept of an ecotype is that it does not require us to invent a name for a particular population but does suggest that we identify its habitat and ecological adaptations. The existence of ecotypes has at least two implications for population structure in bacteria:

'Selective sweeps'

A favourable mutation that spreads to fixation in an asexual population will make that population homogeneous for the genotype in which it first arose. If there is recombination, the homogenizing effect is reduced proportionately. If the population is subdivided into ecotypes, the mutant will be fixed only in the ecotype in which it first arose (unless selection is very strong) and must then 'wait' until horizontally transferred into another ecotype. Ecotypic structure, therefore, has the effect of maintaining genetic variability despite selective sweeps, so long as there is some recombination (Maynard Smith 1991).

Adaptive divergence

Cohan (1994) has investigated theoretically the genetic divergence between ecotypes. An interesting finding is that two populations, able to exchange genes, can diverge genetically at loci adapting them to different habitats, yet maintain shared variability at selectively neutral loci.

Strains

Gupta *et al.* (1996) showed that two or more strains that differ at several antigenic loci can coexist despite recombination. That, is, genotypes AB and ab can be maintained in linkage disequilibrium. Simplifying, suppose that AB and ab are the common pathogenic genotypes. The common host phenotypes will then be $A*B*$ and $a*b*$, where $A*$, for example, means 'resistant to infection by an A pathogen'. Then AB pathogens can infect $a*b*$ hosts, and ab pathogens can infect $A*B*$ hosts, but recombinant pathogens Ab and aB can infect no hosts (assuming that a single antigen is sufficient for rejection of a pathogen). This is an example of linkage disequilibrium being maintained by epistatic fitness interactions, but unusual in that fitnesses are frequency dependent: thus a pathogen population consisting of Ab and aB only would also be stable.

Thus we can expect to find strains differing at more than one locus. This raises difficulties of interpretation. How do we distinguish between the stable maintenance of strains, as proposed by Gupta *et al.* (1996), and the temporary abundance of one or more strains which have recently increased dramatically in numbers under selection ('epidemic' structure; Maynard Smith *et al.* 1993)? The answer is that in the case of epidemic structure we would expect many neutral loci also to be in linkage disequilibrium due to hitch-hiking: if strains are stably maintained by frequency-dependent selection, neutral loci will be closer to linkage equilibrium.

DIVERSIFYING SELECTION

The surface of the cell is often subject to attack from a hostile and variable environment. For pathogens, this attack comes primarily from the host immune system, but all bacteria must evade viral parasites and protozoan predators. Figure 16.1 shows the average amino acid divergence of homologous genes of *E. coli* and *S. typhimurium* for five groups of cytoplasmic proteins and for a group of surface proteins such as flagellae, porins, and pilins (data from Whittam 1995). The average divergence in

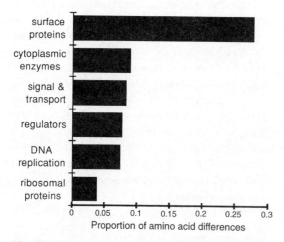

Fig. 16.1 The mean percentage of amino acid differences for six classes of bacterial genes (data from Whittam 1995).

amino acid sequence of cytoplasmically located proteins is 0.07 per cent, and of surface-exposed proteins is 0.28 per cent. Presumably the rapid rate of evolution of surface-exposed proteins reflects the action of diversifying selection, as well as relaxed selection on the cytoplasmic proteins (Whittam 1995).

More specifically, positive selection for diversity has been shown to be a major force acting on the surface-exposed regions of porins (*porB*) of *N. gonorrhoea* (Smith *et al.* 1995c). These highly immunogenic proteins function as pores allowing the passage of small molecules through the outer membrane. A model of the structure of these proteins predicts that eight regions are surface-exposed loops. Most of the variation in the porins is confined to these loops, and amino acid substitutions per site in the loop regions were at least 10 times commoner than synonymous substitutions. These observations suggest a gene that has been positively selected for diversity, a conclusion supported by epidemiological evidence: both transmission probability and the length of the infective state of *N. gonorrhoea* can be enhanced by amino acid changes in the loop region (Ison 1988; Plummer *et al.* 1989).

A more subtle effect of diversifying selection is the apparent increase in the rate of recombination of genes coding for cell surface components in bacterial species that are otherwise highly clonal, such as *Salmonella*. However, the highly diverse flagella filament genes frequently transfer between strains. This can be explained if strains bearing novel recombinant flagellin genes are selected (Nelson and Selander 1994). That is, a strain expressing a brand new flagellin gene is more likely to rise to high frequency in the population than a recombinant of a housekeeping gene. In this way the apparent recombination rate of flagellin genes has been enhanced, although the actual rate of transfer between strains is probably no higher than for housekeeping genes.

Bacteria have evolved many molecular mechanisms actively to increase the variation of gene products that directly interact with the environment. These mechanisms share the following properties: the variation is generated randomly; they generate variation at a relatively high frequency; they act at specific loci; and they involve intragenomic chromosomal changes. Known as phase variable systems (Saunders 1995) or 'programmed DNA rearrangements' (Borst and Greaves 1987), they have recently been referred to as 'contingency genes' (Moxon *et al.* 1994). As well as rapidly generating polymorphism in the surface layers of bacteria, contingency genes may also control the level of expression or specificity of many bacterial systems, including host cell interactions, chemotaxis, and motility (Saunders 1995).

SOME OUTSTANDING PROBLEMS

Why differences in recombination rate?

Why do bacteria vary from the panmictic to the virtually clonal? Are there ecological correlates? For example, are there differences between free-living and endosymbiotic bacteria, or between pathogens and commensals, or between bacteria with and without means of distant dispersal? Competence for transformation is found both in pathogens like *Haemophilus*, *Streptococcus*, and *Neisseria* and in soil bacteria like *Bacillus*. An association between competence for transformation and a low value of I_A is not universal: for example, electrophoretic data on *Haemophilus* indicate a clonal structure (Musser *et al.* 1990). There is still doubt about the selective forces responsible for the evolution of transformation: Redfield (1993) and Michod *et al.* (1988) have offered explanations that have nothing to do with horizontal transfer. The short answer to the question above is that we do not know.

What is meant by the 'rate' of recombination, and how should we measure it?

In organisms with meiotic sex, the rate of recombination means the probability that two genes inherited from the same parent will be transmitted to different offspring. It can readily be measured in the laboratory. No comparable definition is possible for bacteria, even in principle. The effects of recombination on population structure depend, at least, on the following factors:

1. Given that two bacterial strains are simultaneously present, how does the probability of horizontal transfer depend on the genetic distance between them and on the environmental conditions, such as population density?

2. How large a fragment is typically transferred (see McKane and Milkman 1995; Zawadski and Cohan 1995)?

3. How often do bacteria differing in genotype by varying amounts actually meet? How often does *N. gonorrhoea* meet *N. meningitidis*?

In practice, following Guttman and Dykhuizen (1994), we suggest that the parameter we need to know is: (Probability, per unit time, that a chromosomal site will change by recombination) / (Probability per unit time that it will change by mutation).

This parameter needs a name. How can it be estimated? The best bet seems to be to investigate, by computer simulation, the effects of varying the parameter on the genetic structure of populations—for example, I_A and the homoplasy ratio.

A GENE'S-EYE VIEW OF BACTERIAL EVOLUTION

In a strictly clonal population, a gene can increase in frequency under selection only if it increases the fitness of the cell in which it finds itself: there is no room for 'selfishness'. The prevalence of horizontal transfer forces us to ask whether the properties of genetic elements exist because of their effects on cell fitness, or whether we must also allow for their ability to increase by horizontal transfer. This has always been clear for plasmids and phage, but we must now ask the question also for chromosomal genes.

Two recent suggestions have been made along these lines. Kusano *et al.* (1995) have suggested that type II restriction-modification systems are selfish elements. In brief, such an element consists of two tightly linked genes, one coding for an endonuclease cutting DNA at a specific sequence, and the other protecting that sequence by methylating it. The usual interpretation of such elements has been that they protect the bacterium against phage infection: the bacterial chromosome is protected by methylation, but foreign DNA is not. Kusano et al. point out, first, that such protection is very short-lived (see Korona et al. (1993), who argue that bacterial immunity to phage infection depends mainly on changes in surface receptor sites), and second, that if protection was the correct explanation, there would be no reason for the two genes to be linked. They suggest instead that these are selfish elements, ensuring their own maintenance and spread by a mechanism analogous to that ensuring the spread of meiotic drive genes in higher organisms.

A second proposal along similar lines (Lawrence and Roth 1996) is that the linkage of functionally related genes into operons exists, not because it facilitates coregulation, but because it facilitates cotransfer. Their idea is that essential genes (e.g. those needed for DNA synthesis) are usually not linked in operons, whereas genes performing functions that are only occasionally called for (e.g. utilizing lactose) do exist in operons. The comparative data support the idea (although there are snags: ribosomal proteins are in operons). The idea is also supported by the fact that genes in operons often show signs of recent introduction into the bacterial chromosome, in possessing an unusual G + C content.

A final example of a problem that requires a gene's eye view is the evolution of the mutation rate (Taddei et al. 1997). Bacterial populations that have been exposed to directional selection in the laboratory often turn out to carry mutator genes (Lenski and Travisano 1994). Genes slightly raising the mutation rate are not uncommon in natural populations of E. coli (Matic et al., unpublished manuscript). Such observations make sense, for a population carrying a mutator gene will adapt more rapidly to new circumstances. But any attempt to model the process has to allow both for the advantages and disadvantages that a mutator gene acquires simply through being linked to the mutations it causes.

17

WHOLE-GENOME ANALYSES OF PATHOGENS

E. Richard Moxon

Automated DNA sequencing has revolutionized biology by making available the immense fund of information contained in the genome of all organisms. In retrospect, it is surprising that it took so long to apply the rich potential of the available technology to whole-genome sequencing of microbes. Jacques Monod quipped that 'what is true for *Escherichia coli* is true for the elephant', yet despite the enormous interest, funding, and progress on the human genome project, sequencing the genome of *E. coli* languished for many years owing to the lack of public or private funding.

The turning point was the completion of the *Haemophilus influenzae* genome, 1.8 megabases (Mb) of double-stranded, circular DNA, by The Institute for Genome Research (TIGR). This achievement took the scientific world by surprise and elicited great excitement as its potential utility to public health and wealth was grasped. Many had not known that the sequencing of the complete genome of *H. influenzae* was in progress, and even microbiologists had underestimated the impact of this milestone. Within 2 years, the potential availability and utility of the complete genomes of several microbes (bacteria, archaea, fungi, and parasites) had attracted great interest. There are now firm plans to sequence 50 or so microbial genomes of outstanding public health and industrial importance within the next few years, with financial backing provided by government, major charities, and the commercial sector.

In this chapter, I have two aims. The first is to provide some perspective on the sequencing of whole genomes of bacteria because this has had major impact on biology. While it has fundamentally changed theoretical and experimental approaches to microbiology, including

infectious diseases, the implications go much further. In the second part of the chapter, I present some ideas on how whole-genome sequences can be used to identify the prevalence of sequence motifs involved in hypermutation, antigenic variation, and their role in host adaptation of pathogens.

WHOLE-GENOME SEQUENCING OF MICROBES: A REVOLUTIONARY TOOL

New methods were one of the interesting aspects of the sequencing of the *H. influenzae* genome. Most current genome projects construct a physical map in which overlapping cloned fragments are ordered and then sequenced. Venter *et al.* (1996) argued that this assembly-first approach is not necessary and that sequencing a library of randomly sheared fragments (shot-gun approach), coupled with computer programs to do the assembly, might be preferable. Although a shot-gun approach had been taken by Sanger (Sanger *et al.* 1977, 1982) to complete the sequence of phage lambda (48 000 bp in size), the use of this strategy to tackle a genome of almost 2 million nucleotide pairs (2 Mb) was a bold step that relied heavily upon bioinformatics to achieve assembly of the random sequences. Until the completion of the *H. influenzae* project the largest microbial genome that had been completely sequenced was that of cytomegalovirus (229 000 bp) (Bankier *et al.* 1991). The assembled, annotated genome of 1 830 137 bp of *H. influenzae*, strain Rd, was completed in about 1 year and duly acclaimed as a landmark (Fleischmann *et al.* 1995) (Table 17.1).

Table 17.1 Whole-genome sequencing strategy

Stage	Description
Random small-insert and large-insert library construction	Shear genomic DNA randomly to 2 kb and 15 to 20 kb, respectively
Library plating	Verify random nature of library and maximize random selection of small-insert and large-insert clones for template production
High-throughput DNA sequencing	Sequence sufficient number of sequence fragments from both ends for 6× coverage
Assembly	Assemble random sequence fragments and identify repeat regions
Gap closure	Order all contigs (fingerprints, peptide links, clones, PCR) and
Physical gaps	provide templates for closure
Sequence gaps	Complete the genome sequence by primer walking
Editing	Inspect the sequence visually and resolve sequence ambiguities, including frame-shifts
Annotation	Identify and describe all predicted coding regions (putative identifications, starts and stops, role assignments, operons, regulatory regions)

Reprinted with permission from: Fleischmann, R.D., Adams, M.D, White, O., *et al.* 1995. Whole-genome random sequencing and assembly of *Haemophilus influenzae* Rd. Science **269**: 496—512. Copyright 1995, American Association for the Advancement of Science.

The success of the shot-gun approach requires that the library be random; deviations from this are reflected in a proportionate increase in redundancy of the sequenced clones. This in turn indicates the number of physical gaps (i.e. a lack of sequence) for which additional sequencing must be done if the available random sequences are to be assembled into a complete genome. Thus, applying Poisson principles, if the number of nucleotides sequenced in the randomly selected clones corresponds to the number in the genome, the probability is that about 37 per cent of the genome will not be represented. After completing five times the number of nucleotides in the genome, less than 1 per cent of the genome will remain unsequenced. This still amounts to a considerable number of physical and sequence gaps (clones available but not sequenced).

H. influenzae was the first free-living organism to be completely sequenced and thus provided a unique opportunity to probe the absolute requirements for the independent life of a cell.

The utility of information in a microbial genome can be considered under several general categories (Table 17.2). My discussion

Table 17.2 Some categories of information obtained from genomes

- Linear sequence
- Genome organization
- Gene expression
- Evolutionary biology
- Population biology
- Pathogenicity
- Cryptic life forms

emphasizes those that are particularly relevant to 'evolution in health and disease.'

Insights from linear sequence

The *H. influenzae* genome was estimated to contain 1743 open reading frames (putative genes). For these, homologous sequence matches in the database could be found for more than 90 per cent and allowed some indication of function in more than 80 per cent. These figures overestimate the extent to which valid biological functions can be attributed. A more realistic conclusion is that only one-half to two-thirds of these genes could be reasonably assigned functions. These putative genes and their functions, depicted in helpful colour-coded maps, provide an immediately accessible picture of the amount of DNA involved in different functions in *H. influenzae*, for example: 10 per cent to energy metabolism, 17 per cent to transcription and translation, 12 per cent to transport, and 8 per cent to cell envelope proteins. It had been thought that such assignments, anticipated several years previously by Riley (1993), would first be made available for *E. coli*.

Whole-genome sequences are revealing many new genes whose functions are not evident from inspection or use of homology searches. These novel genes (sometimes referred to as orphan genes) represent an exciting challenge and an example of 'the richness of the treasure trove'. Currently, for an investment of about US$1 million, a genome sequence identifies several hundred gene sequences among which are those encoding every virulence factor, every target for therapy, and every molecule that is a potential vaccine or target for host immune clearance mechanisms. The systematic, inevitably selective investigation of biological function represents a massive task. Many have admonished the incautious conclusions drawn by those who assume functions from database searches. Even when such predictions are undertaken with caution (and almost inevitably when they are not), subsequent experiments produce results which surprise or cause previous conclusions to be radically revised or abandoned.

Non-coding DNA in the *H. influenzae* genome also proved particularly interesting. There were 1400 copies of motifs (23 nucleotides) involved in DNA transformation. With this mechanism *H. influenzae* can recognize its own DNA released following spontaneous lysis of sibling or closely related bacterial cells, and then incorporated into the genome of recipient cells by homologous recombination (Smith *et al.* 1995*b*). There were also multiple tandem tetrameric repeats within open reading frames of known or putative virulence factors (Hood *et al.* 1996*a*) (also see below).

Gene expression

Sequence information from a genome opens the door to an understanding of the integration and co-ordination of genotype and phenotype under different conditions, such as the changes in expression occurring when bacteria enter host cells in comparison with their extracellular residence, or differences in gene activity in an insect vector compared with a mammalian host. Such analysis could be undertaken using a library of genes, representative of the whole genome. The nucleotides of every gene sequence can be arrayed on a solid phase and probed with labelled RNA extracted from bacteria isolated and grown under different conditions. A major advance in phenotypic analysis has been made by combining two-dimensional gel electrophoresis with mass spectrometry. The resolving power of the latter, requiring only minute quantities of protein from cells, allows the detection of post-translational modifications so that profiles of cell function can be studied.

Population biology

A key issue is how to use the complete sequence of an index strain to characterize other strains of the named species. Conventional taxonomy in microbiology has relied upon phenotypic characterization which, although useful, can obscure or distort the appreciation of genetic relationships and population structure (see Chapter 16). More than a decade ago, multi-locus enzyme electrophoresis (MLEE) was recruited to investigate population structures of bacteria, focusing particularly on pathogenic species (Selander *et al.* 1987). This indexes genic variation through comparison of polymorphisms in a sample of metabolic (housekeeping) enzymes. It allows large numbers of organisms to be characterized and, using appropriate statistical analyses, allows strong inferences to be drawn concerning the extent of horizontal gene transfer and intergenomic recombination among strains of a species.

With the advent of automated sequencing, the DNA sequence of particular virulence genes can be compared in strains selected from a population framework defined by MLEE. These phylogenetic frameworks have in a few instances been supplemented by more detailed analyses of the interstrain variation in nucleotide sequence data of selected genes (e.g. Kehoe *et al.* 1996). However, only snap-shots exist, for such analysis is still labour intensive. Furthermore, whole-genome sequences provide another level of resolution, for virulence is multifactorial and must be considered in the context of the products of many genes including the molecules that determine tropism, *in vivo* replication, and cytotoxic factors.

Whole genome sequences offer, *per se*, a more informed view of populations. Some major gaps in our knowledge include the extent of the conservation of coding and non-coding nucleotides, the extent of rearrangements and intragenomic mobility, and the extent and rates of recombination. Such data have a very practical application in determining the variation in molecules that are targets for drugs and vaccines.

Comparative genomics

'The availability of complete genome sequences of cellular life forms creates the opportunity to explore the functional content of the genomes of and evolutionary relationships between them at a new qualitative level' (Koonin and Mushegian 1996).

One of the most exciting opportunities opened up by whole-genome sequences lies in comparative biology. Although complete sequences of viral genomes have been available for some years, access to the genomes of cellular life forms makes a real difference. Comparing these genomes should allow us to estimate the minimal set of genes required for the essential functions of a cell and to trace back the gene composition of ancestral organisms. A practical application of this will be the rational attenuation of bacteria and their use as live vaccines or vectors for foreign antigens.

What theoretical biologists can do with whole-genome sequences is suggested by Tatusov *et al.*'s (1996) work on the sequences of *H. influenzae* and *E. coli*. These organisms both belong to the gamma division of purple bacteria, but they have very different lifestyles. *H. influenzae* is an obligate parasite and a major pathogen found in the upper respiratory

tract of humans, the only known natural reservoir. *E. coli*, also a potential pathogen, is a ubiquitous saprophyte found predominantly in the gastrointestinal tracts of a variety of animals, but is also common in inanimate environments. The genome of *E. coli* is more than twice as large as that of *H. influenzae*. In part, this reflects *E. coli*'s capacity to utilize a wide variety of nutrients as energy sources. This is reflected in several families of permeases and transcriptional activators in *E. coli*, each of which is represented by only single proteins in *H. influenzae*.

Pathogenicity

Whole-genome sequences of pathogenic microbes provide a catalogue of the genes responsible for every virulence factor, drug target, and vaccine candidate. How might this information be used? Hood *et al.* (1996*b*) used the complete genome sequence of *H. influenzae* to identify and characterize the multiple genes involved in lipopolysaccharide (LPS) biosynthesis. LPS is a critical structural and functional component of the cell envelope of a whole range of bacterial genera. In *H. influenzae*, LPS is involved in every stage of the pathogenesis of serious infections such as meningitis. It modulates interactions with respiratory epithelial cells, plays a role in the invasive process whereby bacteria translocate from the respiratory tract to the blood, impairs clearance of bacteria from the blood, and affects inflammation and tissue injury. LPS is a highly complex glycolipid whose biosynthesis requires more than 30 genes for synthesis of the precursor units, which include fatty acids, sugars, and side-chain substituents such as phosphoryl choline. These are synthesized and assembled initially in the cytoplasm, transported outwards, and assembled on the cell surface.

Only a few of the relevant genes were known before the availability of the whole-genome sequence, but once this was available, more than 25 additional genes were identified. The most powerful method was the use of previously published sequences deposited in the general database, particularly those known to be involved in lipopolysaccharide synthesis in other organisms. These 'database probes' were used to search the *H. influenzae* genome for homologous sequences. When matches were found within open reading frames, the candidate LPS genes were cloned and characterized. By making mutations in each of these genes, we could determine whether they were involved in LPS biosynthesis, for mutational changes in the LPS molecule could be detected by reactivity with monoclonal antibodies, gel fractionation patterns, and electrospray mass-spectrometry to confirm their role. Virulence studies in an experimental animal infection model provided an estimate of the minimal LPS structure required for intravascular dissemination. We have extended our investigations of LPS structure to other strains and have confirmed the extent of conservation of LPS-related genes across all LPS types. This provides a rational start to evaluating the candidacy of LPS for broad-range vaccine development.

REPETITIVE DNA, HYPERMUTABILITY, AND PATHOGEN EVOLUTION

Having indicated the revolutionary impact of whole-genome sequencing in analysing virulence factors, I now suggest that this information will also ignite interest in what has been called theoretical evolutionary genomics (Koonin and Mushegian 1996). The information in whole-genome sequences will change our ability to generate hypotheses about the evolutionary origins of health and disease, and our understanding of pathogenic microbes and the evolution of virulence will experience a conceptual transition.

Repetitive DNA mediates phenotypic variation

Iterations of short (1–6 nucleotides) tandemly oriented DNA motifs, of the kind that are highly characteristic of eukaryotic genomes, were not thought to be a conspicuous feature of prokaryotes until a relatively short time ago. Over the past decade, many examples of homopolymeric tracts and tandem arrays of

multiple short repeat motifs (2–5 nucleotides) have been reported in bacteria (Table 17.3). Most of these have been described in pathogens, the first description being that of penta-meric repeats (CTCTT) in a gene encoding an outer membrane protein of *Neisseria gonorrhoea* (Stern *et al.* 1986). In *H. influenzae*, tandem tetranucleotide repeats (CAAT) were identified within the translated region of several genes required for the biosynthesis of lipopolysaccharide (Weiser *et al.* 1989).

These iterations of nucleotides are important because they provide a mechanism for varying the expression of genes, as follows: tandem arrays of nucleotide repeats are prone to mispair as the complementary strands, annealed by classical Watson–Crick base-pairing, are separated spontaneously (as DNA 'breaths') or during transcription or replication. As the strands separate and the complementary base-pairing is temporarily disrupted, slippage may occur so that one strand moves relative to the other in either the 5' or 3' direction. This results in one or more repeat units being mispaired when the strands realign (Fig. 17.1). On the next round of replication, mismatch repair mechanisms correct the incompatibility in base-pairing and one or more repeats is gained or lost, depending on the direction of slippage. However, loss or gain of a repeat unit (say one copy of the tetra-nucleotide 5'-CAAT-3', to take one example) will cause a frame-shift, for translation depends upon having the correct triplet codons in relationship to a translational start codon (for example and usually, ATG). Thus, if there are n repeat units and this number of repeats is in frame for translation, $n - 1$ or $n + 1$ repeats, resulting from slippage, would cause a frame-shift and either an altered or truncated peptide depending on whether the frame-shift introduced a stop codon or merely altered the triplet code. Thus, in this example, repetitive DNA promotes variable expression of genes through its effect on translation. The phenotypic correlate of this genetic instability (referred to as slip-strand mispairing) (Streisinger *et al.* 1966; Levinson and Gutman 1987) is the variable expression of the encoded molecule.

Table 17.3 Examples of genetic mechanisms generating random phenotypic variation at high frequency in specific loci

Mechanism	Organism	Determinant	Host
Gene conversion (homologous recombination)	*Borrelia hermisii*	Lipoprotein (*vmp*)	Human
	Bacillus thuringiensis	Toxin (*crylA*)	Lepidoptera
	Haemophilus influenzae	Capsule (*capB*)	Human
Site-specific recombination	*Escherichia coli* Incl1	Fimbriae (*fimA*)	Human
	Plasmid R721 (*E. coli*)	Pilin (*pilV*)	Human
	Moraxella bovis	Pilin (*tfp*)	Cattle
Oligonucleotide repeats	*Haemophilus influenzae*	Lipopolysaccharide (*lic* loci)	Human
	Neisseria gonnorrhoea	Fimbriae (*hifA, hifB*)	Human
		Opacity proteins (*opa*)	Human
	Pseudomonas solanacearum	EPS, virulence determinants (*phcA*)	Plants, especially solanaceous crops
Homopolymeric tracts	*Neisseria meningitidis*	Opacity proteins (*opc*)	Human
	Mycoplasma hyorhinis	Lipoprotein (*vlp*)	Swine
	Bordetella pertussis	Fimbriae	Human
Site-specific transposition	*Pseudomonas atlantica*	Extracellular polysaccharide	Not applicable
Site-specific methylation	*Escherichia coli*	Pilin (*papA*)	Human
Unknown	*Pseudomonas tolaasii*	Symbiotic efficiency	Mushroom
	Rhizobium phaseoli	Virulence, chemotaxis, attachment	Leguminous plants

Adapted from Moxon *et al.* Copyright 1994, Current Biology Ltd.

What sort of gene products are affected? Most are surface-exposed molecules important for interacting with the host. These functionally important genes are not essential for viability, but their context-dependent expression may confer a fitness benefit depending on the stage of the infection. For example, expression of an adhesin during colonization may confer an advantage, but if inflammatory cells accumulate, perhaps at a different site and at a later stage in the infection, an adhesin may result in loss of fitness if it promotes intimate contact and ingestion of the bacterial cell by polymorphonuclear leucocytes or macrophages ('professional' phagocytic cells). Thus, the capacity to switch molecules on and off, a reversible process, offers potential advantages to microbial survival.

Antigenic variation in pathogenic microbes was recognized long ago. Phase variation of the flagella antigens of *Salmonella* was first described by Andrews in 1922; the molecular basis was elucidated much later (Lederberg and Iino 1956; Simon *et al.* 1980). Other celebrated examples are the variant surface antigens of trypanosomes (Borst and Rudenko 1994; Cross 1996) and malaria parasites (Roberts *et al.* 1992). The capacity to switch surface molecules on and off through a variety of genetic mechanisms, iterative DNA being only one example, is well known in a variety of microbes (for reviews see Robertson and Meyer 1992; Moxon *et al.* 1994; Deitsch *et al.* 1997). Phase variation provides a mechanism of phenotypic variation by which microbes can adapt to the differing microenvironments of the host and evade host immune responses (Robertson and Meyer 1992; Moxon *et al.* 1994).

A complete genome sequence (*H. influenzae*) in which novel DNA repeats mediate phenotypic variation

Iterative DNA might provide clues to functionally important sequences because of their likely association with functionally important,

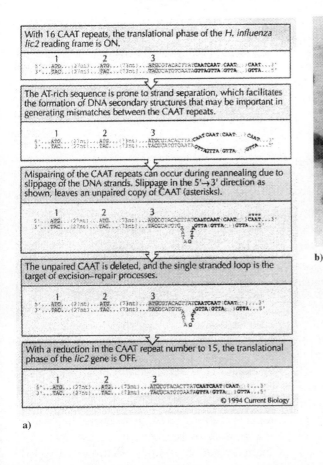

a)

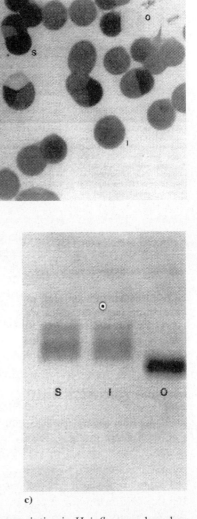

b)

c)

Fig. 17.1 (a) Proposed mutational mechanism responsible for LPS phase variation in *H. influenzae*, based on slipped-strand mipairing. The DNA sequence of the 5′ end of the *H. influenzae lic2* gene, required for the biosynthesis of the disaccharide Galα(1–4)Galβ in the LPS core, is shown. Phase variation of the digalacto-side is mediated through changes in the number of copies of tandem repeats of CAAT (bold); 16 copies of the tetranucleotide sequence are shown. Potential translational start codons (boxed ATG) for a long open read-ing frame are shown, and are positioned such that ATG 1 and 2 are in frame when *lic2* has 16 copies of CAAT, ATG 3 is in frame if *lic2* has 17 copies of CAAT, and there is no ATG in frame if there are 15 copies of CAAT. (Fig. 17.1(a) from Moxon *et al.* Copyright 1994, Current Biology Ltd.)

(b) Individual colonies of RM 7004 immunoblotted with monoclonal antibody (4C4) specific for a lipopolysaccharide epitope. Colonies binding 4C4 strongly (S), to an intermediate extent (I), or to an undetectable extent (O) are indicated. Sectored colonies which express more than one phenotype within the same colony are present.

(c) This shows SDS–PAGE electrophoresis (silver staining) of differing migration patterns of lipopoly-saccharide extracted from S and I colonies (similar glycoforms) compared with O (lower molecular weight glycoforms compared to S and I). (Fig. 17.1(b and c) reprinted from Roche *et al.* Copyright 1994, with kind permission from Elsevier Science—NL, Sara Burgerhaartstraat 25, 1055 KV Amsterdam, The Netherlands.

host-interactive molecules. We tested this idea by searching the whole-genome sequence of *H. influenzae* strain Rd to identify tandem oligonucleotide repeat sequences. Prior to the search we proposed the following hypotheses. Iterative DNA would:

(1) identify genes of importance to interactions with the host;

(2) be located within open reading frames (genes) or within sequences (e.g. promoters) important to transcription;

(3) displace spontaneous changes in the number of repeat units through slipped-strand mispairing (Streisinger *et al.* 1966; Levinson and Gutman 1987);

(4) and code for molecules that display a phase variable (switching) phenotype.

The search identified nine novel as well as three previously recognized loci with multiple (range 6–36, mean 22) tandem tetranucleotide repeats (Hood *et al.* 1996*a*: Table 17.4). All tetramers were located within putative open reading frames in the Rd strain. Additional studies in two epidemiologically different *H. influenzae* type b strains showed that these same genes containing multiple tetramers were also present, indicating the generality of the finding for the species. However, the number of repeats was shown to vary both within the same strain (we did polymerase chain reactions (PCR) on organisms from a frozen culture of Rd that had been stored in our laboratory for several years to compare the numbers of repeats found in it with those of findings of Fleischmann) and among epidemiologically different strains. Moreover, the particular nucleotide composition of the tetramers associated with the same gene, but in different strains, could vary. As one example, instead of the AGTC tetranucleotides found in one locus of strain Rd, a different strain had multiple copies of AGCC.

As for possible functions, database searches indicated that four of the genes were homologues of haemoglobin-binding proteins of *Neisseria*, each located in different positions of the same genome (strain Rd). Iron is an essen-

tial nutrient of bacteria, but its availability to microbes in free form is negligible since most is sequestered or bound to protein. Thus, the efficiency with which bacteria scavenge iron is important for virulence (Weinberg 1984). What might explain both the redundancy (four copies) and the potential for phase variation of these proteins? One possibility is that the genes mediate binding to different sources of protein-bound iron, such as lactoferrin, transferrin, or haemoglobin, a finding in keeping with the multiplicity of iron-sequestering proteins (Genco and Desai 1996). Another possibility is that, given the importance of their function, multiple alleles might afford some insurance against the recognition of any one protein by the immune system, since phenotypic switching would allow a novel variant to be selected.

A homologue of a gene involved in LPS biosynthesis in *Neisseria* was identified and, for this gene, our investigations provided proof in principle for all facets of our hypothesis. In *H. influenzae*, this gene is also involved in LPS biosynthesis, a known virulence factor; the repeats are located in the open reading frame; the number of repeat units varies; and these changes mediate phase variation of the LPS phenotype (Hood *et al.* 1996*a*). These findings indicate the new power of whole-genome sequences in identifying virulence genes. They also raise a variety of intriguing questions about the role of repetitive DNA.

Intra- and interstrain variabilities in repetitive DNA affect the evolution of virulence

Comparisons of the findings in Rd and other strains reveal that tetranucleotide repeats, involving multiple loci, are a general feature of *H. influenzae*. In addition to intrastrain variation—slippage results in phenotypic variation—there is substantial interstrain variation in the number of tetrameric repeats found in particular loci. For example, the interstrain variation in the number of tetramers within *lic1* and *lic2* loci is shown in Fig. 17.2 (High *et al.* 1996). These findings raise several intriguing issues. Do the tract lengths influence the

Table 17.4 Characteristics of tetrameric repeats and associated loci identified through searching the *Haemophilus influenzae* strain Rd genome sequence

Tetrameric repeat	Homologue	(Genus)[a]	Function	Homology[a] (BLAST)	Number of repeats[b]				Open reading frame[c]
					Rd	RM118	RM153	RM7004	
CAAT	Lic1	*Haemophilus*	LPS biosynthesis	–	17	ND	ND	30	ATG ...9 bp... ATG ...5 bp... (CAAT)17 ...895 bp... TAA
CAAT	Lic2	*Haemophilus*	LPS biosynthesis	–	22	ND	17	16	ATG ...11 bp... (CAAT)22 ...698 bp... TAG
CAAT	Lic 3	*Haemophilus*	LPS biosynthesis	–	32	ND	ND	22	ATG ...36 bp... ATG ...76 bp... (CAAT)32 ...878bp... TAG
GCAA	YadA	*Yersinia*	Adhesin	$7.0e^{-20}$	25	24	15	23	ATG ...15 bp... (GCAA)25 ...751 bp... TAA
GACA	LgtC	*Neisseria*	Glycosyl transferase	$2.9e^{-37}$	22	ND	22	32	AT (GACA)22 ...899 bp... TAA
CAAC	Haemoglobin receptor	*Neisseria*	Iron binding	$7.4e^{-84}$	36	36	22	22	ATG ...49 bp... ATG ...12 bp... (CAAC)36 ...3029bp... TAA
CAAC	Haemoglobin receptor	*Neisseria*	Iron binding	$1.8e^{-86}$	20	21	23	23	ATG ...24 bp... ATG ...5 bp... (CAAC)20 ...2841 bp... TAA
CAAC	Haemoglobin receptor	*Neisseria*	Iron binding	$5.0e^{-85}$	18	20	[d]	[d]	ATG ...25 bp... (CAAC)18 ...2839 bp... TAA
CAAC	Haemoglobin receptor	*Neisseria*	Iron binding	$1.3e^{-87}$	20	ND	ND	ND	ATG ...12 bp... ATG 5 bp... (CAAC)20 ...3029 bp... TAG
CAAC	No homology				15	14	8	7	ATG ...16 bp... ATG ...3bp... (CAAC)15 ...550 bp... TAA
AGTC	Methyl-transferase	*Salmonella*	Host restriction/ modification	$4.7e^{-30}$	32	26	30[e]	3[e]	ATG ...49 bp... (AGTC)32 ...1710 bp... TGA
TTTA	32.9 kD protein	*Bacillus*	Unknown	$4.7e^{-05}$	6	ND	ND	ND	ATG ...130 bp... (TTTA)6 ...686 bp... TGA

[a] Designation corresponds to the highest value match after a BLAST search of the repeat-associated open reading frame against the combined GenBank/EMBL databanks.
[b] The number of repeats found for each associated locus in the genome sequence of strain Rd is given in bold; the other values are those obtained after sequencing of cloned PCR products or direct sequencing of PCR products from the culture collection strains RM118, RM153, and RM7004.
[c] Open reading frames are shown with the spacing between the repeat region and the nearest ATG translational start codons and the predicted translational stop codons.
[d] None present. Hybridization with a [CAAC]₅ oligonucleotide probe showed only three associated loci in RM153 and RM7004, consistent with the lack of a PCR product for this locus.
[e] Repeats were AGCC, not AGTC.
ND, not determined.
From Hood *et al.* 1996*a*. Copyright 1996, National Academy of Sciences, USA.

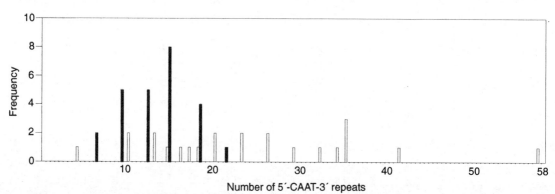

Fig. 17.2 A histogram depicting the relative frequencies at which different numbers of 5'-CAAT-3' repeats occur in both *lic1A* and *lic2A* in epidemiologically distinct strains of *H. influenzae*. Solid bars and open bars represent the number of 5'-CAAT-3' repeats in *lic1A* and *lic2A*, respectively. (From High *et al.* Copyright 1996. Blackwell Science Ltd.)

rate of phase variation? To what extent do tract lengths vary over time in individual strains? What are the effects of a change in tract length in one locus (gene) on other loci? While answers to these questions are not yet known, I propose that the relatively rapid evolution of these hypermutable loci affects commensal and virulence behaviour. To sum up, the iterative DNA found at multiple loci in some bacterial genomes provides a mechanism for phenotypic switching of surface-exposed molecules. We refer to these hypermutable genetic sequences as contingency loci (Moxon *et al.* 1994).

Contingency loci in gene-for-gene arms races

The term contingency focuses attention on concepts that need further elaboration. Microbes must balance the trade-off between mutation, a potential source of useful variation with which to explore solutions to unpredictable aspects of the host environment and without which the microbe may face extinction, while minimizing deleterious effects on fitness. Pathogenic bacteria (but also viruses, fungi, and parasites) have several mechanisms which increase the mutation rates of certain genes or sets of genes and the phenotypes they affect (Robertson and Meyer 1992; Moxon *et al.* 1994; Nowak and Bangham 1996; Deitsch *et al.* 1997). These genetic mechanisms include polymerase slip-

page in iterative DNA, gene conversion, genomic rearrangements, and point mutations.

It is important to recall the distinction between genetic mechanisms of phenotypic variation and those of classical gene regulation. Examples of the latter are catabolite repression or the signalling systems associated with histidine protein kinases (Stock 1990). While gene regulation plays an important role in phenotypic variation, it characteristically affects most of the cells of a population and is therefore not a source of phenotypic variety within a population of organisms at one time in one microenvironment. The position of contingency loci in the genome has been programmed by evolution, but mutation through slippage is stochastic in time (Robertson and Meyer 1992; Moxon *et al.* 1994). The mutability and phenotypic variability characteristic of a few key contingency genes, representing only a small proportion of the genome, can influence antigenicity, motility, chemotaxis, attachment to host cells, acquisition of nutrients, and sensitivity to antibiotics while minimizing the deleterious effect that high mutation rates would impose on housekeeping functions of the genome (Fig. 17.3).

Phenotypic variation and within-host evolution of bacterial populations

Gene regulation and antigenic variation provide contrasting strategies of adaptation, the

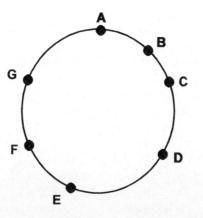

Fig. 17.3 Contingency loci. A simplified bacterial genome, double-stranded circular DNA, is depicted with several hypermutable contingency loci [●] which result in altered activity of genes. The genetic mechanisms include slipped-strand mispairing of rDNA, gene conversion, genomic rearrangements and point mutations. These loci mediate changes in phenotype which occur at high frequency (about 1:100–1000 bacteria per generation) and typically affect the activity of genes which interact with the environment.

The power of the combinatorial effects on phenotype endowed by the independent (stochastic) occurrence of switching at multiple loci makes for an interesting paradigm. For example, one might consider the contingency loci A, B . . . G (controlling phenotypes a, b . . . g) that can swtich to A′, B′ . . . G′ (phenotypes a′, b′ . . . g′), each being binary, reversible, and operating at a rate of 10^{-2} per bacterium per generation. Switching of any one locus occurs independently of any other. Let us suppose that phenotype a′, b, c, d′, e, f′, g is optimal for colonization of a particular anatomical site (S) of, say, the mucosal surface of the nasopharyngeal epithelim of host H. Long-term carriage is favoured by invasive entry of the bacteria to take up residence within the epithelial cells of host H. The new microenvironment (S′) selects for the optimal phenotype a, b, c′, d, e, f′, g. Since this required three independent switching events, the probability of the switch (from a′, b, c, d′, e, f′, g→a, b, c, d′, e, f′, g) is 10^{-6} per bacterium, i.e. a relatively rare event. In time, bacteria from host H may be transmitted to host H′, a new microenvironment in a genetically different host in which the optimal phenotype may be selected from any one of the 128 possibilities. Note also that the 'turns in Muller's ratchet' can be reversed (Muller 1964).

one by prescription, the other by the 'blind' opportunism of stochastic variation and selection (Robertson and Meyer 1992; Rainey *et al.* 1993; Moxon *et al.* 1994). The penchant of pathogenic microbes for antigenic variation is unsurprising when viewed in the context of the within-host pathogenesis (Levin and Bull 1994) and between-host transmission that make up the infectious process. Because of their small size and short generation time, microbial populations vastly outnumber those of their host. Thus, since the absolute number of mutations arising depends on the number of genomes present, far more mutations arise in the bacterial population than in the host population, not only per host generation but even

per bacterial generation. Microbes often produce clonal populations within hosts. In experimental infections caused by *Salmonella typhimurium* and *H. influenzae*, the entire population of infecting bacteria may originate from a single bacterial clone (Meynell 1957; Moxon and Murphy 1978). Host defence mechanisms are continually evolving, both within the lifetime of vertebrate hosts, whose immune system functions according to evolutionary principles on a generation time similar to that of microbes, and through the normal evolution of host populations. It is therefore plausible that microbial clones generate phenotypic diversity as a strategy for adapting to the varying microenvironments, the repertoire of polymorphisms, and the immune clearance mechanisms which characterize their hosts.

For microbes to persist in a host population, the average number of new infections caused by a single infectious host (R_0) must exceed unity (Anderson and May 1982). But, hosts in a sexually reproducing and outcrossing host population differ genetically—especially, for coevolutionary reasons, in the loci that interact with pathogens and parasites. To survive in a new, genetically different host, microbes must be able to adapt rapidly to the new host microenvironment in which polymorphisms in receptors as well as differences in specific and non-specific host defences may pose a precipitous challenge. Here again, mechanisms producing diversity at host-interactive loci within a clonal population could improve clonal survival probability. I suggest therefore that the activities of contingency loci make an important contribution to the value of R_0.

Contingency loci, mutators, and the environment

Would contingency loci be even more effective if they responded only to certain environments, so that their effects could be recruited only when advantageous, such as in times of stress? That cells should be at their most mutable when conditions are unfavourable is appealing; stress-induced mutagenesis is not a new concept (MacLintock 1993).

Changes in osmolarity, temperature, pH, and availability of nutrients are known to affect the level of DNA supercoiling, which affects transcriptional efficiency (Higgins and Dorman 1990). Changes in supercoiling are particularly likely to influence regions of short tandem repeats, facilitating strand separation, exposure of single-stranded DNA, and changes in orientation of sequence and alternative DNA structures. Many contingency genes are found in regions of DNA especially rich in AT nucleotides (High et al. 1993), which are more prone to strand separation. Accumulating evidence supports a role for global regulation of gene expression through changes in supercoiling (Dorman and Higgins 1987).

It has been known for some time that muta-tion rates within a subset of individual cells of a clonal population can be significantly elevated (Treffers et al. 1954; Miyake 1960). These cells are called mutators. The mutator phenotype itself results from mutations in one or more of several genes, including rpoS (Zambrano et al. 1993) and those involved in methyl-directed mismatch repair (Siegel and Bryson 1967). During prolonged incubation, rpoS mutants replace the original population under the strong selection imposed by starvation. Thus, the frequency of mutations in these populations increases through the spread of mutator genes during starvation. Lenski and his colleagues found that 3 of 12 E. coli cultures, continuously propagated over 10000 generations under glucose-limiting conditions, had a mutator phenotype: the defective alleles were mutS and uvrD (Sniegowski et al. 1997).

Through increased mutation and normal recombination (Taddei et al. 1997), mutator phenotypes could benefit pathogen clones by increasing variation that would help to explore host phenotypic landscapes, escape immune surveillance, or elude the action of antibiotics (Thaler 1994). Hypermutability through lesions in mismatch repair would also increase the rate of mutations in contingency loci, and mutators and contingency loci acting in tandem could combine to produce a degree of biological promiscuity that might underlie the emergence of outbreak strains (clones). However, such strains would also have a high probability of extinction unless population-wide and individual locus mutator activity could be kept within limits.

Contingency loci, environmental influences, and directed mutation

A central tenet of molecular biology asserts independence of genetic information from events occurring outside or even inside the cell: 'The very nature and structure of the genetic code and the way it is transcribed implies that no information from outside can ever penetrate the inheritable genetic message' (Judson 1979). Thus, the conventional view is that the environment selects among pre-existing variants; mutations arise without regard for their utility. However, it could be advantageous for organisms to evolve mechanisms through which the environment could increase or decrease the mutation rate that generates the phenotypic variation upon which selection acts.

Might contingency loci be a source of mutations which arise when, and even because, they are advantageous? Cairns et al. (1988) pointed out that the classical Luria–Delbrück experiments were not designed nor were they able to detect mutants that arose during or following the imposition of lethal selection. Data were presented which, they suggested, provided some evidence to suggest that 'cells may have mechanisms for choosing which mutations will occur'. This class of mutations has been given various names, including directed, adaptive, non-random, and selection induced.

In the ensuing years, these and later data have elicited spirited, at times heated, controversy involving evolutionary biologists and bacterial geneticists who have debated whether or not the occurrence of 'directed mutations' was real and, if so, the molecular mechanisms giving rise to them. (Foster 1993; Lenski and Mittler 1993). What does the behaviour of a 'stationary phase' bacterium (E. coli) plated on to simple laboratory media have that is relevant to the biology of infectious disease and the evolution of commensal and virulence behaviour? A link between the so-called directed mutations in E. coli, hypermutation in contingency loci, and the mutators of pathogenic microbes is that the mechanism involves frame shifts in small nucleotide repeats or homopolymeric tracts, probably through polymerase slippage (Foster and Trimarchi 1994; Rosenberg et al. 1994; Mao et al. 1997). (The FC40 story is more complicated and possibly a special case: Bridges 1997.)

Do these ideas challenge the fundamentals of Darwinian theory that mutations are random with respect to their selective utility? There is no convincing evidence that external signals can instruct bacterial cells to make specific mutations—the most unorthodox of the proposed mechanisms invoked to explain directed mutations (Cairns et al. 1988). How-

ever, variants generated by hypermutable sequences or cells (contingency loci or mutators) have apparently evolved to provide flexibility in the activity of some genes of clones or clonal populations. Indeed, the notion that 'directed mutations' might appear to result from environmental instruction is testimony to the power of selection acting on an event (mutation) which is stochastic in time but programmed (evolved) in its assignment to genomic locations (contingency loci), to specific genomes (mutators), or, most powerfully, to the combination of the two (Moxon and Thaler 1997). Thus, the combination of 'contingency loci' with mutators offers a neo-Darwinian explanation for 'directed mutations' in which there is no requirement for any novel molecular mechanisms or reverse flow of information.

Contingency loci and mutators: general implications

Do the effects of contingency loci and mutators apply more broadly to the evolution of health and disease? The pathobiology of tumour formation is considered to be a multistage process, but somatic and germ cell mutation rates, especially where as many as five or six mutations are required for the development of fully fledged tumour cells, are inadequate to satisfy theories based on this model. Mutator clones have been considered as one factor in the ontogeny of cancer (Mao et al. 1997). If a mutator clone arises relatively early in tumour progression, then its genetic instability would be permissive for the generation of a cancer (Loeb 1991), although it still requires that a somatic cell loses the function of the other gene copy so that the resulting cell, lacking mismatch repair, is a mutator and leads to cancer. It has recently been shown that mutagenesis combined with a single selection could result in a population of haploid cells with a majority of mutators (Mao et al. 1997).

The inherited form of human non-polyposis colon cancer results from a defect in the human counterpart to the bacterial mismatch repair system, creating a mutator phenotype (Fishel et al. 1993). Also, several human genetic disorders are caused by trinucleotide repeat expansion, including fragile X, myotonic dystrophy, spinobulbar muscular atrophy, Huntington's disease, and spinocerebella ataxia. Each of these disorders involves cells of the neuromuscular system, all but myotonic dystrophy directly implicating neurones. Although the expanded number of repeats share no common mechanism to perturb gene function, these diseases all stem from allelic diversity generated by a high rate of *de novo* mutation to new length alleles in the germ line.

The following summary paragraph, from a recent review of triplet repeat sequences associated with human diseases, suggests that others (including King *et al.* 1997) have been thinking along similar lines.

Finally, it is possible that the events involved in variation of the repeat sequences at these loci are part of the normal developmental system designed to introduce variation into DNA in various tissues. The precedents from the immune system are quite clear; DNA rearranges to create diversity in immunoglobulins and receptors. These events may be the first glimpses of another mechanism for introducing variability into DNA sequence in organ systems or organisms. Since this variation is found at both the germ cell and somatic levels, it is tempting to speculate that these events could be directed by evolutionary pressures. ... it is not unreasonable to imagine cases of selection for some of these types of trinucleotide mutations, maintaining the machinery involved in generating the variation

Nelson (1993)

SUMMARY

The availability of whole-genome sequences of bacteria has fundamentally changed microbiology and is stimulating novel applications that will benefit the epidemiology, pathobiology, diagnosis, and prevention of infectious diseases. About a dozen whole-genome sequences of bacteria have been completed; sequencing the chromosomes of eukaryotic parasites such as *Plasmodium falciparum* is now planned. Increasingly, investigators will have available a catalogue of gene sequences that include every virulence factor, every potential drug target, and every vaccine candidate for a particular pathogen. The explosion of sequence data also has major implications for population biologists, ecologists, and evolutionary biologists. To illustrate the utility of whole-genome sequences, I describe studies in which the complete *H. influenzae* genome was searched for repetitive DNA. We had hypothesized that repetitive DNA, which is subject to polymerase slippage and is therefore relatively unstable (hypermutable), might be associated with virulence genes. This results in variable activity of genes, many of which code for surface molecules that interact with the host environment. These variations facilitate exploration of the different phenotypic landscapes encountered by microbes, within and between their natural hosts. The evolution and selection of hypermutability, both that conferred by iterative DNA in a restricted class of loci and that arising as mutations of mismatch repair genes, offer fitness advantages by increasing the probability that descendants of individual cells can escape some of the unpredictable challenges that threaten their extinction. These mechanisms increase the probability of the emergence of pathogenic clones that cause disease. In microbes, such virulent clones may cause sporadic, endemic, or epidemic infections. These mechanisms also operate in host cells themselves, examples being certain cancers and degenerative diseases of neurones. Many diseases of higher organisms may be by-products of the evolution of more efficient mechanisms of evolving (Wills 1989). The availability of whole-genome sequences of bacteria, archaea, and eukarya offers an opportunity to examine the prevalence and functional implications of the evolution of iterative DNA and its potential for dynamic mutations, investigations that could reinforce, in the context of disease pathogenesis, Dobzhansky's assertion that 'nothing in biology makes sense except in the light of evolution'.

18

WHAT CAN EVOLUTIONARY BIOLOGY CONTRIBUTE TO UNDERSTANDING VIRULENCE?

Andrew F. Read (rapporteur), Peter Aaby, Rustom Antia, Dieter Ebert, Paul W. Ewald, Sunetra Gupta, Edward C. Holmes, Akira Sasaki, Dennis C. Shields, François Taddei, and E. Richard Moxon (chair)

For host–parasite evolution, which involves complex interactions between potentially coevolving genomes in complex environments, the phrase 'intuitively obvious' is wholly inappropriate.

Work on the evolutionary biology of virulence aims to understand how evolutionary processes shape patterns of disease within and between populations. Consider, for example, the members of the protozoan genus *Plasmodium* that infect humans. Mortality rates per infection are relatively low and differ widely among the four species. People infected with these parasites can be asymptomatic or have mild, severe, or cerebral malaria. Specific signs and symptoms include fever, anaemia, and encephalopathy (due to blockage of blood vessels in the brain). What is responsible for this variation?

More generally, why are some symbionts commensal and others virulent? Why do mortality rates following infection vary? What causes qualitative and quantitative variation in disease symptoms? In this report, we assess how an evolutionary perspective can help answer those questions, and how that contribution might be improved. The first two sections cover general methodological issues. The next two sections cover areas where evolutionary biology can contribute to virulence research: analyses of diversity and cost–benefit (trade-off) models. We finish with a list of research questions.

DEFINITIONS

The term virulence has a tendency to generate controversy, much of it semantic. 'Virulence' is used in several conceptually distinct ways. Some authors equate it with health or disease severity, others with host or parasite fitness, and others with the ability of a parasite to grow on host tissue. This diversity, which parallels lay use, has to be acknowledged, and authors need to make their definition of virulence explicit. We use virulence to mean *the harm done to hosts following infection*; a definition that captures the essence of what we are trying to explain. Here, 'harm' means different things to different people, and that is intentional: it makes the virulence concept coherent across the range of disciplines using the term. To a clinician, harm may involve specific symptoms, signs, and pathologies; to a population biologist, harm is a reduction in host fitness, or some component of it. The phenomena of interest to clinicians and to population biologists may be correlated, but they need not be. We do not see that any differences matter: semantic ambiguity vanishes once the phenomenon under study is made explicit. Thus, we may be seeking to explain variation in rate of host mortality, fever severity, or levels of anaemia. More complex definitions may also be appropriate (e.g. LD_{50}, the inoculating dose which kills 50 per cent of hosts). Which is chosen will be context dependent, and an evolutionary approach is applicable to any of these. Questions like 'Is fever healthy?' are easily dealt with in a specific context: given that a host is infected, is fever adaptive (beneficial) for the host? Does it reduce the harm to the host that would otherwise have occurred (Ewald 1980)?

Importantly, virulence is a property of a host–parasite interaction; mutuality is of the essence. Thus, the function of molecules of a pathogen which are involved in pathogenesis (virulence factors) is contingent upon the host context.

EXPLANATIONS

Variation in virulence is generated by several factors. Purely ecological factors, such as inoculating dose and history of exposure, can be very important, and molecules encoded by both host and parasite can be critical. Thus several causal factors can be invoked to explain any disease phenomena (see Box 18.1 for an example). Determining their relative importance, often a daunting task, is critical for understanding the evolutionary processes involved in disease pathology. The evolutionary consequences of intervention strategies depend on whether relevant genetic variation is found in hosts, pathogens, both, or neither. The response of virulence determinants to selection depends on the extent to which genetic effects are swamped by environmental noise. Distinguishing between multiple causal factors, their interactions, and their associated factors is a formidable challenge when interpreting field and experimental data; even carefully controlled manipulations of a single factor can perturb several. Careful formulation of hypotheses and competing predictions is critical.

Confusion and disagreement is often generated by a failure to distinguish between what evolutionary biologists (e.g. Bulmer 1994) call proximate and ultimate explanations. Proximate explanations refer to immediate causal mechanisms (e.g. the hypothesis that cerebral malaria is caused by a phenotype encoded by particular parasite molecules); ultimate explanations invoke evolutionary processes (that phenotype enhances parasite survival and is thus favoured by selection). The use of a value-laden term like 'ultimate' was considered unfortunate by some of us; in no sense can either type of explanation be seen as superior. Which is more useful depends on the question being asked; a complete understanding of virulence must involve both. Determining which genetic constraints to incorporate in evolutionary models requires mechanistic detail; when evolutionary models fail to match with data, misunderstandings of underlying mechanisms may be an important reason.

Infectious disease research has had many successes without evolutionary biology. Why does a full understanding of disease virulence involve evolutionary explanation? Because natural selection and genetic drift can affect disease severity. Genes determining virulence are under strong selection because they often affect host or pathogen fitness, and we know there is genetic variation on which selection can act. Virulence polymorphisms are common, virulence determinants are frequently encoded by mobile genetic elements, and artificial selection can maintain or reduce virulence (for a review see Read 1994).

Thus medical research on virulence, which has largely concentrated on proximate explanations, can be informed and extended by understanding evolutionary processes. An evolutionary approach contributes more than an intellectually complete understanding of virulence, important as that is. The tools of evolutionary biology can be used to analyse proximate mechanisms, and evolutionary models can predict some consequences of changing conditions, allowing us to avoid circumstances which promote the evolution of increased virulence and to identify intervention strategies which lead to reductions in virulence. The next two sections illustrate and support those claims.

Box 18.1: Multiple explanations of disease patterns in local epidemics: genetics or ecology (epidemiology)?

It has often been observed that the level of virulence increases when an infection is introduced into densely clustered host groups, such as schools, hospitals, refugee camps, or families. For example, the case mortality arising during measles epidemics in refugee camps has been observed to increase from 5 to 50 per cent over a period of 3 to 4 months (Reder 1918). Likewise, the virulence of measles transmitted within households in West Africa increased three- to fourfold between the primary and secondary cases, while the spread of the disease in the general population was not associated with an increase in virulence (Aaby 1988). Several factors may contribute to these observations.

1. During the epidemic spread of a pathogen within a local host group, the inoculating dose is likely to increase. Virulence often increases with dose.
2. Parasite virulence may be host genotype specific. Rapidly evolving microparasites have the potential to adapt to the host genotype. If the local host population has a low genetic diversity, evolving pathogens may increase in virulence with each infection through their hosts.
3. Within-host competition of different strains (or mutants) can lead to an increase of virulence because more rapidly growing (and presumably more virulent) strains will outcompete slower growing strains (van Baalen and Sabelis 1995a). The frequency of multiple infections may increase with increasing prevalence of infection, for example, during epidemics.
4. Life-history theory suggests that in growing populations early reproduction is favoured over late reproduction. In parasites, the first transmission stages released after the start of an infection contribute most to the rate of parasite spread. Thus, parasites with a high within-host growth rate, which release more transmission stages early in an infection, will outcompete parasites which grow slower. If within-host growth rate is positively correlated with virulence, more virulent parasites will be favoured in spreading epidemics (but not under equilibrium conditions). In this case, the parasite-induced death of the host has a substantially reduced impact on parasite fitness.

Controlled experimental studies in animal models can discriminate between these alternatives. Epidemiological evidence may also help. For example, the pattern of measles virulence found in refugee camps, military barracks, and children's homes, where the density of individuals and transmission rate is high, is similar to that observed within families, which argues against the second alternative above (Aaby 1988). Instead, increases in the inoculating dose may explain the higher levels of mortality observed.

DIVERSITY

A major advance in biomedical science in recent years has been the ability to characterize pathogens genetically at the level of the nucleotide sequence. Many studies of genetic diversity in pathogens have simply sought to classify variation without asking how this variation might explain different disease outcomes. Understanding the processes that structure pathogen populations is therefore one of the key examples in which evolutionary thinking can benefit the health sciences: phenotypic variation matters and we gain important insights by understanding it. Here we discuss three ways in which evolutionary analyses of diversity can contribute to virulence research.

Organizing variation

It is often easy to be overwhelmed by variability. The first step is to organize variation so that one can identify the causes and consequences of it. Using a particular genetic or serological marker just because it is possible to do so may reveal little about biologically relevant diversity and biologically relevant population structure, but evolutionary considerations can give a sense to diversity that is theory driven rather than technique led. For example, parasites in a population can be functionally classified with respect to any property, the usefulness of which depends on the biological relevance of the property used and its stability through time. Loci that affect transmission order genetic diversity from an epidemiological perspective, avoiding the need to consider the entire genomic constitution of all parasites in a population. Thus, whether a particular serological classification of a parasite population is interesting to an epidemiologist depends on the degree to which the serotype-specific immune responses are capable of affecting the transmission of a given serotype. An application of this approach to population structure is given in Box 18.2.

Phylogenetic analyses

Phylogenetic trees, which depict the genealogical relationships of DNA sequences, represent a powerful tool for understanding sequence data (see Chapter 15). More than just a visual classification scheme, phylogenetic trees can reveal underlying evolutionary processes, for the pattern of evolutionary relationships between sequences depicted in the phylogeny

is shaped by historical processes, such as natural selection, genetic drift, and mode of population growth, all of which have important implications for disease. Molecular phylogenies have many applications to understanding pathogen variation. Whatever the question, the basic approach is the same: a phylogenetic tree linking the pathogen sequences in the sample is reconstructed, then a correlation is made (using appropriate statistical methods) between the phylogeny and other biological variables of interest, such as host organism, virulence, geographical location, time of origin, or mode of transmission (see Chapter 15).

Phylogenies illuminate both proximate explanations of virulence (e.g. hepatitis B; P.L. Bollyky *et al.*, unpublished) and the evolutionary processes that produce it. For example, if virulence is costly to the pathogen, then selection will reduce it. If this assumption holds and we find virulent pathogens, they will probably be continually regenerated through mutation and so will appear at the tips of the phylogenetic tree. This seems to be the case

---

Box 18.2: Population structure in malaria parasites

The malaria parasite *Plasmodium falciparum* can be categorized on the basis of host immune responses to a variety of polymorphic antigens, but only one such antigen—the variant surface antigen (VSA)—is known to elicit a protective effect against clinical symptoms, and hence possibly against parasite transmission. VSA serotypes can probably be used to recognize relevant variants of *P. falciparum* (Gupta *et al.* 1994*a*), while serotypes corresponding to some other antigen, such as the S-antigen, are not useful for categorizing parasites with respect to transmission. VSAs are also known to be involved in the process of cytoadherence (the sticking of infected blood cells to epithelium), which has been implicated as a virulence factor in the parasite, suggesting that VSA serotypes may correspond to a discrete virulence spectrum. Simple models indicate that a stable discrete set of parasites that do not share alleles or variants at the loci that influence transmission will cocirculate within a host population in a manner analogous to totally unrelated pathogens such as mumps and measles (Gupta *et al.* 1996). Thus even though they may exchange genetic material, VSA serotypes may effectively constitute discrete morphs which behave like *independently transmitted strains*.

These analytical results can explain a number of epidemiological observations regarding the virulence of *P. falciparum*, such as the differential trends that have been noted in protection against mild and severe disease among carriers of the haemoglobin S variant (sickle), for certain HLA associations, and also with regard to the impact of impregnated bed nets (Gupta *et al.* 1994*b*; Gupta and Hill 1995). This theoretical framework may also be used to address issues of intervention such as the epidemiological consequences of using vaccines that only partially cover the spectrum of strains circulating within a community (Gupta *et al.* 1997). More generally, the selection of vaccine elements should be based on an assessment of the current population structure of the parasites circulating within a community.

with syncytium-inducing (SI) strains of HIV-1, which appear to be more virulent and tend to be generated during the later years of HIV-1 infections. In contrast, if virulence benefits the pathogen, then the virulent genotypes will probably occupy deeper positions on the phylogenetic tree because they are actively maintained in populations by natural selection.

Phylogenetic trees are not without problems. The key issue in recovering correct phylogenies and making reliable inferences from sequence data is having a large and unbiased sample of lineages. Another factor complicating the construction of molecular phylogenies is recombination. In the future it will be necessary either to detect recombinant sequences and remove them from the sample or to devise evolutionary models which have the probability of recombination built in.

Finally, the evolutionary processes that have shaped pathogen populations are often too complex to be able to detect with the phylogenetic methods currently available. For example, it is possible in principle to distinguish between genetic drift and natural selection by the distinctive footprints each can leave in the structure of phylogenetic trees (Takahata and Nei 1990). But analysis is greatly complicated if more than one evolutionary process has been in operation, as is often the case, and if these processes occurred so long ago that their signal has been lost. Furthermore, although it is usually possible to choose between natural selection and genetic drift, it is hard to choose between some forms of balancing selection, especially overdominant and frequency-dependent selection. Clearly, more and better methods are needed to reveal the subtleties of molecular evolution. Despite these caveats, phylogenetic trees represent important aids that guide us through the forest of genetic variation. They will continue to help us understand the evolution of pathogen virulence.

Diversity-generating mechanisms

Pathogens reproduce and evolve in potentially hostile and changing environments in which host polymorphisms both within and between species may vary considerably and in which the acquired immune responses of host populations may change rapidly. The variety and fluctuations of host environments create substantial and dynamic selection pressures favouring adaptive variation within pathogen populations over evolutionary time. We expect evolutionary biology to make major contributions to understanding the causes and consequences of this variation. How is pathogen survival optimized in the face of challenge by the imposition of intentional (drugs, vaccines) or unintentional (novel interspecific contacts, international travel) factors? What governs the rate at which virulent variants are generated? *Understanding the evolutionary mechanisms that generate diversity is of immense public health importance.*

Variations in pathogen phenotype may reflect changes in gene expression (physiological switches) which are governed by classical changes in gene regulation and two-component sensor-transducing signalling systems (e.g. histidine protein kinases). In addition, several genetic mechanisms can produce novel and adaptive variants. These include point mutations, insertions, and deletions in protein-coding and non-coding nucleic acid sequences, transpositions, recombination between the DNA of existing strains, and gain of genes from other strains or species. Most such novel mutants are deleterious and are eliminated; a small proportion, acquired singly or accumulated over time, confer an advantage. Of particular interest are mutants that change the interaction between microbe and host, some of which confer pathogenic potential.

Pathogens that can exchange genetic material can rapidly acquire new genes from other strains or species. Extra-chromosomal elements (plasmids) carrying antibiotic resistance or toxins move rapidly across a wide range of bacterial species. Efficiency in acquiring foreign DNA probably limits the time taken for a bacterial species to evolve resistance to antibiotics; we are not aware of any evidence so far that this ability is increased in pathogenic bacteria (see Chapter 16). Some bacteria, including *Salmonella*, have a clonal population structure indicating either that they rarely recombine with related strains or that such recombinants are rarely advantageous, while others appear to undergo frequent recombination. It is not yet clear whether the extent of recombination is associated with pathogenicity. Limits on recombination include the molecular ability to do so (which may be determined by pathogen-encoded factors; Figlerowicz *et al.* 1997) and the ecological opportunity to meet (e.g. within the same organism, anatomical site, or cell).

The speed with which a pathogen species can evolve a novel variant to evade a novel host defence is determined in part by the number of molecular mutational steps between susceptible and resistant forms of the pathogen protein. To evolve away from an epitope recognized by antibodies present in the host population may require as little as one nucleotide substitution in the pathogen genome. To evolve the ability to use an alternative host surface protein to bind and enter a target cell may require many changes in the pathogen surface proteins. The other critical factor is the rate of acquisition of adaptive variants in the pathogen population, which is roughly proportional to the mutation rate, to the population size, and to the intensity of selection on the variants. Too high a mutation rate will have deleterious consequences, because many mutations lower fitness. Therefore, the observed rate of mutation in a pathogen is probably adjusted by selection to an intermediate rate determined by the advantages of acquiring new variants and the disadvantage of producing fewer infectious offspring per pathogen replication. In a pathogen with a small genome, a much higher rate of mutation is permissible, since there are fewer sites in the genome to be damaged. Thus, there is a general tendency for mutation rate to be inversely proportional to genome size (Drake 1991, 1993).

Having a mutation rate which is constant over all genes, and constant over evolutionary time, is clearly not in the best interests of the pathogen. Ideally, mutations would be limited to the genes in which adaptive variants are likely to arise and to time periods when there is a requirement for novel variants. Both theory

and data suggest that mutation rates evolve and that the mechanisms governing mutation rates may be environmentally induced. Models of infinite populations of pathogens indicate that mutation rate can escalate in a host–parasite arm race (Haraguchi and Sasaki 1996). Simulating finite clonal populations under directional selection has shown that mutators can be selected for, allowing faster adaptation (Taddei *et al.* 1997).

One strategy adopted by many pathogens, including *Trypanosoma*, *Plasmodium*, and *Borrelia*, is to switch on alternative coat-protein genes with different immunological specificity. These switches can involve several genetic mechanisms, including gene conversion between multiple copies of homologous genes within the genome or insertions and deletions in the control region of these genes. Altered transcription or frame-shifts in genes of some pathogenic bacteria, such as *Haemophilus influenzae* and *Neisseria meningitidis*, arise at high frequency through insertion or deletions of short runs of repeated DNA. These mutations switch surface molecules on or off (see Chapter 17), and when there are many of these hypermutable loci, substantial phenotypic variation can be generated from independent, stochastic switching of expression of different genes. Increased mutation rates of genes that interact with the host environment might characterize other pathogens; although there are many other examples of antigenic variation, either there are few data or it has been difficult to determine the relative contributions of selection and mutation.

Retroviruses encode their own polymerase (reverse transcriptase), and the fidelity of this enzyme is probably determined by selection for an intermediate mutation rate. The substitution rates of retroviruses per year are known to vary considerably, but it is not clear whether this reflects differences in the mutation rate per replication, as they are also influenced by the numbers of genome replications per year, which vary considerably. The lower basal mutation rate of DNA-based micro-organisms relative to RNA-based ones is partly due to proof-reading mechanisms such as mismatch repair. However, in pathogenic and commensal *Escherichia coli* or *Salmonella*, up to 1 per cent of the strains lack this function (LeClerc *et al.* 1996; Matic *et al.* 1997). This defect increases both point mutations and horizontal gene transfer 100- to 1000-fold (Radman *et al.* 1995).

Interestingly, deficiency in human mismatch repair genes is associated with tumourigenesis, suggesting that the mutation rate selected during the clonal expansion is not as low as mechanistic constraints would allow, contrary to widespread beliefs (Kimura 1967; Leigh 1973; Drake 1991).

Long-term experimental studies of evolution in a bacterial population under weak selection have shown that mutators can be selected for in some populations (Sniegowski *et al.* 1997). These results suggest that mutator bacteria might play an important role *in vivo* in generating more virulent variants. Furthermore, when bacterial clones are submitted to strong selective pressure, including antibiotics, mutators are selected for in a few days (Mao *et al.* 1997). These *in vitro* data suggest the alarming possibility that antibiotic treatments might select for mutators which in turn would increase the production rate of virulent mutants. If virulent mutants are otherwise relatively rare, this could increase average levels of virulence.

COST–BENEFIT ANALYSES

One of the most productive areas of evolutionary biology has been the development of optimality approaches to understanding trait evolution and adaptation (Parker and Maynard Smith 1991). This approach attempts to understand organism design by assessing the fitness costs and benefits of particular traits; it performs a microeconomic analysis to predict or explain the outcome of natural selection. Optimization has been particularly successful in understanding animal behaviour (Krebs and Davies 1997), biomechanics (Alexander 1997), genetic systems (Charnov 1982), and life histories (Stearns 1992), and has made successful quantitative predictions about infectious dis-

ease (Read *et al.* 1995). Central to it is the idea of trade-offs between the competing demands of different fitness components for limited resources.

Several authors have used optimality analyses to explain patterns of virulence (Ewald 1980, 1994; Levin and Pimentel 1981; Anderson and May 1982). They adopt a parasite-centred approach and consider how parasites could maximize their reproductive success (for reviews see Bull 1994; Frank 1996; and Chapter 14). The optimality approach also applies to host responses to infection, where natural selection should balance the costs of responding (resources, immunopathology) against the parasite-caused costs of not responding. The parasite-centred cost–benefit approach, which treats virulence as an unavoidable consequence of parasite adaptation, has generated considerable enthusiasm about the application of evolutionary biology to medicine (Williams and Nesse 1991; Ewald 1994; Futuyma 1995).

It has also generated controversy, not least at this meeting. Views ranged from agnostic to convert. Most considered the ideas intuitively appealing, and all agreed on the urgent need for more empirical work. In principle, specific hypotheses are easily tested—they have the benefit of being vulnerable to falsification. To illustrate the issues involved, we deal with one of these hypotheses: Ewald's view that virulence, which can affect host immobility, is a selective consequence of transmission mode (Ewald 1994). We then turn briefly to vaccination strategies to illustrate the sort of practical suggestions a cost–benefit framework can generate.

Costs and benefits of host immobility

The argument is as follows. Host illness is a potential liability for the pathogen. When pathogens rely on the mobility of their current host to reach uninfected individuals, the illness caused by intense exploitation of the host reduces the potential for transmission. These costs of host exploitation are to be weighed against the competitive benefits that the pathogen accrues from the exploitation. High pathogen virulence can be evolutionarily stable if the costs incurred by parasites from host damage are particularly low or the benefits obtained by the parasites from exploitation are particularly high. Thus if host immobilization has little negative effect on transmission, then pathogen variants that exploit the host so intensely that the host is immobilized may accrue the fitness benefits of exploitation while suffering a relatively small decrement in transmission. Parasites that do not rely on host mobility for transmission should therefore evolve to be more virulent than those that do rely on host mobility.

From this hypothesis, Ewald made several predictions. One is that diarrhoeal disease organisms transmitted by water should evolve to relatively high levels of virulence because effective transmission can occur when infected hosts are immobilized: attendants remove and wash contaminated clothing and bed sheets, contaminating water supplies which then can infect large numbers of uninfected individuals. In support of this hypothesis, Ewald argued there is a positive relationship among diarrhoeal pathogens between mortality rates and the proportion of disease outbreaks involving water-borne pathogens (Ewald 1991).

At the meeting, both the hypothesis and the data used to support it were the subject of considerable debate. Concern about the hypothesis centred on two related aspects: its verbal rather than mathematical formulation, which some of us argued leads to ambiguity, and its generality. In the context of specific diseases, everyone had their pet complexities which might confound the hypothesis (e.g. host immune status, host evolution, other trade-offs which might be more important, the effects of intrahost competition: cf. van Baalen and Sabelis 1995*b*). There is a clear need for more formal modelling that would allow the consequences of various specific assumptions to be investigated and the potential impact of other factors to be assessed.

The controversy about the supporting data was typical of tests involving cross-population or interspecific comparisons elsewhere in evolutionary biology (see Chapter 14): the correlational nature of the data means there will always be potentially confounding variables, and the biological (as opposed to statistical) significance of the associations are hard to determine. In the present context, there are the additional problems of choosing a virulence measure (surrogates, with associated caveats, will often be necessary), especially measures appropriate for comparative analyses (e.g. how to keep both denominator and numerator comparable across diseases when estimating mortality rates). Many potentially confounding variables seem more likely to produce noise rather than artefactual associations, but several were discussed which may generate the predicted correlations for reasons unrelated to the hypothesis. In principle, most of these could be evaluated empirically.

There was general agreement that comparative analyses represent just the first stage in the process of hypothesis testing, but there was little agreement about how compelling the existing evidence is, by how much the comparative data could be improved, and whether that was a profitable path to follow. Other approaches may be more useful. Laboratory experiments on animal model systems can offer a way to alter one variable at a time, but even there, caution needs to exercised (see Chapter 14). Field experiments can also be conducted to determine whether changes in the transmission variables influence parasite virulence. There is always the need to distinguish competing hypotheses.

Can one use existing theory to make public health recommendations? There was considerable nervousness on this issue, and little consensus was reached. Most of us felt that the ideas need more empirical evaluation before useful recommendations can be made. We noted that interventions which have been predicted to induce pathogen evolution beneficial to human health (e.g. sanitation) are often beneficial for other reasons. In these cases, there is little risk of harm and the interventions could constitute experimental tests of the evolutionary hypotheses; if they accrue, the evolutionary benefits would increase the cost-effectiveness of such strategies. However, unsuspected mechanisms and trade-offs might result in evolution towards *increased* virulence in response to intervention strategies which have short-term ecological benefits (e.g. fewer parasites). The likelihood of such undesirable evolution remains an open and critically important question relevant to all large-scale interventions.

Evolutionary effects of vaccines on virulence

We were less divided about the public health possibilities of altering the evolution of virulence using vaccination. Vaccines are currently developed and chosen with little consideration of their evolutionary consequences, but in some cases, it may be fruitful to choose antigens on grounds other than protective efficacy.

The virulence-antigen strategy suggests that policy makers question the tendency to use vaccines that inhibit the broadest possible spectrum of existing variants (Ewald 1996). It proposes instead that antigens be restricted, wherever possible, to those that make otherwise mild organisms harmful to their conspecifics. Such virulence-antigen vaccines should provide exceptionally strong long-term control because they disproportionately suppress the virulent variants, leaving behind mild variants which can augment disease control by competing with any remaining virulent variants. When the mild forms share cross-reactive antigens with the severe forms, the immunity generated against the mild variants should allow mild forms to be used like a free live vaccine that will protect those who are not vaccinated and those who develop insufficient immunity to the administered vaccine. These effects may be particularly important when vaccine coverage temporarily wanes, a change that has occurred often in recent decades even in countries with high-quality health care (Hewlett 1990; Pichichero *et al.* 1992). Evolutionary reductions in the virulence of circulating organisms will increase the chances that mild organisms will cause infections during such lapses. Also, if virulent organisms arise by mutation or enter from other areas, they will be less able to spread because of the increased resistance attributable to mild natural infections. The standard diphtheria vaccine is an example of an apparently effective virulence-antigen vaccine (Ewald 1996).

Intervention outcomes: eradication versus coexistence

Many medical interventions which have been thought to cause evolutionary reductions in virulence, such as alterations in transmission patterns and the use of virulence-antigen vaccines, would leave populations with mild pathogens. Although this living arrangement might seem less than favourable, these waters are uncharted. The value of such coexistence could be worse than eradication, but it might be better, especially if the remaining pathogens are mild.

Measles immunization offers an illustration of how the continued cycling of pathogens might be advantageous. With measles, the advantage could be generated from modification of the immune response to prevent acute mortality and severe disease. There are two kinds of reasons for this. First, recent epidemiological studies from West Africa have suggested that mild measles infection and measles immunization may be beneficial in the long term, providing protection against causes of childhood mortality other than measles infection (Aaby *et al.* 1995, 1996). Hence, it might be beneficial to maintain stimulation with measles virus by vaccination and natural infection rather than attempting to eradicate measles virus and stop immunization.

Second, if measles were nearly perfectly controlled but could not be totally eradicated, this would have major implications for the maintenance of immunity levels in the population. With no re-exposure to the virus, antibody levels would decline and many pregnant women would become susceptible. Those who had lower antibody levels would transfer less antibody to their offspring (Maldonado *et al.* 1995), implying that future infants might become susceptible to measles in the first months of life. Measles during pregnancy and in the first months of life can both produce severe infection (Aaby *et al.* 1986).

These problems would be particularly important in developing countries with a high birth rate where the majority of women between 18 and 30 are either pregnant or breastfeeding. To maintain control of severe and fatal measles, it might be necessary to provide multiple doses of measles immunization throughout life, a strategy that is potentially very costly. Measles re-immunization is much less effective in stimulating high antibody levels than the natural disease. Relying only on vaccines to maintain immunity in a near eradication situation would require repeated doses. Alternatively, a less protective vaccine could prevent acute mortality, permit the circulation of virus thereby gaining from the benefit of restimulation, and at the same time allow for re-exposure and maintenance of protective antibody levels among older people, pregnant women, and young infants. In a situation where measles could not be completely eradicated, there might be risks of outbreaks of severe infection, and the immunization strategy would have to be designed to protect the known high-risk groups.

RESEARCH QUESTIONS

Several research questions emerged from our discussions. Generating these proved the least contentious part of our deliberations; that we readily agreed on areas of ignorance speaks volumes about how much is unknown.

1. What selective factors act on virulence? How important are within-host competition and host demography (density, mortality rates)?

2. Over what time-scale do we expect virulence to evolve in human populations?

3. To what extent is coevolution involved in virulence?

4. Does coevolution with pathogens occur in human populations?

5. How important is selection acting on virulence determinants encoded by host genes relative to those encoded by parasite genes?

6. How can natural selection be detected in host and pathogen molecular sequences?

7. What are the selective factors responsible for the origin and maintenance of polymorphisms of genes involved in host–pathogen interactions?

8. How does host genetic diversity alter pathogen evolution?

9. Can we predict which virulent pathogens will emerge in human populations?

10. Does selection affect mutation rates? Does the rate of mutation predict the probability that virulent pathogen mutants will arise?

11. Can control strategies affect the evolution of virulence?

12. Can evolutionary attenuation of virulence using the virulence-antigen strategy make an important contribution to public health?

13. Under what conditions can infections have beneficial effects?

14. What is the relative importance of ecology and genetics (and their interaction) in determining virulence of specific pathogens?

15. What are the evolutionary and epidemiological consequences of shifts in average age of infection?

16. Can evolutionary biology help to anticipate future epidemiological patterns of infectious disease?

17. Can evolutionary biology help to understand variation in disease severity in individual patients?

CONCLUSION

Variation in the virulence of infectious disease clearly begs evolutionary understanding. We hope the above discussion has shown that an evolutionary perspective can contribute to elucidating proximate mechanisms responsible for virulence. It also has the exciting potential to predict the future direction of virulence evolution, making it possible to evaluate the long-term consequences of large-scale intervention strategies. Importantly, natural selection can result in increases and decreases in virulence; avoiding the former and promoting the latter should be an aim of biomedicine.

We are well aware of the largely theoretical basis of what little we know and how weakly the theory is supported, for the most part, by rigorous empirical analysis. There is relatively little to convince sceptical clinicians and hard-pressed funding agencies. Indeed, in the immediate future, considerable ingenuity will be needed to attract funding to generate the discoveries which will justify funding. We do not doubt that the effort will be worthwhile. Benefits for human health seem likely; we have given examples.

However, the field is very new and much remains unexplored. For example, coevolution (arms races with reciprocity of genetic interactions and coadaptation) undoubtedly occurs at a molecular level. But does it have clinical implications? There are likely to be many niches of interaction, and we have no theoretical understanding of which are likely to be more or less useful in designing intervention strategies, such as vaccine targets. Is the identification of host genes involved in resistance (for instance, HLA alleles) a useful pointer to virulence factors in parasites? Coevolution may be less relevant in the human health context than it is, for example, in plant pathology, because the rapid somatic evolution of higher vertebrate immune systems has the potential to buffer hosts from coevolution. There may be little intergenerational host genetic change even if parasites are undergoing rapid evolution. Such questions, and their implications, are ripe for exploration.

SUMMARY

Patterns of disease within and between populations are conventionally explained in terms of underlying mechanisms. A complete understanding of virulence also requires explanation in terms of the evolutionary processes responsible. We discuss several ways in which an evolutionary perspective can contribute to research on virulence.

1. It can provide a theoretical basis for recognizing the relevant genetic diversity and population structure.

2. Phylogenetic analysis (mapping disease phenotypes on to trees) can pinpoint the mechanistic basis of virulence; phylogenetic information can also help to determine underlying evolutionary processes.

3. Much of evolutionary biology concerns the mechanisms that generate diversity and how selection acts on them. Understanding the frequencies and rates that characterize the evolution of virulence and the evolutionary responses to vaccines are of immense public health importance.

4. Analyses of the costs and benefits for fitness of the traits that cause virulence can suggest how to induce pathogen evolution beneficial to human health, and when it might be necessary to prepare for, or attempt to avoid, the evolution of less benign pathogens.

For each case, we gave examples and discuss the theoretical and empirical requirements for further progress. We finished with a list of research questions; there is an urgent need for empirical work.

Part V: Non-infectious and degenerative disease

19

EVOLUTION, SENESCENCE, AND HEALTH IN OLD AGE

Thomas B.L. Kirkwood, George M. Martin, and Linda Partridge

The mechanisms underlying human senescence are so complex (Medvedev 1990; Kirkwood and Franceschi 1992; Martin 1992) that even recently some scientists have dismissed ageing as a field not deserving immediate attention (e.g. Wolpert 1992). But human senescence cannot be ignored. Its effects are increasing as a result of demographic change, and the negative impact of age-related disease on quality of life is growing in consequence. Fortunately, these demographic changes are occurring at a time when (i) new theory is providing a sharper focus on the key questions which need to be addressed and (ii) new techniques are providing the experimental means to investigate complex, multicausal phenomena at the requisite level of detail (Kirkwood 1996).

A major contribution to our understanding of senescence has come from evolutionary theory (Medawar 1952; Williams 1957; Hamilton 1966; Kirkwood 1977, 1985; Charlesworth 1980, 1994; Finch 1990; Rose 1991; Partridge and Barton 1993; Holliday 1995). Evolutionary questions about senescence are asked not so much to satisfy curiosity about events that occurred long ago as to understand the genetic factors affecting longevity and health in old age now, and how these factors interact with the environment. Evolutionary insights tell us what kind of process ageing is and guide the experimental study of its mechanisms. Understanding how the human life history evolved sheds light on the environmental and constitutional factors that produce different styles of ageing and on interventions that can improve the quality of life in old age.

AGEING AND ITS MEASUREMENT

We use the words 'ageing' and 'senescence' interchangeably, to refer to the biological declines in structure and function that unfold gradually and progressively in humans and other animals after reaching adulthood. Mortality rates in humans rise approximately exponentially with increasing chronological age, noted first by Gompertz (1825); a similar pattern has been observed in other mammals (see Sacher 1978; Finch 1990). For this reason ageing has commonly been defined as a 'progressive, generalized impairment of function resulting in an increasing probability of death' (Maynard Smith 1962). This definition reflects the primary interest of physicians in the ageing process. However, in an evolutionary sense, senescence may be defined more broadly as a decline in future fitness affected by reductions in fertility as well as survival (Partridge and Barton 1996). Age-associated decline in fertility, especially in women, can itself present as a medical problem to which evolutionary thinking can contribute (see below and Chapter 10).

The rate of increase in human mortality appears to decelerate among centenarians (Smith 1994), but it is not known whether this reflects (i) genetic heterogeneity within the population, (ii) particularly assiduous care of the most elderly, or (iii) intrinsic biological processes. Heterogeneity is probably at least part of the explanation if, as seems likely, centenarians represent a robust subset of the population (Vaupel *et al.* 1979; Schächter *et al.*

1993). As a cohort ages, the frailer individuals die sooner, and the mortality rate of the survivors is based on a fraction of the population who may have started their adult lives with intrinsically greater capacity for survival. Non-exponential mortality patterns have been described in other species, for example, the medfly *Ceratitis capitata* (Carey *et al.* 1992), the fruitfly *Drosophila melanogaster* (Curtsinger *et al.* 1992), and the nematode *Caenorhabditis elegans* (Brooks *et al.* 1994). For these reasons, the Gompertz model of exponentially increasing mortality rates, once regarded by some almost as a law of ageing, is now seen by most as a convenient description of the effects of ageing on age-specific mortality rates in instances where it fits well.

Many important diseases of late life show approximately exponential increases in age-specific incidence. Examples include Alzheimer's disease, carcinoma of the prostate, and carcinoma of the colon. However, these should not be taken as evidence that the underlying physiological processes necessarily follow exponential kinetics. Some physiological parameters show accelerating declines but others are linear. Cross-sectional studies reveal great variability in both the slopes and patterns of the changes observed with age.

Many of the observable features of ageing in animals are consistent with the idea that ageing is a result of damage. Damage can arise at all levels within the organism, ranging from macroscopic tissue and organ damage (e.g. wounding, burns, infections), to a continual background level of molecular damage (e.g. oxidation by free radicals, mutations, protein denaturation). The inevitability of damage has led to the suggestion that ageing itself is inevitable though the time-dependent accumulation of damage, or 'wear and tear'. But ageing is not inevitable, because some species do not age (see Comfort 1979; Finch 1990). Furthermore, the germ line of all extant species must in a sense be immortal, otherwise these species would die out. Yet germ cells are also subject to damage; thus their immortality is evidence that maintenance and repair processes, perhaps aided by selection at the cell

level, can prevent the accumulation of damage (Kirkwood and Cremer 1982). If ageing is not inevitable we need to ask why it is not eliminated by natural selection. This brings us to the evolutionary theories.

EVOLUTIONARY THEORIES OF AGEING AND THEIR IMPLICATIONS

Evolutionary theories of senescence explain why ageing occurs, in spite of its clearly deleterious impact on Darwinian fitness. Because senescence is so obviously deleterious for the individual, several attempts have been made to explain its evolution in terms of an advantage to the population as a whole. Typically, it is suggested that senescence functions as a form of population control to prevent overcrowding. These theories are given little credence today, first because there is scant evidence that animal numbers in the wild are regulated by senescence, most deaths occurring at younger ages from extrinsic causes such as predation, and second because they invoke group selection, which is a weak evolutionary force compared with selection on individuals to contribute genetically to the next generation (Williams 1966). Nevertheless, these ideas periodically resurface (e.g. Kenyon 1996).

Most credence is now given to evolutionary theories which are 'non-adaptive' in the sense that they do not suggest senescence confers any fitness benefit of itself, and they recognize that senescence may indeed be harmful (Kirkwood and Cremer 1982). All the non-adaptive theories follow from the reduced strength of natural selection on older organisms (Medawar 1952; Hamilton 1966; Charlesworth 1980, 1994).

Even if senescence did not happen, death would, because of disease, predation, and accidents. Any new mutation that reduces the fitness of the young will be expressed in most of the individuals that bore the mutation at birth and will therefore be strongly opposed by natural selection. In contrast, an otherwise similar mutation that affects only older individuals will be expressed only in that subset of carriers that survives to the age of expression. Natural selection will therefore be less effective at removing deleterious mutations with later ages of effect, and these mutations will reach a higher frequency under mutation–selection balance: this is the 'mutation accumulation' theory of ageing (Medawar 1952). In addition, some genes may confer fitness benefits in early life but produce disadvantages later: this is the 'pleiotropy' or 'trade-off' theory of ageing (Williams 1957). The 'disposable soma' theory (Kirkwood 1977, 1981; Kirkwood and Holliday 1979) predicts a particular form of trade-off and provides a link between evolutionary and physiological aspects of senescence by recognizing the evolutionary importance of the allocation of metabolic resources among activities of growth, somatic maintenance, and reproduction. Increasing maintenance promotes the survival and longevity of the organism but only at the expense of significant metabolic investments that could otherwise be used for greater reproductive effort. Formal models demonstrate that the optimum allocation strategy results in a smaller investment in somatic maintenance than would be required for an indefinite life span (e.g. Kirkwood and Holliday 1986; Kirkwood and Rose 1991; Abrams and Ludwig 1995). The disposable soma theory predicts higher levels of maintenance in somatic cells of longer-lived species, for which there is growing evidence.

The evolutionary theories predict that many genes will contribute to senescence and thus suggest that uncovering the genetic basis of senescence will be a complex task requiring a combination of approaches, such as quantitative trait locus (QTL) mapping, methods for analysis of polygenic diseases, transgenic animal models, and so on.

Highly polygenic traits tend to have a major environmental component. This is likely to be of great importance in industrial human populations, which have undergone drastic environmental change over a few generations (see Chapter 22). We need to think in term of genes, environment, gene–environment inter-actions, and the interactions between the effects of environment at different ages. Environmental effects are proving less reversible than was once thought; for example, there is some evidence that the level of fetal nutrition may influence health in later life (see Chapter 21).

The evolutionary theories also tell us that senescence is not a programmed process like development, with a closely regulated machinery of hierarchical genetic control. No genes evolved to cause ageing. Rather, genes active during development determine the resistance of the soma to damage, genes that determine rates of activity and reproduction influence the impact of damage, and genes active in various forms of repair determine the extent to which damage is reversed. We therefore do not expect ageing to yield readily to the kinds of genetic analysis that have been so successful in developmental biology.

An implication of the trade-off theory of ageing is that senescence may be influenced by changing the balance between good and bad effects. Indeed, the trade-off principle should be borne in mind when considering interventions in human ageing. Interventions that might increase longevity or postpone a late-age disease may have side-effects mediated by trade-offs. In addition, because the accumulation of damage affecting different vital processes is expected to proceed at approximately similar rates, amelioration of one problem may simply expose the effects of another. These evolutionary deductions are not a counsel of despair; it is abundantly clear from work on experimental organisms that the rate of ageing can be changed by environmental and genetic interventions. But it is important to proceed in full awareness of the multifactorial nature of ageing.

MODEL SYSTEMS FOR TESTING THE EVOLUTIONARY THEORIES

Conserved mechanisms of genetic control of development across huge evolutionary distances have been revealed in recent years, and it is natural to ask if a similar set of homologies will apply to events at the other end of life. The idea that there are indeed such 'public' mechanisms of ageing (Martin *et al.* 1996) is attractive, and has motivated much of the gerontological research carried out on a wide range of model organisms, including rodents, flies, nematodes, yeast, and filamentous fungi. No model system will adequately represent all aspects of human ageing, but they can act as an important source of hypotheses about its nature and mechanisms.

Model organisms have been used to ask if there is genetic variation for the rate of ageing in natural populations, and to explore the possibility of genetic and environmental intervention in the ageing process. For practical reasons, most work to date has focused on the fruitfly *D. melanogaster* and the nematode *C. elegans*, both of which combine advantages of short lifespan, low maintenance costs, and excellent background genetics.

Drosophila has a 10-day development period from egg to adult and an adult life span of about 30–40 days. The fly is a random-bred, diploid organism. Its populations therefore harbour high levels of standing genetic variation, in which regard it resembles humans. One contribution of *Drosophila* to our understanding of ageing has come from the use of this quantitative genetic variability to produce lines of flies with different lifespans.

Genetic variation for rate of ageing has been demonstrated, mainly by the use of artificial selection on age at reproduction. 'Young' lines of flies are propagated in each generation from adults that have just emerged from pupae, while 'old' lines are propagated from adults that have already been reproducing for some time. Compared with the 'young' lines, the 'old' lines are therefore selected for increased lifespan because, to contribute to the next generation, they must survive to the time of egg collection. Greater longevity of 'old' lines has consistently been found in these studies (e.g. Luckinbill *et al.* 1984; Rose 1984; Partridge and Fowler 1992; Roper *et al.* 1993). In an alternative approach, direct selection for increased lifespan has been accomplished using an experimental design that exploited the effect of temperature on *Drosophila* longevity (Zwaan *et al.* 1995). In this experiment, the progeny from each pair of breeding flies were divided into two groups, one of which was maintained at 15°C and the other at 29°C. At 29°C maximum lifespan was less than 60 days, whereas at 15°C mean lifespan was greater than 120 days. This meant that lifespans could be measured on the groups maintained at 29°C and the longest-lived families selected, at a time when the sibling groups kept at 15°C were still reproductively viable.

If trade-offs are important in ageing, then we would expect the long-lived flies resulting from artificial selection experiments to pay for their late-life advantage at some earlier point in their life history; and indeed, the long-lived flies were shown to have lower fecundity in early adult life (e.g. Luckinbill *et al.* 1984; Rose 1984; Zwaan *et al.* 1995). Several studies have also reported an increase in the duration of the pre-adult period in 'old' lines (e.g. Partridge and Fowler 1992; Chippindale *et al.* 1994, 1996). However, these findings have not been universal. Indeed, experiments have revealed the crucial importance of both the environment during selection and testing, and the exact selection protocol (Partridge and Fowler 1992; Roper *et al.* 1993; Chippindale *et al.* 1994, 1996; Leroi *et al.* 1994*a, b*).

Drosophila has also been used to test the contribution of mutation accumulation to ageing. Demonstrating the effects of mutation accumulation is not straightforward. One line of evidence has come from experiments where selection on older flies has been relaxed, allowing deleterious mutations to accumulate more easily (e.g. Mueller 1987). Another comes from studying the increase with age in additive genetic variance for survival rates and levels of inbreeding depression (Hughes and Charlesworth 1994; Charlesworth and Hughes 1996; Promislow *et al.* 1996; Tatar *et al.* 1996). As yet we cannot assess the relative importance of mutation accumulation and trade-offs in the evolution of senescence in *Drosophila*.

An alternative approach to the selection experiments is to recover mutations that affect lifespan. Mutagenesis for any trait related to fitness is problematic in *Drosophila*. New mutations, which are nearly always at least partially recessive, have to be made homozygous before they are scored, and in the process large parts of the genome become inbred, leading to confounding effects on lifespan from inbreeding depression. These difficulties are avoided in the nematode *C. elegans* which is a self-fertilizing hermaphrodite. Most worms are homozygous at most loci and do not show inbreeding depression for lifespan.

C. elegans has a short generation time (about 20 days) during which it progresses through a series of immature larval stages to adulthood and then reproduces. If population density is high, causing low nutrient levels, the larva enters an alternative developmental pathway which gives rise to a third-instar 'dauer' larva with arrested development, and which is resistant to various forms of stress (Riddle 1988). Mutations in several genes have been found to extend hermaphrodite lifespan (Table 19.1). Many of the genes are in the signalling pathways controlling the formation of the dauer larva (*age-1* and the *daf-* genes), and the mutants show increased resistance to various forms of environmental stress, such as heat (Lithgow 1996). In addition, lifespan is extended by mutations in genes that affect timing, including developmental timing (*clk-* mutants and *gro-1*). In *Drosophila*, extension of lifespan has been associated with growth and reduced fertility of the young adult. The costs (if any) that are associated with the *C. elegans* mutants that extend lifespan have not yet been identified.

Table 19.1 Genes for which mutations have been recovered that extend the adult lifespan of hermaphrodite *C. elegans* (see Lithgow (1996) and Martin *et al.* (1996) for extended reviews)

Gene	Other information and sources
age-1	Dauer constitutive at 27°C. Encodes a homologue of mammalian phosphatidylinositol-3-OH-kinase catalytic subunits (Klass and Hirsh 1976; Friedman and Johnson 1988a, b; Malone *et al.* 1996; Morris *et al.* 1996)
spe-26	Isolated as a male sterile. Homology to actin-binding proteins (van Voorhies 1992; Varkey *et al.* 1996)
daf-2	Dauer constitutive at 25°C (Dorman et al. 1995; Larsen *et al.* 1995). Encodes an insulin receptor-like gene (Kimura *et al.* 1997)
daf-12	Interacts with daf-2 to extend lifespan (Larsen *et al.* 1995)
daf-23	Dauer constitutive at 25°C (Larsen *et al.* 1995)
daf-28	Dauer constitutive. Does not affect recovery from dauer (Malone *et al.* 1996)
clk-1	Maternal effect. Extended pre-adult development period. Homologous to yeast metabolic regulator Cat5p/Coq7p (Wong *et al.* 1995; Lakowski and Hekimi 1996; Ewbank et al.1997)
clk-2	Maternal effect. Extended pre-adult development period (Lakowski and Hekimi 1996)
clk-3	Maternal effect. Extended pre-adult development period (Lakowski and Hekimi 1996)
gro-1	Maternal effect. Extended pre-adult development period. Common pathway with *clk-* genes (Lakowski and Hekimi 1996)

EVOLUTIONARY ASPECTS OF THE PATHOBIOLOGY OF HUMAN SENESCENCE

We know far more about the phenotypes relevant to ageing in our own species than we do for any other species (see Table 19.2), and research on the genetic basis of human diseases is accelerating. As longevity increases, research focuses increasingly on diseases that arise in older people, diseases that are likely to be polygenic in nature and to involve significant gene–environment interactions. Important examples are cardiovascular disorders, which are currently the major causes of morbidity and mortality (Martin 1993), and dementias of the Alzheimer type (DAT), which until recently were largely ignored as a cause of death but have become the fourth leading

Table19.2 Comparatively prevalent pathological processes associated with the senescent phenotype of *Homo sapiens sapiens* (modified from Martin 1993)

Body system	Phenotype
Cardiovascular	Atherosclerosis, arteriolosclerosis, medial calcinosis, basement membrane thickening of capillaries; hypertension; increased susceptibility to thromboembolism; myocardial lipofuscinosis; myocardial hypertrophy with interstitial fibrosis, valvular fibrosis and valvular calcification (calcific aortic stenosis)
Central nervous	β-Amyloid depositions; lipofuscin depositions; neuritic plaques; neurofibrillary tangles; gliosis; diminished synaptic density; regional neuronal loss, leading to disorders such as Parkinson's disease and Alzheimer's disease
Peripheral nervous	Segmental demyelination with decreased nerve conduction velocity
Special senses	Loss of visual accommodation; ocular cataracts; senile macular degeneration; loss of high frequency auditory acuity; loss of olfactory acuity
Respiratory	Chronic obstructive pulmonary disease; interstitial fibrosis; decreased vital capacity
Renal	Glomerulosclerosis and loss of nephron units; interstitial fibrosis
Male reproductive	Decreased spermatogenesis; hyalinization of seminiferous tubules; benign prostatic hyperplasia; adenocarcinoma of the prostate
Female reproductive	Depletion of ovarian primordial follicles; ovarian stromal cell hyperplasia; endometrial atrophy and hyperplasia; endometrial carcinoma, carcinoma of breast and ovary, 'fibroid' tumours of the myometrium (leiomyomas); vaginal atrophy
Musculoskeletal	Skeletal muscle atrophy and interstitial fibrosis; osteoporosis, osteoarthritis
Haematopoietic	Anaemias (including pernicious anaemia); chronic lymphocytic leukaemia; chronic myelogenous leukaemia, myelodysplastic syndromes; myelofibrosis; myoclonal gammopathies and multiple myeloma; polycythaemia vera
Endocrine	Interstitial fibrosis of thyroid; hypercortisolaemia; asynchrony of growth hormone release; amyloid depositions in β-cells of pancreatic islets; non-insulin-dependent diabetes mellitus
Gastrointestinal	Colonic polyps and adenocarcinoma; diverticulosis of colon; gastric anal atrophy; fatty infiltration and brown atrophy of pancreas; adenocarcinoma of pancreas; cholelithiasis; periodontal disease
Integumentary	Epidermal atrophy; pigmentary alterations; basophilic alteration of collagen; senile elastosis; senile keratoses (especially in sun-exposed skin); regional atrophy and hypertrophy of adipocytes; thinning and greying of hair of scalp and body

cause of death in the United States (Blass 1985). We consider atherosclerosis and DAT (and related dementing illnesses of the elderly) as model systems for genetic studies of such diseases and briefly examine how evolutionary theory might contribute to their investigation. We also consider Werner syndrome as an example of a single gene mutation that appears to accelerate several aspects of ageing.

Research in atherogenesis, the formation on the walls of arteries of raised, fibrotic, lipid-containing lesions, is well advanced and serves as a good model for the evolution of a complex disease with polygenic and environmental effects. The leading risk factor for the disease is age. A study of 19 ethnic groups demonstrated approximately linear but variable rates of increase, with age, in how rapidly the intimal surfaces of ageing aortas and coronary arteries develop raised, fibrotic, lipid-containing atheromas (Eggen and Solberg 1968). The great majority of deaths from atherosclerosis (typically via myocardial infarctions) occur after the age of 45 years: this is a disorder which has largely escaped the force of natural selection.

Evolutionary theory predicts two classes of gene action which might contribute to the pathogenesis of late-life disorders such as atherosclerosis: namely, trade-offs and late-acting deleterious mutations. There is some evidence to support both types of gene action, although the atherosclerosis research community has not yet discussed its findings from this perspective. A trade-off is suggested by recent evidence that mice deficient in macrophage-scavenging functions are more resistant to atherosclerosis. One theory of atherogenesis invokes a major role for arterial wall damage mediated by post-translationally modified (particularly oxidized) low-density lipoproteins (Brown and Goldstein 1983; Aviram 1996; Steinberg 1997), which are picked up by macrophages, the body's 'professional' scavengers (Krieger *et al.* 1993). Macrophages are thought to be the major sources of the lipid-laden foam cells that appear in the early 'fatty streak' stage of atherogenesis. In wild-type mice, spontaneous atherosclerosis is rare, but mice homozygous for null mutations at the apolipoprotein E locus develop marked hyperlipidaemia and progressive atherosclerosis (Palinski *et al.* 1994). However, when these mice also bear mutations in a macrophage scavenger receptor gene, they are more resistant to atherosclerosis but become highly susceptible to infection by certain microbial pathogens (Suzuki *et al.* 1997). Thus, atherosclerosis may develop, in part, as a by-product of the evolution of superior mechanisms for dealing with infectious agents.

How might late-acting deleterious mutations affect atherogenesis? Aberrations in the embryonic development of the vascular wall, which may have a genetic basis, may result in foci of intimal cells capable of multiplying, eventually leading to atherosclerotic lesions (Schwartz *et al.* 1995*a, b*; Majesky and Schwartz 1997). However, it is also possible that this process involves somatic rather than germ-line mutations.

By far the most prevalent dementing illness of the elderly in developed countries is dementia of the Alzheimer type (DAT), characterized by numerous neuritic plaques containing aggregates of β-amyloid peptides, deposits of β-amyloid within the walls of cerebral blood vessels, neurofibrillary tangles (consisting of unusual phosphorylated forms of the microtubule-associated protein tau), granulovacuolar

degeneration of hippocampal cells, decrease in neocortical synaptic density, and regional loss of neurones. Confirmation of DAT at autopsy is required, because clinical diagnosis can only indicate possible or probable DAT; but, even at autopsy, the semiquantitative criteria currently in use are relatively arbitrary (Mirra *et al.* 1991). The lesions in DAT are not qualitatively distinct from what one can observe in the brains of older subjects without dementia. This fact, and the evidence for some degree of cognitive decline in essentially all subjects of very advanced ages, suggests that DAT is an interesting model to study genetic and evolutionary factors that may have a general relevance to ageing.

A large number of genetic loci, combined with gene–environment interactions, undoubtedly contribute to the variable rates of decline of neural function among ageing human subjects. A crude estimate can be made from Table 10 of Martin (1978), in which a systematic search was made in McKusick's *Mendelian inheritance in man* (McKusick 1975) for genetic loci, variation at which could result in abiotrophic degenerative or proliferative disorders of potential relevance to the pathobiology of ageing, including dementias, certain relevant types of neurodegeneration, or both. From a total of 2336 entries, 55 loci relevant to the dementias and related pathologies (2.35 per cent) were identified. Assuming a total of 100 000 human genes, this suggests that variation at up to 2350 genetic loci could modulate such phenotypes. In 1978, McKusick listed only one DAT locus, based upon evidence for the Mendelian segregation of an autosomal dominant gene. We now know of at least three autosomal dominant genes (*APP, PS1, PS2*) that can result in early-onset familial DAT. At least one polymorphic locus (*APOE*) can influence the age of onset of the common 'sporadic' late-onset forms of the disease. This information is summarized in Table 19.3.

The first evidence of an important role for a genetic susceptibility factor in DAT came from the study of patients with Down syndrome (trisomy 21). Essentially all such individuals surviving to around the age of 40 years exhibit, at autopsy, all of the neuropathological stigmata

Table 19.3 Mutant, polymorphic, or wild-type loci shown to modulate age of onset of Alzheimer's disease (the order in which the loci are listed reflect the chronologies of their discoveries; additional loci must exist)

Chromosome band	Genetic locus	Formal genetics	Protein products	Prevalence (US, Europe, Japan)	Age of onset (years)	Role in amyloidogenesis
? 21q21– 21q22	?*APP*	Gene dosage (trisomy 21)	?βPP isoforms	~ 1.5/1000 newborns[a]	< 40	βPP Aβ
21q21.3– 21q22.05	*APP* (*AD2*)	Autosomal dominant	βPP isoforms and Aβ species	~ 20–30 families	~ 50–60	Aβ· Aβ42
19q13.2	*APOE*	Three-allele polymorphism	Apolipoprotein E	ε3 > ε4 > ε2 allele[b]	~ 60–90	Aβ
14q24.3	*PS1* (*AD3*)	Autosomal dominant	Presenilin 1	~ 120–130 families	~ 30–60	Aβ42
1q31–1q42	*PS2* (*AD4*)	Autosomal dominant	Presenilin 2	~ 10 families	~ 50–70	Aβ42

βPP = β-amyloid precursor protein; Aβ = Aβ-peptide; Aβ42 = 42-amino acid variant (long form) of Aβ-peptide.
[a] The prevalence of Down syndrome births varies with the age distribution of parents (sharply increased prevalence with advanced maternal age) (Wald and Kennard 1996).
[b] Allele frequencies vary among human populations (ε4, the allele associated with earlier onsets of Alzheimer's disease, is comparatively common in New Guinea and Nigeria and comparatively uncommon in Asian populations) (Gerdes *et al.* 1992). Reviews and relevant references are cited in Hardy (1996*a, b*),Tsuang and Faraone (1996), Pericak-Vance and Haines (1995).

of DAT, including numerous neuritic plaques containing aggregates of β-amyloid peptides, deposits of β-amyloid within the walls of cerebral blood vessels, neurofibrillary tangles (consisting of unusual phosphorylated forms of the microtubule-associated protein tau), granulovacuolar degeneration of hippocampal cells, decrease in neocortical synaptic density, and regional loss of neurones. (Clinical evidence of a progressive dementing illness may be ambiguous, however, given the mental retardation due to Down syndrome itself.) Given this evidence of DAT susceptibility of Down syndrome subjects and the discovery that the amyloid precursor protein, responsible for the genesis of the β-amyloid aggregates, maps to chromosome 21, it was not long before families were identified in which DAT cosegregated with mutations at the β-amyloid precursor protein locus (*APP*) on chromosome 21. However, only about 20 to 30 such families have so far been discovered around the world. Far more common (but still only a minute proportion of all DAT cases) are mutations at a locus on chromosome 14 that has become

known as the presenilin 1 locus (*PS1*). These mutations lead to exceptionally early onset of DAT, including some individuals in their early thirties. The *PS1* mutations led to a search for homologous sequences responsible for the disease in other families. In a very short period, cosegregating mutations were identified at a homologous locus (*PS2*) on chromosome 1. Most of these families are ethnic Volga Germans and exhibit a genetic founder effect.

A major question is how the various mutations discussed above will elucidate mechanisms underlying the pathogenesis of the vastly more prevalent 'sporadic' forms of the disease, which reach prevalence estimates of between 28 and 47 per cent of community-dwelling American subjects over the age of 85 years (Evans *et al.* 1989; Bachman *et al.* 1992). The major centres specializing in genetic investigations of DAT all have concluded that there must be other loci modulating the rates of development of DAT, for there are several pedigrees in which wild-type sequences have been found for each of the three known major dominant genes listed in Table 19.3. The role of the *APOE* polymorphism in the genesis of β-amyloid or in other aspects of the pathogenesis of DAT is unknown. More extensive deposits of β-amyloid have been found in the brain parenchyma and in the cerebral blood vessels of individuals with the *APOE ε4* allele, who are known to have, on average, an earlier age of onset than individuals with the more common *ε3* allele or the rare *ε2* allele (associated with later onset of DAT). The *ε4* allele, however, is neither necessary nor sufficient for the development of DAT. It is entirely possible that in the non-familial forms of late-onset DAT, the deposits of β-amyloid are a result rather than a cause of cell injury. Thus, the autosomal dominant mutations so far found in early-onset familial DAT may be relevant for certain 'private' mechanisms of senescence in the nervous system. In this regard, the *APOE* story looms as being particularly important, as polymorphic variations are more likely to be of relevance to how most of us age and could be more relevant to 'public' mechanisms of cognitive ageing. Thus, just as we saw for the case of atherosclerosis, DAT may provide us with examples of both of the classes of gene action predicted by the evolutionary theory of ageing (Charlesworth 1996). *APP*, *PS1*, and *PS2* appear to be examples of late-acting mutations whose prevalence may be the result of mutation–selection balance. The *APOE* polymorphism may be an example of balancing selection with trade-offs, in which certain alleles, under certain environmental conditions, have beneficial effects on early life-history parameters but deleterious effects late in life. The challenge is to discover how such presumptive trade-offs evolved and to elucidate their physiological and pathophysiological consequences.

The evolutionary theory of ageing predicts a highly polygenic basis of senescent phenotypes and of lifespan. Are there spontaneous mutations in man that challenge this prediction? Martin (1978) addressed this question and concluded that there are no known 'global' progeroid mutations in our species—that is, single gene mutations that accelerate all of the commonly observed features of human senescence. Several mutations, however, appear to accelerate the onsets and rates of development of many aspects of the senescent phenotype ('segmental' progeroid syndromes). The most striking example is the Werner syndrome, the features of which include short stature, premature greying and thinning of hair, bilateral ocular cataracts, osteoporosis, non-insulin-dependent diabetes mellitus, skin atrophy and ulceration, regional atrophy of subcutaneous fat, multiple benign and malignant neoplasms, gonadal atrophy, and various forms of arteriosclerosis, atherosclerosis, arteriolosclerosis, and medial calcinosis (Epstein *et al.* 1966). Subjects with Werner syndrome die at a median age of 47 years, typically from a myocardial infarction resulting from severe atherosclerosis which, histologically, does not differ from the atherosclerosis one observes in the general population (Epstein *et al.* 1966). Werner syndrome shows a number of phenotypic discordances with 'usual' ageing, however, including the distribution of the osteoporosis (predominantly involving the long bones of the extremities), the spectrum of neoplasia (predominantly involving mesenchymal cells), and the peculiar skin ulcerations, which are associated with soft tissue calcifications.

Recently, the gene *WRN*, whose mutation causes Werner syndrome, was shown by sequence homology to be a member of the RecQ family of helicases (Yu *et al.* 1996). Helicases serve to unwind DNA or RNA double helices, an essential early step in DNA repair, transcription, recombination, and chromosomal segregation. A helicase mutation could explain a number of previous observations indicating that somatic cells from subjects with Werner syndrome exhibit a mutator phenotype, with a particular propensity to undergo intragenic deletions and chromosomal deletions, inversions, and translocations (Hoehn *et al.* 1975; Salk *et al.* 1981; Fukuchi *et al.* 1989, 1990; Runger *et al.* 1994). All currently known mutations at the *WRN* locus appear to act as nulls (Oshima *et al.* 1996; Yu *et al.* 1996, 1997; J. Oshima and C.M. Martin, unpublished). Given the well-documented impact of all known mutations on fertility, and the rarity of Werner syndrome (homozygotes: 1–22/million), it is clear that we are dealing with a set of mutations which could not have escaped the force of natural selection.

An intriguing new finding suggests a broader link between *WRN* polymorphism and cardiovascular disease. In a preliminary study, Ye *et al.* (1997) found that the risk of myocardial infarction among Japanese subjects with at least one Arg allele at a biallelic polymorphic site within the *WRN* coding region was only about one-third (0.36; 95% confidence interval: 0.15–0.82) that of age-matched controls with the more common Cys allele. Could the Arg allele confer an enhanced catalytic efficiency of the *WRN* helicase? This would seem unlikely for the following reasons. First, as we have noted above, current evidence indicates that all spontaneously occurring *WRN* mutations are nulls. Thus, heterozygous carriers can be expected to have approximately one-half of the wild-type level of enzyme activity; but we know that such carriers can live well into their eighties without any of the stigmata of the Werner syndrome. Polymorphism at the *WRN* locus is likely to have much more subtle effects on enzyme activity than null mutations, and therefore any such changes in catalytic efficiency are unlikely to explain the variable susceptibilities to myocardial infarction. Perhaps the observed coding region polymorphism is acting as a surrogate for a structural alteration in a regulatory domain that can result in modulation of the speed with which this gene product can participate in repair to damage of the vascular wall.

REPRODUCTIVE SENESCENCE

Humans in developed countries typically initiate reproduction in early adulthood, voluntarily limit its extent, and have the children fledged by the time their parents are middle aged. There follows an often substantial postreproductive period. It is easy to forget how atypical these patterns are in the animal world, and how little relevance some of them probably have for the circumstances under which the human life history itself evolved. In most species, postreproductive individuals are very rare in nature (Austad 1994). A number of important consequences follow.

Postreproductive humans, unlike flies or nematodes, can influence the survival and fecundity of their dependent offspring and other relatives, by providing assistance with parental care, gathering of food, protection, and the benefits of acquired wisdom. Postreproductive survival is therefore, potentially at least, not selectively neutral in humans. The postreproductive period is much more clearly marked in women than in men by menopause. It seems clear that menopause would not be adaptive in the absence of ageing. This has led to the suggestion that menopause is an artefact of the benign environment inhabited by woman in developed countries, since many female mammals that do not show a postreproductive period in nature do so when brought into captivity (Austad 1994). Alternatively, the declining reproductive prospects associated with ageing could make it advantageous to abandon personal reproduction and to divert attention to existing dependent offspring and, possibly, to other relatives (Medawar 1952; Williams 1957), especially if mortality associated with childbearing increases because of somatic senescence (Kirkwood and Holliday 1986). Two models have suggested that the amount of help that mothers would have to provide in order to make it worth their while to forego personal reproduction is unrealistically high (Hill and Hurtado 1991; Rogers 1993). However, neither of these models took into account the decline in fertility that would occur if the mothers attempted to continue reproduction, or the possibility that the costs of continued reproduction could lead to an increased probability of death of mothers with dependent offspring. Taking account of these factors provides some support for the adaptive value of menopause (D.P. Shanley and T.B.L. Kirkwood unpublished).

PROSPECTS FOR IMPROVED HEALTH IN OLD AGE

The spectacular growth in the fraction of the human population over the age of 85 years in the developed countries will be paralleled in time in the developing countries. The question of improved health in old age will be one of the most important social, psychological, economic, political, and medical issues of the twenty-first century. An evolutionary approach suggests that the most effective interventions are likely to be preventative and may need to be taken long before the health problems of old age become manifest.

The highly polygenic nature of ageing means that single gene interventions are unlikely to be of much utility. Selective breeding to favour late-life vigour is of course out of the question. It is a moot point whether current secular trends of delayed childbirth in some segments of developed societies may eventually reproduce the *Drosophila* experiments of indirect selection for enhanced lifespans, but this will not be known for many generations! In any case, many other factors are likely to shape the future of the human life history. Environmental differences between individuals may be at least as important as genotypic differences in determining styles of ageing. Environmental interventions should be geared to extending health span. An encouraging recent analysis suggests that there have been significant declines in chronic disability prevalence rates among the elderly population of the United States over the period 1982 to 1994 (Manton *et al.* 1997). This is consistent with a 'compression' of morbidity into the last years of life, and it will be important to understand the factors that have produced this change.

One very interesting question with evolutionary significance is whether dietary restriction, which has long been known to extend lifespan in rodents and retard late-life pathologies (Weindruch and Walford 1988), may act similarly in humans. Dietary restriction appears to increase lifespan in several other short-lived animals (Weindruch 1996). The mechanism is as yet unknown, but it appears to involve a general upregulation of somatic maintenance functions, such as the levels of antioxidants and stress proteins. There may be some analogy with the phenomenon of dauer larva formation in *C. elegans* where food shortage induces a process which also upregulates maintenance functions, although in this case by a more clearly defined developmental switch. The extension of lifespan in response to dietary restriction in rodents may reflect an adaptive response whereby breeding stops and maintenance is increased during times of food shortage in nature (Harrison and Archer 1988; Holliday 1989; Masoro and Austad 1996). Shanley and Kirkwood (1998) have modelled the ecological and physiological factors involved in such an adaptive response and found, for mice, that some conditions favour its evolution. It will be interesting to use a similar model to test whether similar effects would be predicted in humans.

The leading current hypothesis about the mechanism of ageing is the 'free-radical' theory, which proposes that ageing results from damage to macromolecules caused by endogenously derived reactive oxygen species (reviewed by Martin *et al.* 1996). The hypothesis is far from proved, however. Given the predictions of the evolutionary theory of ageing for multiple mechanisms of senescence, it would seem highly unlikely that such a single global mechanism could explain all aspects of the senescent phenotype. The 'network model' of Kowald and Kirkwood (1996) illustrates the value of considering multiple, interacting mechanisms within a single framework.

An area of potentially viable intervention could come from more detailed research on a phase of life that could be referred to as 'sageing', a postreproductive period during which several adaptive behavioural and physiological mechanisms, used only intermittently during earlier phases of the life cycle, come into play more regularly, or constitutively, as compensations for age-related decrements in structure and function. An interesting example is the phenomenon of neuronal 'sprouting'—a growth of the dendritic arborization with an associated enhancement of regional synaptic density of neurones in response to the age-related or disease-related loss of neighbouring neurones (Flood and Coleman 1986). Trophic factors and their receptors must participate in such adaptive responses and could conceivably serve as the basis for pharmacological intervention.

For the medical geneticist, the most exciting prospects are the possibilities of tailoring a programme of preventive medicine to the particular strengths and weaknesses of subjects with known genotypes. For instance, individuals homozygous for the *APOE ε4* allele might gain particular benefit from early prophylactic treatment with non-steroidal anti-inflammatories, such as ibuprofen. Similarly, women concerned about osteoporosis might attach particular weight to family history in deciding whether to receive oestrogen replacement therapy. Such strategies would need to take careful account of ethical considerations of genetic confidentiality, including the extreme view that genetic diagnoses ought to be excluded from the medical record. Such individualized attention is seen by some to be in conflict with the economic imperative to reduce the proportion of the gross national product devoted to health care, but this view may be short-sighted ('an ounce of prevention being worth a pound of cure'). Given the current emphasis on cost-effectiveness, evolutionary biologists suggest that research and intervention be focused upon 'public' mechanisms of ageing. This would involve more support for research on the late-life effects of genetic polymorphisms.

SUMMARY

The global impact of human senescence has grown dramatically as a result of increased life expectancy arising from reductions in early mortality. Many more people are now living long enough to experience the infirmity, disability, and disease that comes with old age. Research on mechanisms of ageing is a priority, but these mechanisms are complex. Evolutionary theory is an important guide.

The evolutionary theories of ageing indicate that senescence is not a programmed process like development but is a consequence of the declining force of natural selection acting on later ages in the life history. One idea is simply that late-acting deleterious mutations accumulate under mutation–selection balance because selection against them is weak. Another suggestion is that trade-offs are responsible; that is, there exist pleiotropic genes which confer fitness benefits early in life but are harmful later on. Genes that control levels of somatic maintenance and repair are examples of genes that involve trade-offs, because maintenance uses resources that might otherwise be used for reproduction. The disposable soma theory predicts that the optimal investment in somatic maintenance is less than would be required for indefinite longevity and thus accumulation of damage is the principal cause of ageing.

Model organisms have been used to test the evolutionary theories. In fruitflies, it has been shown that genetic variation in natural populations can be used to select for increased lifespans, and that trade-offs and mutation accumulation may both be important. In nematodes, mutations have been recovered that extend lifespan, often linked with increased resistance to somatic stress. Numerous genes are being discovered in humans that are associated with common age-related diseases, such as cardiovascular disease and dementias. The prevalence and actions of these genes need to be understood in the context of the evolutionary theories of ageing. Evolutionary theory can also help to explain how the reproductive system is affected by ageing, especially in the case of menopause.

20

AN EVOLUTIONARY PERSPECTIVE ON THE GENETIC ARCHITECTURE OF SUSCEPTIBILITY TO CARDIOVASCULAR DISEASE

Sharon L.R. Kardia, Jari Stengård, and Alan Templeton

In Westernized industrialized societies, diseases of the heart rank as the number one cause of morbidity and mortality in the adult population (Higgins and Luepker 1989; Murray and Lopez 1997). More than 6 million people alive today in the United States have a history of symptomatic cardiovascular disease and approximately 1 500 000 Americans will experience a myocardial infarction this year as estimated by the American Heart Association (1994). In general, cardiovascular disease represents a very broad range of clinical presentations. For instance, many individuals with cardiovascular disease go undetected until a fatal event. Excluding angina pectoris, 18 per cent of all myocardial infarctions in males and 24 per cent in females present with sudden death as the first and only symptom (Kannel and Schatzin 1985). Many other individuals, approximately 25 per cent of cases of cardiovascular disease in the Framingham study, manifest disease through silent myocardial infarctions characterized by changes on periodic electrocardiographic examinations but no other clinical symptoms (Kannel and Abbott 1984). These studies highlight the breadth of the biological variation among those who have this common disease. Most genetic studies of cardiovascular disease focus on the quantitative measures of cardiovascular health that have a continuous unimodal distribution between the very healthy and those who die of cardiovascular disease. This strategy recognizes that clinical symptoms, or disease endpoints, represent an arbitrary threshold which dichotomizes the population at large into 'normal' or 'diseased'

categories that ignore the continuum of variation associated with cardiovascular health and disease.

It is now widely accepted that the cardiovascular disease process begins in childhood, and perhaps even in infancy or *in utero* (see Chapter 21), and develops rapidly between the ages of 15 and 34 years (Berenson 1985; PDAY Research Group 1993). Cardiovascular disease is a consequence of the age- and gender-specific interactions of many genetic and environmental factors that influence variation in the characteristics of the anatomical, biochemical, and physiological systems that influence cardiovascular health. Atherosclerosis, a major cause of cardiovascular disease, is currently thought to be initiated from vascular injury which then precipitates monocyte recruitment, macrophage formation and lysis, and subsequent lipid deposition into a developing atherosclerotic plaque. Vascular injury is also accompanied by platelet aggregation followed by the migration and proliferation of vascular smooth muscle cells which influence the synthesis of an extracellular matrix that contributes to the progression of atherosclerosis. Badimon *et al.* (1993) give an excellent overview of the wealth of details known about the hundreds of intermediate biochemical, cellular, and vascular traits that influence the initiation, progression, and severity of cardiovascular disease. The majority of these traits are quantitative and have a continuous distribution among individuals in the population at large.

In the past, much research has focused on measures which have come to be known as 'traditional risk factors,' specifically, plasma cholesterol level, hypertension, and smoking, and 'additional risk factors' such as lack of exercise, body size, and gender. However, variation in these factors explains only a fraction of the observed cases in the United States population (Crouse 1984). Also, these risk factors do not explain all of the population differences in cardiovascular disease prevalence (Higgins and Luepker 1989; Murray and Lopez 1997). Interindividual variation in genetic factors, when considered as independent of other factors, are thought to explain as much of the variation in cardiovascular disease risk, or associated intermediate biochemical and physiological risk factors, within and between populations as explained by exposures to environmental factors considered alone (Sing and Moll 1989). However, we emphasize that no single gene is responsible for the majority of the variation in cardiovascular disease risk (see e.g. Goldstein *et al.* 1995). Rather, it is the joint distribution of variations in multiple genetic factors with variations in multiple environmental factors that is responsible for the distribution of susceptibility to cardiovascular disease in the population at large (Sing *et al.* 1996). Although variations in a large number of genetic and environmental factors are known to be involved, variations in only a subset of these factors will be associated with cardiovascular disease in a particular individual, a particular family, or a particular population. Moreover, it is likely that the synergistic effects of different combinations of genetic and environmental variations in different individuals, different families, or different populations will make a greater contribution to determining the distribution of risk than the separate effects of variations in either genetic or environmental factors.

Given the very recent history of the high prevalence of cardiovascular disease in Westernized societies, a fair question is how can evolutionary principles help us to understand the distribution of this common disease within and among populations? There are some who believe that it is not necessary to include information about evolution of the genome when studying the role that genomic information can play in understanding the aetiology of, or in predicting, the common diseases (Morton 1997). In contrast, we believe that evolutionary principles may be very important for understanding susceptibility to late age of onset diseases because evolutionary forces have shaped the genetic architecture of interindividual variation in the continuously distributed intermediate anatomical, biochemical, and physiological traits that are associated with the continuum from health to disease. However, we emphasize that the role of evolutionary history in shaping the contemporary distribution of causes of the common multifactorial diseases is probably very complex, has not been well studied, and may be impossible to resolve because the selective forces may no longer exist. Consequently, instead of trying to identify the specific evolutionary mechanisms that may be associated with the cardiovascular disease epidemic in this century, we propose that evolutionary concepts and principles be used to inform our assumptions and to guide our biological and statistical modelling of the distribution of susceptibility to cardiovascular disease within and among populations. New information about genetic variation interpreted according to evolutionary principles has fundamentally altered the way we view the nature of genetic variation in the human species (Weiss 1996). This new information is seldom included in the study of the genetics of human disease. Cladistic analysis (Templeton *et al.* 1987; Sing *et al.* 1992), which utilizes an evolutionary view of genetic variation to design studies to untangle the complex genetic architecture of susceptibility to cardiovascular disease, is one of the best examples of a synergism between evolutionary concepts and the analysis of variation in susceptibility to cardiovascular disease. Below we expand on the notion that evolutionary theories and principles can influence the design of studies, the choice of analytical methods, and the interpretation of results of studies designed to investigate the genetic architecture of susceptibility to cardiovascular disease.

A DEFINITION OF GENETIC ARCHITECTURE

The genetic architecture of a trait is defined by the number of genes involved, their genomic positions, the number of alleles per gene, the relative frequencies of these alleles, the organization of allelic variations into genotypic variation, and the relationships between genotype variation and phenotypic variation. The relationships between genotype variation and trait variation involve dominance, epistasis, genotype by environment interactions, and age- and gender-dependent penetrance. Studies of the genetic architecture of the anatomical, biochemical, or physiological traits that are associated with the onset, progression, and severity of disease are central to understanding the genetic architecture of susceptibility to a common disease like cardiovascular disease (Sing *et al.* 1988, 1992). Although not yet standard in a definition of genetic architecture of disease susceptibility, we also consider the influence of genetic variation on phenotypic plasticity of each intermediate trait and on the functional relationships between traits over the life cycle to be an important component to our understanding of the genetic architecture of cardiovascular disease susceptibility (Reilly *et al.* 1992, 1994).

One reason to study the genetic architecture of intermediate anatomical, biochemical, and physiological traits is to improve our ability to predict risk of cardiovascular disease for an individual, a family, or a population. The objective is to determine which genes and which allelic variations improve prediction beyond the information provided by the traditional risk factors. A second reason for understanding genetic architecture is the possibility for insights into the disease aetiology that can guide efforts to develop treatments or identify those who would benefit most from preventative or therapeutic interventions. The ability to use genetic information to identify subsets of individuals with increased susceptibility to cardiovascular disease, to understand the distribution of cardiovascular disease aetiologies, and to tailor treatments to susceptible sub-groups depends on (i) an understanding of the distribution of functional genotypic variations among individuals, families, and populations and (ii) knowing the extent to which the impact of a particular allelic variation on risk of cardiovascular disease is dependent on the contexts defined by other genetic and environmental factors. As we point out throughout this chapter, interpreting the results of studies designed to reveal the genetic architecture of cardiovascular disease susceptibility can be greatly influenced by the evolutionary perspective, or lack there of, that is employed by the investigator.

EVOLUTIONARY MODELS FOR THE GENETIC ARCHITECTURE OF COMMON CHRONIC DISEASES

Early in the twentieth century, two sharply divergent viewpoints arose within evolutionary genetics about the nature of genetic architecture and the implications of genetic architecture for the origin of adaptive transitions. These two positions were exemplified by the contrasting viewpoints of R.A. Fisher and Sewall Wright. Both men ultimately introduced a visual metaphor to communicate their ideas about the mechanisms of adaptation and the resulting genetic architecture of adaptive traits.

Fisher's metaphor involved the concept of an adaptive 'target' (Fisher 1930). In this metaphor, the degree of adaptiveness of an organism can be represented by its closeness to some fixed point (the optimal adaptive type) in a multidimensional space. Let the organism's current phenotype be represented by a second point A. Then all points that are closer to the optimal type can be represented by a hypersphere whose centre is at the optimum and whose radius is the distance between the optimum and point A. When dealing with two dimensions, this becomes a simple circle or 'target'. Fisher then represented mutations as random vectors coming from the point A, with these mutational vectors being both random in direction and magnitude. Fisher then calcu-lated the probability of such a random vector having its endpoint within the adaptive target, that is, closer to the optimum than the initial point A. He showed that this probability is very small when the magnitude of the vector is large, but as the magnitude decreases, the probability of a random mutation resulting in an adaptively favourable change increases. Fisher concluded from this exercise that muta-tions having very small effects will be the primary raw materials of adaptive change, and mutations having large effects only rarely would play a role in adaptive evolution. More-over, he argued that because the genetic background is constantly changing in large random-mating populations, only mutations that have a consistent advantageous effect regardless of genetic background will be utilized by natural selection to build, over time, the genetic architecture of adaptive traits. Hence, adaptation occurs by the accumulation of many small, independent steps toward an optimum resulting in a genetic architecture characterized by a large number of loci scat-tered throughout the genome, by each locus having functional alleles with small phenotypic effects, and by little or no epistasis among loci.

In contrast, Sewall Wright, who worked in developmental and physiological genetics before moving into evolutionary genetics, introduced the metaphor of the 'adaptive topography' (Wright 1932). In this metaphor, the average fitness of a population is plotted above 'the field of gene combinations' to define a landscape which Wright regarded as being very rugged, with multiple peaks and valleys. Like Fisher, Wright regarded genetic architecture as generally involving multiple loci, but beyond that common point Wright's views were sharply divergent from Fisher's. If one takes a transect from one of the genetic axes in Wright's adaptive topography, the tran-sect in general will slice through many differ-ent peaks and valleys of the rugged adaptive landscape. This means that the effect of any particular allelic variation at a single locus is highly dependent upon its context. The same allele could have large or small phenotype

effects, and these could vary from being bene-ficial to being deleterious, depending on how it is combined into genotypes with alleles of other loci involved in the genetic architecture of the adaptive trait.

Wright's metaphor also incorporates great potential for epistasis and pleiotropy, which he regarded as a 'universal' phenomena. Wright (1934) emphasized the important role of gene-by-environment interactions in determining causation and the dynamics of gene expression and gene-by-gene interaction effects through-out development and the ageing process. In summary, Wright argued for a dynamic inter-active network of traits influenced by many genes and environmental exposures that traverses the hierarchy from genome type to phenotype rather than the non-interactive mapping of many small gene (and environmen-tal) effects to a phenotype that was advocated by Fisher. Fisher's metaphor was built on agent-based causation and Wright's metaphor was built on variation in relationships among agents as the primary causation of phenotypic variation. A schematic summary of Fisher's and Wright's metaphors compared with the Mendel and the Kimura, neutral allele, models is given in Fig. 20.1, which is a modified version of a figure from Wright (1982). The Mendel mapping function in Fig. 20.1 represents a model where variation in a single gene is responsible for variation in a trait. The Kimura mapping function represents a model where there is allelic variation in many genes but it is not related to the phenotypic variation of interest.

Because the nature of the genetic architec-ture of traits that influence health has pro-found implications for the meaning of adaptive evolution and its application to medicine, it is important to address the question of whether the Fisherian or Wrightian views best approxi-mates reality. In the next section we point out how different approaches to studying genetic architecture are based on different views of reality and therefore yield different insights into the genetics of cardiovascular disease susceptibility.

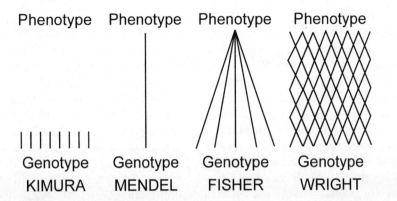

Fig. 20.1 Four classic models of genetic architecture.

THE GENETIC ARCHITECTURE OF SUSCEPTIBILITY TO CARDIOVASCULAR DISEASE

In this section we give an overview of insights about the genetic architecture of susceptibility to cardiovascular disease that have become apparent from using both top-down and bottom-up analytical strategies. We also draw attention to how Fisher's and Wright's views of evolution lead to different predictions about the nature of the genetic architecture of cardiovascular disease and juxtapose their multigenic views with a Mendelian view of common diseases. First, we begin with a brief overview of the top-down and bottom-up approaches that have been used to gain insights into the genetic architecture of common diseases.

Top-down versus bottom-up analytical strategies

For most of this century the top-down approaches were the only methods available to make inferences about the influence of un-measured genetic variation on trait variation. The top-down strategy uses phenotype data on related individuals, collected as pedigrees, sibs, or twins, to make inferences about whether unmeasured genetic variation is associated with phenotypic variation. One top-down strategy, linkage analysis, is used to localize gene regions on chromosomes that are associated with phenotypic variation. Linkage analysis has been of somewhat limited utility to the search for genes associated with common diseases be-cause it relies on a single-gene model of geno-type–phenotype relationships and common diseases do not segregate as simple Mendelian diseases. Therefore for common disease re-search, linkage analysis has only been success-fully used to identify genes associated with simply inherited versions of cardiovascular disease (e.g. familial hypercholesterolaemia) which are rare in the population at large. Because the pedigrees used in these studies are often ascertained through affected individuals and selected to be enriched for affected mem-bers, they identify only a subset of alleles segregating in a population and cannot be used to estimate the probability of having a particu-lar phenotype given a particular genotype. Population-based bottom-up strategies are necessary for estimating the relative frequency and penetrance of a particular susceptibility allele in the population at large. More recently, new linkage methods have been developed to identify quantitative trait loci in population-based samples of sib-pairs. Research is under-way to apply these methods to the intermediate traits that are associated with cardiovascular disease susceptibility (e.g. Fornage *et al.*, in press). In general, the main contribution of the linkage top-down approach is the localization of new genes.

A second top-down approach, often called the biometrical approach, has been used by quantitative genetics to test the null hypothesis that unmeasured genetic variation (either polygenic and/or major gene) inferred from familial relationships was associated with variation in quantitative intermediate traits (Sing *et al.* 1988). This analytical approach summarizes unmeasured genetic effects into a few parameters such as heritabilities or estimates of genotypic means associated with an inferred major gene. However, the biometrical approach does not yield information in terms of the actual genes involved or their locations. This approach has traditionally used a Fisherian view of genetic architecture that ignores or conceals biological interactions (Cheverud and Routman 1995) and therefore provides limited information on genetic architecture except in broad terms of whether there is evidence for additive genetic effects.

In contrast to the top-down approaches, the bottom-up approach starts with a candidate gene whose gene product is involved in the disease process, characterizes allelic variations in the gene, and then estimates how that genetic variation is associated with phenotypic variation in the population at large. The majority of bottom-up studies have examined only one gene at a time and have characterized only a minor fraction of the allelic variation in these genes. Multigenic bottom-up studies, which both Fisherian and Wrightian models would suggest are needed to understand genetic architecture, are much more difficult to conduct because of the large sample sizes that are needed to estimate genotype–phenotype relationships since the number of genotypes increases exponentially with the number of loci considered and the number of potential interactions between loci is also prohibitively large.

Number of genes involved and their genomic positions

Both Fisher and Wright would predict that a large number of genes influence the continuous variation in cardiovascular disease susceptibility that is associated with quantitative variation in anatomical, biochemical, and physiological traits implicated in the disease process. Using a top-down linkage approach, multiple genes have been identified that are

associated with extreme lipid phenotypes, such as familial hypercholesterolaemia (Breslow 1988; Goldstein *et al.* 1995). However, the majority of candidate genes have been identified from molecular studies of the genes that are involved in coding molecules that are associated with the intermediate traits involved in the disease process. Specifically, most of these candidate cardiovascular disease susceptibility genes have been identified because of their role in lipid metabolism, carbohydrate metabolism, blood pressure regulation, and haemostasis. In Table 20.1 and Fig. 20.2, we show a selected list of these candidate genes to illustrate the type of genes that are currently being studied and their map location. This list is unavoidably biased because research in this field using both the top-down and bottom-up approaches has concentrated most heavily on lipid metabolism and haemostasis. Other classes of candidate genes not listed here include the genes that code for components of the cardiac and smooth muscles, DNA regulatory genes, and many other genes that are yet to be identified. In summary, the number of genes that are involved in the genetic architecture is very large and they are scattered around the genome, as both Fisher and Wright would have predicted.

Allelic variation in candidate genes

Because of technical and financial limitations and the large number of genes involved, we are only at the very beginning of our understanding of the DNA sequence variations in each candidate gene, their locations (i.e. the loci involved), the number of variants per locus, their relative frequencies, and how they form haplotypes (i.e. alleles defined by variation at multiple loci). In general, the terms 'allele' and 'haplotype' are used interchangeably, except when referring to the variations at a single locus, when 'allele' is used exclusively. Both the top-down and bottom-up approaches have yielded only small glimpses into the allelic variation in candidate genes. For example, the top-down linkage approach has helped in the location of rare alleles associated with simply inherited versions of common disease but has

Table 20.1 List of candidate susceptibility genes

Lipid metabolism	Carbohydrate metabolism	Blood pressure regulation	Homeostasis
Apolipoprotein AI (Apo AI)	Insulin (INS)	Renin (REN)	Fibrinogen-α (FGA)
Apolipoprotein AII (Apo II)	Insulin-like growth factor I (IGFI)	Angiotensinogen (AGT)	Fibrinogen-β (FGB)
Apolipoprotein AIV (Apo AIV)	Insulin-like growth factor II (IGFII)	Angiotensinogen receptor (AGTR)	Fibrinogen-γ (FGG)
Apolipoprotein B (Apo B)	Insulin-like receptor (INSR)	Angiotensin-converting enzyme (ACE)	Factor VII
Apolipoprotein CI (Apo C)	Glucose transporter I (GLUT1)	Glucocorticoid receptor (GCR)	Factor VIII
Apolipoprotein CII (Apo CII)	Glucose transporter II (GLUT2)	Kallikrein I (KLK)	Factor IX
Apolipoprotein CIII (Apo CIII)	Glucose transporter III (GLUT3)	Neuropeptide Y (NPY)	Factor X
Apolipoprotein D (Apo D)	Glucose transporter IV (GLUT4)	Proenkephalin (PENK)	Factor XI
Apolipoprotein E (Apo E)	Glucose transporter V (GLUT5)	Pro-opimelanin (PMOC)	Factor XII
Hepatic lipase (LIPC)	Pyruvate kinase (PK)	FMS-related tyrosine kinase (FRT)	Factor XIII (Factor XIIIb)
Lipoprotein lipase (LPL)	Glucokinase (GCK)	Mineralcorticoid receptor (MLR)	C-reactive protein (CREP)
Lecithinacyl transferase (LCAT)	6-phosphogluconate dehydrase (PGD)	Arginine vasopressin (AVP)	Thrombin (THR)
Cholesterol ester transfer protein (CETP)	Insulin-like gf binding protein (IGFBP)	Endothelin (EDNI)	Phospholipase C (PLC)
Low-density lipoprotein receptor (LDLR)		Growth hormone (GH)	Platelet-derived growth factors (PDGF)
HMG coreductase (HMGR)		α-Cardiac Actin gene (ACTC)	Antithrombin (AT)
HMG cosynthetase (HMGS)			Plasminogen (PLG)
Very low-density lipoprotein receptor (VLDLR)			Plasminogen activator (PLAT)
Apolipoprotein (a) (Apo (a))			Plasminogen activator inhibitor (PAI)
Apolipoprotein M (Apo M)			Haptoglobin (HP)
LDL-receptor-related protein (LRP1)			Von Willebrand Factor (F8VWF)

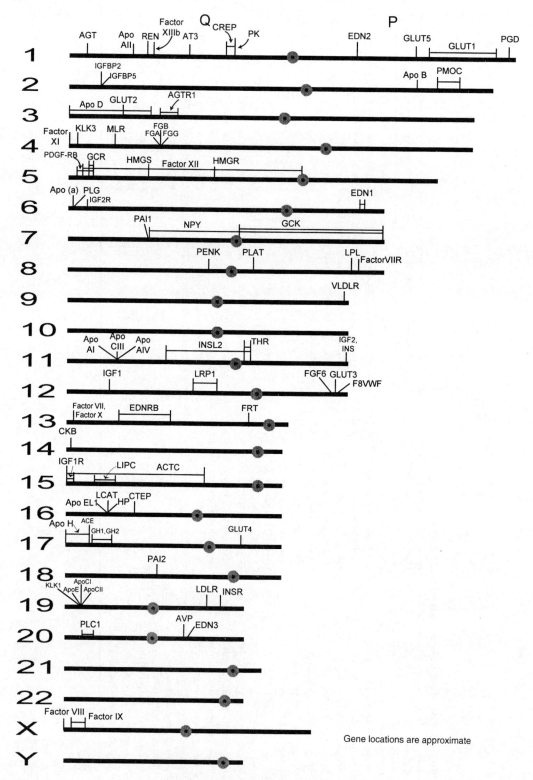

Fig. 20.2 Genomic location of candidate susceptibility genes for cardiovascular disease.

not been successful in identifying common alleles with functional effects. On the other hand, the bottom-up approach has been limited by the type of allelic variation that is technically possible to measure on relatively large samples which are needed for the study of genotype–phenotype relationships in the population at large. Although protein isoforms were originally the primary indicators of genetic variation in a gene region, more recently, variation in candidate genes has been measured almost exclusively through restriction fragment length polymorphisms (RFLPs)—that is, polymorphisms within restriction enzyme sites within a gene region. In many cases, these polymorphism have no obvious functional significance. Currently, the increased efficiency in genotyping multiple RFLPs per gene or in DNA sequencing (Hobbs *et al.* 1990) has provided data that have revised our conceptualization of genetic variation and its role in determining risk of cardiovascular disease. Many of these conceptual changes are directly related to how evolutionary and population processes give rise to the genetic variation observed in today's populations (Weiss 1996).

There are three characteristics of genetic variation that reflect its evolutionary history. First, most mutations within a gene are unique. The ability to examine haplotypes composed from many polymorphic loci within a gene has shown that most mutations arise only once in the allelic history of a population and that subsequent mutations then fall on the existing allelic background within a population (e.g. Haviland *et al.* 1995). Second, each gene is expected to have a large number of alleles. For example, genetic studies of the low-density lipoprotein receptor associated with familial hypercholesterolaemia have shown that hundreds of alleles can be associated with a particular phenotype (Hobbs *et al.* 1990). This is especially important in biomedical research because the methodological challenge of sorting through 20, 50, or 100 different alleles to find the subset of alleles with phenotypic effects has yet to be addressed. Third, haplotype variation is hierarchical (cladistic), geographically structured, and history dependent.

Generally, after one mutation has occurred on a given chromosome, some generations will usually pass before another mutation occurs at a nearby site on a descendant chromosome carrying the first mutation. Because of this hierarchical and history-dependent process, the current allelic variation in a population is a trace of the allelic history within that population. The most frequent alleles are most likely to be the oldest alleles while the rarer alleles are more likely to be the most recent alleles (Castelloe and Templeton 1994). Likewise, younger, rarer alleles are more likely to be geographically localized whereas older, more frequent alleles are more likely to be found in many populations or across many continents (Weiss 1996). Using multiple RFLPs within candidate genes, the cladistic nature of variation in the *Apo AI-CII-AIV* gene region (Haviland *et al.* 1995), LDLR gene region (Haviland *et al.* 1997), and lipoprotein lipase gene region (Kardia *et al.* 1996) has been demonstrated. As described below, cladistic analysis (Templeton *et al.* 1987) uses this information about evolutionary history of haplotypes to guide the study of genotype–phenotype relationships and to identify haplotypes carrying potential functional mutations.

In spite of our growing appreciation of how little we know about the number of loci that are variable per gene, or the relative frequencies of the associated haplotypes, there have been a few instances where it has been thought that the functional alleles within a candidate gene have been identified and the frequency distribution of that allelic variation studied. For example, even though we do not know about the allelic variation of the *Apo E* gene at the DNA level, we do know that in nearly every population there are three common protein isoforms, designated E2, E3, and E4. These isoforms are associated with three common haplotypes, designated $\varepsilon2$, $\varepsilon3$, and $\varepsilon4$, that encode amino acid variations at the 112 and 158 positions of this 299-amino acid protein (Utermann *et al.* 1980). Specifically, the $\varepsilon2$ allele encodes Cys112 ... Cys158, while the $\varepsilon3$ allele encodes Cys112 ... Arg158, and the $\varepsilon4$ encodes Arg112 ... Arg158. Studies over the past 10 years have established that these three common alleles are present in representative samples of over 50 populations (Davignon *et al.* 1988; Gerdes *et al.* 1992). In every case the relative frequency of the ε3 allele is the largest, ranging from 0.70 to 0.85. Although this feature appears to be an invariant property of the *Apo E* allelic system, the differences in allele frequencies between Caucasian, Japanese, Chinese, African, New Guinean, Mexican, and Malaysian populations are statistically significant (Davignon *et al.* 1988; Hallman *et al.* 1991; Hanis *et al.* 1991). Of particular interest is the observation that the populations of southern Europe have *Apo E* allele frequencies similar to Japanese populations, whereas Caucasian populations in northern Europe, Canada, and United States that have significantly higher rates of cardiovascular disease mortality have consistently higher ε4 alleles frequencies. In general, there is an association between ε4 allele frequency and the risk of cardiovascular disease both within populations (Davignon 1993; Stengård *et al.* 1995) and between populations (Sing and Moll 1989; Gerdes *et al.* 1992).

Genotype–phenotype relationship: marginal allelic effects

One of the major areas of debate about the Fisherian and Wrightian models for genetic architecture is whether selection on the marginal effect of an allele is the primary evolutionary force shaping the genetic architecture or whether the context-dependent effects of an allele, plus other evolutionary forces (e.g. restricted gene flow, restricted population size), would strongly interact with selective forces to form the genetic architecture. The marginal effect of an allele is a statistical measure of the impact of that particular allele averaged over the phenotypes of all individuals who have a genotype that includes that allele (Sing and Davignon 1985; Sing *et al.* 1996). The genetic architecture of diseases such as cardiovascular disease is expected to involve only a small number of genes with large marginal allelic effects and a large number of genes with small marginal allelic effects (Weiss 1995; Sing *et al.* 1996).

In general, our understanding about the marginal effects of allelic variation in candidate genes is relatively poor. First, many of the rare alleles with large effects identified through top-down studies have not been measured and studied in large population-based samples. Consequently, we do not know the frequency and phenotypes of individuals in the population at large that may carry such an allele but do not have an extreme phenotype. Second, because of the limited ability to identify variation in genes, bottom-up studies of the genetics of cardiovascular disease have consisted primarily of single-locus association studies using RFLPs (e.g. Deeb *et al.* 1986). When significant associations were found, the inference has been that some other unknown DNA sequence variation, in linkage disequilibrium with the RFLP locus, must be the cause of the phenotypically important effect. The challenge, not often even attempted because of technical limitations in DNA sequencing a large number of individuals, was then to find the 'functional' mutation that the RFLP was marking. In cases where amino acid substitutions have been identified (e.g. Apo E, Apo B, Apo AIV, Apo H, lipoprotein lipase), the marginal effects on intermediate traits or risk of cardiovascular disease have been studied but often the results have been inconsistent among studies. Currently, the marginal effects of the common *Apo E* haplotypes on cardiovascular disease risk or its associated intermediate biochemical and physiological traits have been the most widely studied and appear the most consistent among studies (Davignon 1993).

In the majority of studies, the ε4 allele is associated with increased levels of total plasma cholesterol and the ε2 allele is associated with lower levels. In all but a few studies, the ε2 allele has been associated with elevated levels of plasma Apo E and triglycerides whereas the ε4 allele is associated with lower than average levels of these traits. With respect to other biochemical and physiological risk factor traits, such as HDL-cholesterol, blood pressure, or other apolipoprotein levels, the results have been more heterogeneous (e.g. Kaprio *et al.* 1991). During the same time period, numerous studies have investigated the association between ApoE variation and variation in risk of cardiovascular disease (for review see Davignon 1993). In particular, many studies have found strong associations between the relative frequencies of *Apo E* genotypes carrying the ε4 allele and the frequency of cardiovascular disease in case–control studies (Cumming and Robertson 1983; Kuusi *et al.* 1989; Luc *et al.* 1994), in studies of special population subgroups (Laakso *et al.* 1993), and recently in a population-based prospective study (Stengård *et al.* 1995, 1996). These results have raised expectations that the *Apo E* polymorphisms may be a useful 'marker' system for identifying asymptomatic individuals who are at high risk of cardiovascular disease. However, there have also been many studies that have found no associations between *Apo E* polymorphisms and risk of cardiovascular disease (Menzel *et al.* 1983; Lenzen *et al.* 1986; Stuyt *et al.* 1991) or that the relative frequency of the ε4 allele was actually lower in cases than controls (Utermann *et al.* 1984), or that the ε2 (and not the ε4) allele was associated with an increased risk of cardiovascular disease (de Andrade *et al.* 1995; Eichner *et al.* 1993).

Here we remind you that the emphasis on characterizing an allele's marginal effect is a direct consequence of Fisher's model of genetic architecture. In many instances, the search for alleles with consistent marginal effects has influenced the way investigators interpret their results. For instance, when the results from one study of an allele's effect conflict with a previous study, it is not uncommon for the explanation of the differing results to be that the allele either contributes more than what was originally found or less than previously expected. Only in rare instances does the author direct our attention to a more Wrightian interpretation of the result, namely that the allele's effect may be context dependent.

Genotype–phenotype relationship: context-dependent effects of allelic variation

Understanding how the genotype–phenotype relationship changes with variation in environment, variation in other genes, or over time (e.g. as a consequence of development or senescence) is as important, if not more important from an individual's point of view, than understanding an allele's marginal effect since a particular genetic factor may have a large effect in a particular subset of individuals because of its interaction with some other necessary genetic or environmental factor (Sing *et al.* 1996). As the common *Apo E* polymorphisms have become more established as a genetic factor that effects cardiovascular disease susceptibility, the heterogeneous results among studies have brought more attention to how its effects on cardiovascular disease risk or intermediate trait levels vary across contexts indexed by age, gender, body size, smoking status, or other indices of the environment. For example, several studies have established that the effects of *Apo E* alleles are gender specific and dependent on hormone use in females (Hanis *et al.* 1991; Kaprio *et al.* 1991; Reilly *et al.* 1991; Xhignesse *et al.* 1991). Zerba *et al.* (1996) found that the fraction of variance in plasma Apo E that is associated with genotypic variation determined by the three common *Apo E* alleles was a function of age and gender. Studies comparing the influences of the *Apo E* genotypic variation in newborns with 3-year-olds also provide evidence that the *Apo E* allele effects are age dependent (Lehtimäki *et al.* 1994). Reilly *et al.* (1992) demonstrated that *Apo E* genotypic variation influenced the relationships between concomitants (age, weight, waist-to-hip ratio, and smoking) and nine lipid and apolipoprotein traits. Others have shown that the influence of the *Apo E* genotype on plasma risk-factor levels are dependent on variation in other genes such as the low-density lipoprotein receptor (Pedersen and Berg 1989). Overall, many studies suggest that the *Apo E* genotype's influence on biochemical and physiological traits is context dependent— that is, it may change over time, it may be influenced by the cardiovascular disease process, and it may vary among different environmental or genetic backgrounds.

A number of studies have investigated whether the relationship between variation in

Apo E and variation in cardiovascular disease risk is context dependent. Specifically, studies have shown that the Apo E association with cardiovascular disease risk is dependent on age and gender (Eichner *et al.* 1993; Katzel *et al.* 1993; Luc *et al.* 1994). Two studies of special note tested whether the association between *Apo E* genotype and cardiovascular disease risk was different at different intermediate traits levels. Specifically, Reilly *et al.* (1995) demonstrated that although variation among *Apo E* genotypes did not have an average, marginal association with variation in measures of coronary artery calcification, there were significant genotypic differences in the relationships between this measure of cardiovascular disease and established risk factors such as cholesterol, triglycerides, and their associated apolipoproteins. A study by Stengård *et al.* (in press) also recently demonstrated that the relationship between body mass index and the probability of death from coronary heat disease in a 5-year period was dependent on *Apo E* genotype in a population-based cohort of Finnish men aged 65 to 84 who were born between 1900 and 1915. These studies suggest that variation in the *Apo E* genotype influences the relationships between established lipid risk factors and some measures of cardiovascular disease. In more general terms, these studies indicate that the aetiological mapping function between variation in intermediate traits and variation in disease risk can vary among genotypes. This is an important point because Fisher's model of genetic architecture is built on a non-interactive mapping of many small genetic (and environmental) effects to a phenotype, while Wright's model is built on a dynamic, interactive network of traits where many genes and environments influence not only the levels but also the relationships among traits. So, these *Apo E* studies support a more Wrightian view of genetic architecture and could have major implications for the public health goals of using genetic information for prediction of disease risk and design of appropriate interventions. Specifically, a Wrightian genetic architecture would lead to a more individualized (i.e. context dependent) view of medical treatment and

risk assessment. Also, it would suggest that making health care decisions or policies today based on evolutionary adaptations to past environments has diminished validity because there is no single adaptive optimum from which to form medical recommendations for the common diseases.

UTILITY OF AN EVOLUTIONARY PERSPECTIVE

There is much that an evolutionary perspective can add to the understanding and study of the genetic architecture of susceptibility to cardiovascular disease, more than can be alluded to in one section or one chapter. Consequently, in this section, we give a single, well-developed example of how an evolutionary perspective can influence something as fundamental as the method we use to study the influence of genetic variation on phenotypic variation.

The cladistic approach for identifying functional mutations

For most candidate genes the functional mutations are still unknown even though single RFLP association studies have implicated the gene regions as being associated with variation in intermediate risk factors or cardiovascular disease risk. Relatively recently, a new method of analysis, known as cladistic analysis, was developed to take advantage of the phylogenetic structure of allelic variation to organize genotype–phenotype analyses so as to identify haplotypes or haplotype classes that were associated with significant phenotypic effects. The fundamental theory underlying cladistic analysis is that if haplotypes that are close in evolutionary time are associated with significantly different phenotypic effects, then during the evolutionary time period in which they diverged at the DNA marker level, a functional mutation must have occurred that affects the phenotype.

The cladistic approach involves two basic steps. First, the evolutionary relationships among observed haplotypes are estimated using maximum parsimony and the haplotypes are arranged next to their nearest neighbours in a cladogram (i.e. a branching tree structure). Second, the haplotype cladogram is converted into a nested series of clades (branches) by using the nesting rules given in Templeton *et al.* (1987) and Templeton and Sing (1993). In Fig. 20.3 we give an example of a cladogram based on haplotypes constructed from five RFLPs which illustrates the nesting rules. These nesting rules start at the tips of the haplotype network and move one mutational step into the interior, uniting all haplotypes that are connected by this procedure into a '1-step clade' (e.g. 1–1, 1–2, 1–3, 1–4 in Fig. 20.3). The next level of nesting uses the 1-step clades as its units, rather than individual haplotypes. The nesting rules are the same, but result in '2-step clades' this time. This nesting procedure is repeated until a nesting level is reached such that the next higher nesting level would result in only a single category spanning the entire original haplotype network. The resulting nested clades are designated by 'C-N' where 'C' is the nesting level of the clade and 'N' is the number of a particular clade at a given nesting level.

The nesting scheme is important because it provides a basis for a phenotypic analysis that focuses on contrasts between haplotypes that will provide the greatest amount of information about the presence of functional mutations with significant phenotypic effects while also providing a parsimonious use of the available degrees of freedoms. Specifically, the phenotypic analysis begins at the highest level of nesting. In Fig. 20.3, it would be associated with testing for phenotypic differences between 2-step clades (i.e. testing whether alleles in clade 2–1 are significantly different from alleles in clade 2–2). The analysis then turns toward an investigation of phenotypic differences between 1-step clades within each of the 2-step clades (e.g. comparing clade 1–1 and clade 1–2 within clade 2–1). Finally, there is an analysis of the phenotypic differences among haplotypes within 1-step clades (e.g. comparing ++++− and +++++ within clade 1–1). The nested testing scheme enables one to identify places in the cladogram where a

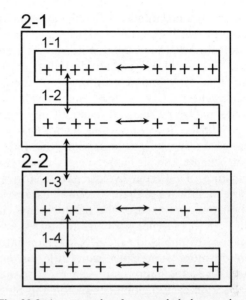

Fig. 20.3 An example of a nested cladogram based on haplotypes using RFLP markers (+ = cut, − = no cut). One-step clades (1–1, 1–2, 1–3, 1–4) group together haplotypes that differ by a single mutation step. Two-step clades (2–1, 2–2) group together haplotypes that are two mutational steps apart.

functional mutation potentially occurred. For example, if the individuals in 1–1 have a significantly higher level of a trait, such as plasma cholesterol, than individuals in 1–2, then this suggests that there was a mutational difference between those two haplotype classes associated with the phenotypic difference. However, we point out that it does not tell us where in the gene region it occurred. Further genetic and molecular studies are needed to identify the actual mutation, but the cladistic analysis identifies the haplotype classes to compare.

The ability of the cladistic method to identify haplotypes that carry potentially functional mutations is directly related to how well the cladogram portrays the evolutionary history of the haplotypes observed in the sample. Two main factors influence how well the cladogram portrays the evolutionary history. The first is recombination. The second is multiple mutational events at the same site that can create a situation in which more than one plausible tree is possible. For most (but not all) small regions of DNA (less than 1 MB), recombination is rare and the magnitude of linkage disequilibrium is primarily determined by the exact chromosomal background upon which a mutation initially occurred. The magnitude of disequilibrium reflects temporal positioning of mutational events over evolutionary time, not spatial positioning over a physical region (Templeton *et al.* 1987). However, when recombination does occurs, it results in loops within the cladogram because a single haplotype can represent a mixture of different evolutionary histories. Consequently, when there is recombination, it is best to either exclude the recombinants (if only a few haplotypes were affected by recombination) or subdivide the DNA region into two or more subregions with little or no internal recombination and then do separate analyses on each region (Templeton and Sing 1993). Some recombination actually strengthens and extends the inferences possible with the cladistic analysis. For example, Templeton *et al.* (1987) used a cladistic analysis of the drosophila alcohol dehydrogenase gene to identify a restriction site change that was associated with a major change in alcohol dehydrogenase activity. A few recombinants had been identified and were excluded from the initial cladistic analysis. By examining the phenotype of one of these recombinants, Templeton *et al.* (1987) concluded that the restriction site change associated with the enzyme activity change was not the major functional mutation, which had to lie in a different region of the DNA. These predictions were confirmed by subsequent molecular studies (Laurie *et al.* 1991).

With respect to cladogram ambiguity because there is more than one parsimonious tree, Templeton *et al.* (1992) have developed algorithms to weight the set of plausible trees using principles of population genetics and coalescent theory in order to construct 90% confidence sets of alternative cladograms. Moreover, the role of the cladogram in the cladistic analysis is to provide a nested design, and the nesting rules are robust to much cladogram ambiguity (Templeton and Sing 1993). When cladogram ambiguity does affect the nested design, one can iterate over all nesting possibilities to insure the robustness of inferences. Alternatively, one can use additional nesting rules that incorporate cladogram ambiguity, but at the price of reducing the resolution of the cladogram (Templeton and Sing 1993).

One of the main advantages of the cladistic analysis is that it identifies those haplotypes that differ phenotypically but that are close in evolutionary history. This closeness in evolutionary history greatly reduces the number of sequence differences between haplotypes that will have to be examined and should reduce the noise present when searching for functional mutations. A recent cladistic analysis of the lipoprotein lipase gene indicated that variation in the LPL gene is associated with a large variety and range of effects on risk factor traits for cardiovascular disease (Kardia *et al.* 1996). Specifically, the cladistic analysis suggested that there were multiple mutations in the LPL gene region, rather than a single mutation, and that these multiple mutations reside on multiple haplotypes which have functional effects on levels of established quantitative risk factor traits for cardiovascular disease (Kardia *et al.* 1996). Cladistic analysis of the *Apo AI-CIII-AIV* gene region (Haviland *et al.* 1995) and the LDLR gene region (Haviland *et al.* 1997) have shown similar types of results that multiple mutations on multiple haplotypes are associated with variation in intermediate trait levels. Another advantage of the cladistic approach is that it can be used iteratively to refine the cladogram by searching for more sequence variation in those regions of the cladogram where haplotypes carrying potential functional mutations have already been identified. A third advantage is that it can be used to identify functional mutations with context-dependent effects since the analysis can be carried out easily in different subsets of a population defined by relevant environmental or genetic strata. Thus, the cladistic approach can be used under either a Fisherian or a Wrightian model of genetic architecture to sort through the large number of alleles per candidate gene to identify those alleles with significant phenotypic effects.

SUMMARY

In the not so distant past, the study of the genetics of common diseases, such as cardiovascular disease, was hampered by a lack of knowledge about candidate genes for the disease. With advances in genetic technologies, the number of candidate genes and candidate DNA sequence variations available for study have risen exponentially. From current studies of candidate genes, we now have three basic insights into the genetic architecture of susceptibility for cardiovascular disease. First, the number of genes involved in determining susceptibility to cardiovascular disease is very large, probably in the hundreds. Second, a large number of alleles for each of these genes is expected in each population. Third, the relationship between genetic variation and phenotypic variation is complex. From both an evolutionary biology and public health perspective the key questions still to be addressed in the population at large are: (i) what is the distribution of allelic variations in each gene, (ii) which alleles are associated with variation in risk of cardiovascular disease, and (iii) what is the impact of context on these effects? The answers to these questions, and how those answers reflect either the Fisherian model, Wrightian model, or some mixture of their models of genetic architecture, has important implications for the success of different public health strategies. If the genetic architecture of cardiovascular disease susceptibility reflects more of a Wrightian view of reality, then a public health strategy that concentrates on identifying and treating high-risk individuals based on their combination of genetic and environmental factors may be more successful than a strategy based on a Fisherian viewpoint that concentrates on lowering risk of cardiovascular disease through population-based environmental risk-factor programmes or genetic screening for a particular cardiovascular disease susceptibility allele. We suggest that evolutionary models provide a valuable perspective on these public health issues and should be taken into consideration when we design our studies, develop methods of analysis, and interpret our results.

ACKNOWLEDGEMENTS

We thank Charles F. Sing for his guidance and helpful comments. Support for this work was provided in part by National Institutes of Health grants HL39107 and HL58240.

21

THE FETAL ORIGINS OF CORONARY HEART DISEASE AND STROKE: EVOLUTIONARY IMPLICATIONS

David J.P. Barker

Coronary heart disease has recently been shown to be associated with small size at birth. In a study of 16 000 men and women born in Hertfordshire, England during 1911–1930, death rates from the disease fell twofold between the upper and lower ends of the birth-weight distribution (Barker *et al.* 1989) (Table 21.1). (Death rates in Table 21.1 and sub-sequent tables are expressed as standardized mortality ratios which take account of the age and sex of people at each birthweight). A study in Sheffield showed that it was people who were small for dates, rather than born prema-turely, who were at increased risk of the disease (Barker *et al.* 1993). The association between low birthweight and coronary heart disease has been confirmed in the United States: among 88 000 women in the Nurses study there was a similar twofold fall in the relative risk of non-fatal coronary heart disease across the range of birthweight (Rich-Edwards *et al.* 1995). Recent findings from Hertfordshire and Sheffield show that death from stroke is also associated with low birthweight (Martyn *et al.* 1996): standard-ized mortality ratios for stroke fell by 40 per cent across the range of birthweight. Birth-weight is, however, a crude summary index of fetal growth and in studies where more detailed measurements of birth size were available, including length and head circumference, body proportions have been found to predict cardio-vascular disease more strongly than birth-weight (Barker *et al.* 1993, Barker 1995, Forsen *et al.* 1997).

These associations are independent of adult lifestyle including smoking, obesity, and socio-economic status. They have led to the hypothesis that cardiovascular disease, that is coronary heart disease and stroke, are 'programmed' *in utero* (Barker 1995). Programming is the pro-cess, well documented in animals, whereby undernutrition and other adverse influences acting during early life permanently change the structure and function of the body. If pregnant animals are undernourished, their offspring show permanent changes which include raised blood pressure and altered lipid and glucose metabolism (Barker 1995). These changes are thought to be adaptive, enabling the fetus to survive and grow; but at the cost of harmful long-term effects. The 'fetal origins hypothesis' proposes that undernutrition of the human fetus leads to responses that permanently alter the physiology and metabolism of the body in ways which lead to cardiovascular disease in adult life (see Chapter 7, on maternal–fetus conflict).

Table 21.1 Standardized mortality ratios from coronary heart disease among 15 726 men and women according to birthweight

Birthweight, pounds (kg)	Standardized mortality ratio	Number of deaths
– 5.5 (2.50)	100	57
– 6.5 (2.95)	81	137
– 7.5 (3.41)	80	298
– 8.5 (3.86)	74	289
– 9.5 (4.31)	55	103
> 9.5 (4.31)	65	57
Total	74	941

Table 21.2 Prevalence of non-insulin-dependent diabetes and impaired glucose tolerance in men aged between 59 and 70 years

Birthweight, pounds (kg)	Number of men	% with impaired glucose tolerance or diabetes	Odds ratio adjusted for body mass index (95% confidence interval)
− 5.5 (2.50)	20	40	6.6 (1.5–28)
− 6.5 (2.95)	47	34	4.8 (1.3–17)
− 7.5 (3.41)	104	31	4.6 (1.4–16)
− 8.5 (3.86)	117	22	2.6 (0.8–8.9)
− 9.5 (4.31)	54	13	1.4 (0.3–5.6)
>9.5 (4.31)	28	14	1.0
Total	370	25	

THE CORRELATES OF POOR FETAL NUTRITION

Fetal growth, diabetes, and hypertension

We are beginning to understand the mechanisms by which cardiovascular disease is programmed. The trends in cardiovascular disease with birthweight have been found to parallel similar trends in its major risk factors, which include non-insulin-dependent diabetes, hypertension, and disordered lipid metabolism and blood coagulation (Barker 1995). These are strong trends. Table 21.2 shows that the prevalence of non-insulin-dependent diabetes falls threefold between men who weighed 2.5 kg or less at birth and those who weighed more than 4.3 kg (Hales *et al.* 1991; Lithell *et al.* 1996). Trends among women are similar.

Obesity in adult life adds to the effect of low birthweight so that the highest prevalence of non-insulin-dependent diabetes is seen in people who were small at birth and obese as adults.

There is evidence that people who have low growth rates *in utero* become resistant to the action of insulin. Insulin resistance is associated with a particular pattern of fetal growth which leads to thinness at birth, measured by a low ponderal index (birthweight/birth length: Rich-Edwards *et al.* 1995). Men and women who had a low birthweight and a low ponderal index tend to be insulin resistant as children and adults (Lithell *et al.* 1996), and they have markedly increased susceptibility to the insulin resistance syndrome, which leads to coronary heart disease (Phillips 1996) (Table 21.3). The

Table 21.3 Prevalence of the insulin resistance syndrome in men aged between 59 and 70 years according to birthweight

Birthweight, pounds (kg)	Number of men	% with insulin resistance syndrome	Odds ratio adjusted for body mass index (95% confidence interval)
− 5.5 (2.50)	20	30	18.0 (2.6–118)
− 6.5 (2.95)	54	19	8.4 (1.5–49)
− 7.5 (3.41)	114	17	8.5 (1.5–46)
− 8.5 (3.86)	123	12	4.9 (0.9–27)
− 9.5 (4.31)	64	6	2.2 (0.3–14)
> 9.5 (4.31)	32	6	1.0
Total	407	14	

Table 21.4 Mean systolic pressure in men and women aged between 60 and 71 years according to birthweight

Birthweight, pounds (kg)	Number of subjects	Systolic blood pressure mmHg (adjusted for sex) ± SE
− 5.5 (2.50)	54	168 ± 3.4
− 6.5 (2.95)	174	165 ± 1.9
− 7.5 (3.41)	403	165 ± 1.2
− 8.5 (3.86)	342	164 ± 1.4
− 9.5 (4.31)	183	160 ± 1.8
> 9.5 (4.31)	72	163 ± 2.9
Total	1228	164 ± 0.7

thin newborn baby lacks muscle as well as fat, and muscle is the main peripheral site of insulin action, which has a key role in stimulating cell division in fetal life. It is thought that at some point in mid–late gestation the thin neonate became undernourished, and that in response its muscles became resistant to insulin. Muscle growth was therefore sacrificed, perhaps to spare brain growth.

Other components of the insulin resistance syndrome, which include raised blood pressure and altered blood lipids, may similarly be persisting effects of adaptations which enabled the fetus to continue growth in the face of a limited nutrient supply and to protect key organs and tissues, amongst which the brain is paramount.

Thirty-two studies world-wide have shown that low birthweight is associated with raised blood pressure in childhood and adult life (Law and Shiell 1996: Table 21.4). This association may reflect persisting loss of elasticity in arteries, or permanent resetting of hormonal axes including the hypothalamic–pituitary–adrenal axis, growth hormone/insulin-like growth factor axis, and the renin–angiotensin system (Edwards *et al.* 1993). Raised blood pressure is the major risk factor for stroke.

Liver growth and cholesterol

In response to undernutrition in late gestation the human fetus appears to divert oxygenated blood away from the trunk to sustain the brain. This adaptation prejudices growth of the liver and other abdominal organs, and leads to a reduced abdominal circumference at birth. A small abdominal circumference predicts persisting abnormalities in systems regulated by the liver, including cholesterol metabolism and blood coagulation. People born with a small abdominal circumference have raised serum cholesterol and plasma fibrinogen concentrations, two risk factors for coronary heart disease (Barker *et al.* 1995: Table 21.5). One interpretation is that restriction of liver growth *in utero* is associated with persisting alteration in liver metabolism. Animal studies demonstrate that liver metabolism is readily programmed by undernutrition in intrauterine life (Hales *et al.* 1996).

Table 21.5 Mean serum cholesterol concentrations according to abdominal circumference at birth in men and women aged between 50 and 53 years

Abdominal circumference, inches (cm)	Number of people	Total cholesterol (mmol/l)	Low-density lipoprotein cholesterol (mmol/l)
− 11.5 (29.2)	53	6.7	4.5
− 12.0 (30.5)	43	6.9	4.6
− 12.5 (31.8)	31	6.8	4.4
− 13.0 (33.0)	45	6.2	4.0
> 13.0 (33.0)	45	6.1	4.0
Total	217	6.5	4.3

EVOLUTIONARY IMPLICATIONS

Genes and fetal adaptations

Studies of size at birth among relatives, together with experiments in animals, suggest that fetal growth is controlled by the intra-uterine environment rather than genetic inheritance (Penrose 1954). For example among half siblings related only through one parent those with the same mother have similar birthweight whereas the birthweights of half siblings with the same father are dissimilar (Morton 1955). Fetal genes must, however, determine fetal cardiovascular and metabolic responses to undernutrition. *In utero* modification of the expression of these genes by nutrient deficiency, acting directly or through hormonal signals, will have lasting effects that include reduced cell numbers and changes in the distribution of cell types, in hormonal feedback, in metabolic activity, and in organ structure (Hales *et al.* 1996). Genes which allow the fetus to adapt successfully to undernutrition are likely to be favoured by natural selection even though they may lead to disease and premature death in postreproductive life (see Chapter 19). Individuals in a population are likely to vary genetically in their ability to maintain homeostasis under environmental challenge. In response to poor nutrient availability *in utero*, some fetuses will fail to make appropriate homeostatic responses and will die; some will make responses that will allow growth to continue at the same rate; others will make homeostatic responses that will ensure survival but lead to reduced growth in selected tissues. This last group will be at risk of coronary heart disease in adult life. Thinness and reduced abdominal circumference at birth represent two forms of disproportionate growth in which different tissues, muscle and liver, are sacrificed to sustain the brain. Perhaps the origins of coronary heart disease partly lie in the large size of the human brain, in comparison with that of other mammals. Adaptive responses that protect the brain do so at exaggerated costs to other tissues.

The coronary heart disease epidemic

Whereas the incidence of stroke has been falling in Western countries for 50 years, the incidence of coronary heart disease has, until recently, been rising. The decline in stroke is consistent with improvements in maternal physique and nutrition leading to improved fetal nutrition. Improved fetal nutrition has been reflected in continuing declines in death rates among newborn babies during this century (Barker 1989). The epidemic of coronary heart disease in the Western world could be explained in two ways. First, adverse effects of fetal adaptations to undernutrition may lead to coronary heart disease only when followed by overnutrition in adult life. It is known that development of insulin resistance *in utero* is more likely to lead to non-insulin-dependent diabetes if it is followed by obesity in adult life (Hales *et al.* 1991; Lithell *et al.* 1996). It seems unlikely, however, that such a model can fully explain the rise in coronary heart disease during this century. An alternative is that continuing improvements in nutrition have brought increases in maternal weight and height, and alterations to diet during pregnancy which have led to qualitative or quantitative alterations in fetal adaptations. The first direct evidence to support this comes from a recent study of men born in Helsinki, Finland (Forsen *et al.* 1997). Those with the highest death rates from coronary heart disease were thin at birth with a small placenta, but their mothers had a high body mass index. The inference is that a baby's metabolism may be permanently impaired if its nutrient supply is limited by a small placenta while its mother's metabolism is altered through obesity. This could be of especial relevance to Asian countries where coronary heart disease rates are now rising rapidly.

In India, death rates from the disease are expected to overtake those due to infectious disease by the year 2010 (Bulatao and Stephens 1992). Already, cardiovascular deaths account for half the deaths occurring under the age of 70 years. These high rates of coronary heart disease in India are not explained by known risk factors including obesity, raised blood pressure, smoking, and raised cholesterol. Coronary heart disease in Indian populations is, however, associated with a particular metabolic profile that is known to be unfavourable, which includes insulin resistance (McKeigue *et al.* 1991). This metabolic profile is often associated with non-insulin-dependent diabetes requiring treatment. Coronary heart disease in India differs in other ways from the Western stereotype. It is more common in lower socioeconomic groups (Stein *et al.* 1996), men with the disease are often thin, and rates in women are similar to those in men, even though women in many parts of India do not smoke.

When people from India migrate to other countries they take their high rates of coronary heart disease with them. This raises the possibility that Indian people have a genetically determined susceptibility to coronary heart disease which is enhanced on exposure to a sedentary lifestyle, high energy intake, and other aspects of Westernization (Williams 1995). The genes responsible for this have not been identified, but it is hypothesized that they conferred a survival advantage to Indian people in past times when food supplies were unreliable and physical work was demanding. A similar speculation has been put forward to account for the high rates of non-insulin-dependent diabetes in Indian and other populations (Neel 1962).

The 'thrifty genotype' hypothesis proposes that the diabetogenic gene or genes persist at a high level because they confer a survival advantage in times of nutritional deprivation. The implications of these speculations are that Indian people will continue to have high rates of coronary heart disease and non-insulin-dependent diabetes unless they return to a more primitive way of life. This conflicts, however, with experience elsewhere in the world, where epidemics of coronary heart disease have been followed by declining rates (Barker 1989) which, though perhaps assisted by health education, are largely unexplained.

The 'thrifty phenotype' hypothesis has been put forward as an alternative to the 'thrifty genotype' (Hales and Barker 1992). It proposes that coronary heart disease and non-insulin-dependent diabetes are the outcome of the fetus being programmed by undernutrition. Preliminary findings from India support this, showing that, similarly to Western countries, people who have coronary heart disease had low birthweight (McKeigue *et al.* 1991).

Transmission of disease through the mother

Non-insulin-dependent diabetes and hypertension tend to be inherited through maternal rather than paternal lines (Alcolado and Alcolado 1991). This is consistent with the 'fetal origins' hypothesis. Remarkable experiments in animals have shown that changes in the glucose concentrations to which a fetus is exposed produce effects through several generations, mimicking genetic transmission (Van Assche and Aerts 1985). When pregnant rats were made mildly diabetic by injection of streptozotocin on the day of mating, their female offspring developed disordered glucose metabolism and gestational diabetes when they became pregnant. In turn, the next generation of females similarly developed disordered glucose metabolism as adults. The changes in the second and third generation were independent of the origins of the father.

CONCLUSION

The search for the environmental causes of coronary heart disease has hitherto focused on a 'destructive' model in which adverse external influences, such as smoking, hasten ageing processes, such as the formation of atheroma and rise in blood pressure. This model is an extension of germ theory. Recent findings, however, suggest that cardiovascular disease may be the delayed cost of successful adaptations which preserved homeostasis and growth *in utero*.

22

THE EVOLUTIONARY CONTEXT OF CHRONIC DEGENERATIVE DISEASES

S. Boyd Eaton and Stanley B. Eaton III

Since 1800 the expectation of life has nearly doubled, from around 40 years to nearly 80 years in developed countries, a demographic phenomenon reflecting both reduced mortality in early life and improving survivorship at all ages, including the very old. Intrinsic human longevity has remained unaltered (Kirkwood 1996), so this revolution in human life expectancy must reflect the effects of environmental improvements in sanitation, housing, caloric availability, medical care, and general living standards. However, as the proportion of elderly has increased, age-related illnesses such as coronary atherosclerosis, cancer, osteoporosis, and diabetes have become paramount health issues. Modern culture, which has doubled life expectancy, appears also to promote the development of these disorders. The sanguine corollary of this argument is that such illnesses are not unavoidable but—to a significant degree —the consequence of novel gene–environment interactions.

Collectively, these illnesses have been called 'afflictions of affluence' or 'diseases of civilization', terms which recognize their relative rarity in preindustrial societies. This rarity is not merely an artefact of lower average age: younger members of affluent societies commonly exhibit vascular pathology, elevated blood pressure, and undesirable body composition not observed in age-matched members of preindustrial cultures. Furthermore, older members of preliterate societies remain largely free of the heart attacks, diabetic complications, and strokes which cause much mortality among their affluent counterparts (Eaton *et al.* 1988).

This comparative freedom of traditional peoples from Western diseases may reflect the similarity of their lifestyles to those of palaeolithic humans. An emerging discipline— evolutionary medicine—builds awareness of humanity's past into a theoretical framework unifying the epidemiology, pathophysiology, genetics, and prevention of chronic degenerative diseases. Neel (1994) has expressed two of the field's cardinal tenets: 'When [our] hunting and gathering ancestors began to take up agriculture some 3000 to 10 000 years ago ... they were already essentially the same biological material as ourselves today', and ' ... there is now little room for argument with the proposal that health ... would be substantially improved by a diet and exercise schedule more like that under which we humans evolved'. In other words, genetic evolution has not kept pace with cultural change so that—in affluent nations more than in traditional cultures—our genes and lives are no longer in harmony. This mismatch contributes to the causation of chronic diseases: understanding its nature can aid in their prevention.

MULTIFACTORIAL DISEASES

Monogenic disorders, whose expression is often independent of environment, are tragic for those involved but comprise only a small fraction of our disease burden and health care costs. Accordingly, the health status of any Western population approximates the interaction between its genetic make-up and prevailing lifestyle factors that affect the initiation and progression of common polygenic patho-physiological processes. This dynamic involves many complex gene–environment interactions modified over time by fetal experience, child-hood development, and advancing age. A population's genetically determined suscept-ibility to any given multifactorial chronic de-generative disorder can be represented by a Gaussian distribution curve, with each individ-ual's inherent resistance or vulnerability deter-mined by his/her place on the curve. Adherents of evolutionary medicine believe that changes in the shape of such genetic susceptibility curves have lagged behind lifestyle transforma-tions since the adoption of agriculture.

HUMAN GENOME

The current human gene pool has been shaped both by neutral mechanisms (drift and founder effects) and by natural selection. While the relative importance and interactive dynamics of these forces are not fully understood, con-sensus exists on the following points.

1. Many genes affect susceptibility to chronic illnesses, from tens to hundreds in most cases (Sing *et al.* 1996).
2. Most genes were selected by forces acting on early life traits, not because of their influence on disease in later life.
3. Most of the current human genome was formed during preagricultural evolution. While allele frequencies have been affected by recent migration, population expansions, and encounters with infectious disease, it is not likely that our susceptibility to chronic disease has changed during the Holocene.
4. Therefore citizens of affluent nations still have essentially the same innate suscept-ibility to chronic disease as did their pre-agricultural ancestors (Neel 1994).

CULTURAL CHANGE

From their first appearance 2.5 million years ago, the stability of stone tool industries is a hallmark of the Lower and Middle Palaeolithic; even early anatomically modern humans failed to manufacture innovative stone tools (Klein 1989). However, about 50 000 years ago a creative revolution began; thereafter new tool types appeared in comparatively rapid succes-sion. This technological development identifies the Upper Palaeolithic and is thought to signal improved cultural capability, for example, the settlement of glacial central Asia (the 'mam-moth steppe') and voyages across open water to Sahul (the ice-age continent joining New Guinea, Australia, and Tasmania). Despite cultural acceleration, most humans remained nomadic hunter–gatherers who used their own muscle power to extract naturally occurring foods from the environment like other animals. Before and during the Pleistocene (2 million years ago to 10 000 years ago), natural selec-tion tailored the human genome to match our ancestors' environmental niches, but since agriculture/animal husbandry and especially since the Industrial Revolution, genetic evolu-tion has been unable to keep pace with cultural changes which have made modern human environments increasingly different from the palaeolithic. Our ancient genes now contend with the realities of Space Age life.

Genetic–cultural discordance has developed unevenly over time. For example, agriculture altered human nutrition from its earliest appearance: intake of animal protein was re-duced, dietary breadth was narrowed and, break-ing totally with primate experience (Milton 1993), grains became the main source of energy. On the other hand, the level of obligatory ex-ertion changed little until the introduction of labour-saving devices, primarily in this century. The time-course of change in other important areas (reproduction, microbiology, psycho-social experience) also varied from case to case. Nevertheless, all are now markedly different from what they were preagriculturally.

LIVING IN THE LATE PALAEOLITHIC

Nutrition

Human diets from 50 000 to 10 000 years ago must have varied with climate, season, and latitude—all of which influenced local availability of plants and animals. Undoubtedly, periods of relative plenty alternated with shortages: there was no universal nutritional pattern. However, subsistence was based largely on uncultivated fruits and vegetables together with wild game; honey, fish, and shellfish made varying, generally minor, contributions and wild grains were only an emergency food. From these premises follow several generalizations (Eaton *et al.* 1997).

1. Typical protein intake was greater than at present, about 30 per cent of daily energy. Game was abundant in the Pleistocene, hence optimal foraging considerations (Bettinger 1991) justify this assertion. For over 50 hunting and gathering groups studied in this century, protein provided 34 per cent of caloric intake (Lee 1968). In most Western diets protein contributes 10–15 per cent of nutrient energy.

2. Fat intake was typically 20–25 per cent of total energy. Game animals are lean, wild plants provide little fat, and there were no separable fats or oils (Eaton and Konner 1985).

3. The diet contained much less saturated fat and had a nearly equal ω-6:ω-3 polyunsaturated fatty acid partition (that in the United States is estimated at 10–25:1) (Eaton 1992).

4. Cholesterol intake would have been similar to that now observed in most Western nations. Even lean meat contains considerable cholesterol.

5. Carbohydrate intake was similar to that in current Western nations but came from vastly different sources. Most current carbohydrate intake is derived from cereals, sugar, and dairy products—all unavailable before agriculture—so remote ancestors consumed three times more fruits and vegetables than do current Westerners (Eaton and Cordain 1997). Fruit and vegetable consumption is inversely correlated with cancer prevalence; a similar correlation has not been demonstrated for cereal grains (Block *et al.* 1992).

6. Dietary fibre intake would have been perhaps five times that of today's Western population (Eaton 1990), and the sources would have differed (vegetables and fruits rather than cereals): Stone Age people had a higher proportion of soluble fibre.

7. Despite liking honey, palaeolithic humans would have obtained far less energy from simple sugars, perhaps 1–2 per compared with 20 per cent for current Americans.

8. Consumption of 'empty calories' (i.e. energy dissociated from micronutrients and essential fatty /amino acids) was far less for palaeolithic humans. Consequently, their intake of vitamins, minerals, and probably phytochemicals was greater, about 1.5 to 5 times as much depending on the nutrient, but nothing like the levels advocated by megavitamin extremists.

9. Much less sodium was consumed: 90% of current intake reflects processing, preparation, and table use of extraneous salt; only 10 per cent is intrinsic to the foods themselves. Like all other free-living terrestrial mammals except recent humans, our ancestors consumed much more potassium than sodium.

10. Preagricultural humans had no milk after weaning. However, because of high average calcium content in wild plants, calcium intake was typically twice ours (Eaton and Nelson 1991).

Physical exertion

Until the appearance of social stratification, *Homo sapiens*, like all other mammals, had to work to eat: food procurement and energy expenditure were directly related. Cro-Magnon Venus figurines, accurately portraying obese women, indicate the existence of elite individuals long before agriculture. With the development of neolithic cultures, hierarchical social structure became more widespread, but for most humans hard work remained life's lot, and it has been estimated that in agricultural societies, total daily energy expenditure (TEE) averaged 12 MJ (approximately 3000 kcal)

(Åstrand 1988). When corrected for body weight, TEE for foragers is similar (Cordain *et al.* 1997). Energy expended as physical activity by hunter–gatherers typically approaches 100 kJ/kg per day (25 kcal/kg per day) for men and 83 kJ/kg per day (19 kcal/kg per day) for women. Corresponding values for sedentary American office workers are: TEE 2000 kcal, physical activity 4.5 kcal/kg per day (men) and TEE 1600 kcal, physical activity 4.3 kcal/kg per day (women).

Thus, on average, late palaeolithic humans expended *four to five times more* energy as daily physical activity than do typical Western-ers. These data apply to adults, but differences in children's activity may be of greater impor-tance (Cooper 1994). During the age period most critical for development of peak bone mass, American children spend 40 hours watching television (Dietz and Gortmaker 1985) and 25 hours sitting in classrooms each week—to say nothing of video games and home computers.

Preagricultural life required both stamina and strength. Recent male hunter–gatherers have maximum oxygen consumption rates of about 52.5 ml/kg per min which contrast with rates in industrial states of about 37.5 ml/kg per min (Cordain *et al.* 1997). Thus the mean aerobic fitness of forager males is about 40 per cent greater. The discrepancy for women is similar—in fact women from Inuit (Eskimo) communities exhibit greater aerobic power than do age-matched Western men (Rode and Shephard 1994).

Preagricultural human bony remains can be used to estimate physical strength: the promi-nence of muscle insertion sites, area of articular surfaces, cortical thicknesses, and the cross-sectional shape of long bone shafts reflect the magnitude of muscular forces exerted on them. A concatenation of these measures suggests an average 17 per cent reduction in physical strength since the Cro-Magnon era (Ruff *et al.* 1993). Osteological differences may under-estimate changes in strength: after only 20 years acculturation, the leg extension force of 20–29-year-old Inuit males declined 37 per cent (Rode and Shephard 1994).

Alcohol and tobacco

No hunter–gatherers studied in this century have been able to manufacture alcoholic drinks (Eaton *et al.* 1988), so palaeoanthropologists believe alcohol consumption in the Palaeo-lithic was rare, infrequent, or altogether unknown. For adult Americans it provides approximately 4.5 per cent of daily energy intake.

Tobacco is indigenous to the Americas, Australia, and a few Pacific Islands. Australian Aborigines do chew tobacco, but for human ancestors living in Eurasia none was avail-able until the fifteenth- and sixteeth-century European voyages of discovery.

Micro-organisms

Preagricultural humans were almost all nomadic, living at low population density, and while there were trading and/or gift-exchanging net-works, communication between distant groups was infrequent. Therefore high-mortality epidemic infectious diseases were rare or un-known (at least in their present form) during the Pleistocene (Motulsky 1960). Nevertheless, micro-organisms probably constrained human population growth throughout our evolution (Groube 1996). The specific agents may have been less lethal ancestors of current 'crowd' diseases (Motulsky 1960), viral diseases marked by latency and recurrence, and ordinarily saprophytic bacteria. Occasional infection by pathogens acquired from animal contact or via arthropod vectors would also have occurred (Fenner 1970). These sporadic, endemic infec-tious encounters probably affected population growth during the Palaeolithic and left evi-dence of their impact in polymorphisms of the major histocompatibility complex and human leucocyte antigen system.

Agriculture, animal husbandry, urban living, and regular long-distance trade changed the dynamics of human–microbial interaction and promoted epidemic infectious diseases. The historical significance of these disorders has been widely appreciated (e.g. Zinser 1935; McNeil 1976) and their relationship (especially that of malaria) to the human genome is well established. The development of immunization and antibiotics represent biomedical counterstrikes in this ongoing conflict whose evolutionary implications are discussed elsewhere in this volume.

Reproduction

Recently studied hunter–gatherer women are the best available surrogates for preagricultural women. Analysis of their life histories, augmented by studies of women in traditional settings (see Chapter 8), yields a picture of palaeolithic reproductive experience that differs greatly from that of today's typical affluent women. For foragers, menarche is later (16 compared with 12.5 years) and first birth earlier (at around age 19) so that the nubility (menarche–first birth) interval is much shorter, say 3 years compared with an average of 12 years for American and European women. Total parity is greater (6 for foragers as opposed to 1.8 for Americans). Women in traditional societies nurse intensively (i.e on demand, not on schedule) for 2 to 3 years. Only 50 per cent of American babies are nursed at all; average duration is barely 3 months. Menopause is earlier for hunter–gatherers (mid-to-late forties) than for Western women (early fifties). These reproductive differences all parallel epidemiologically established risk factors for breast, ovarian, and endometrial cancers; the longer menarche to first birth interval and reduced total number of ovulations (110–160 for foragers compared with 450 for current women who do not use oral contraception) appear especially significant (Eaton *et al.* 1994).

Psychological factors

At least some of the psychiatric disorders discussed in Chapter 23 probably relate to the novel circumstances of life in current Western nations. Many factors traditionally viewed as influencing psychic development and interpersonal behaviour are now radically different from what they were during the Pleistocene (Konner 1988), when important selection on the genetic component of human behaviour

probably occurred (e.g. brain size doubled during this period).

The usual birth spacing pattern for pre-agriculturalists means that children seldom had a non-twin sibling closer than 3 to 4 years in age. Physical contact between infants and older humans (especially mothers) was much more extensive: among the Paraguayan Aché, infants are in physical contact with another person for 90 per cent of their first year (Hurtado *et al.* 1985). Cribs, play pens, car seats, and isolation during night-time sleep have created a new separation: nothing like the former were utilized by foragers, and in all known instances mothers slept with their babies, not separately. As infants became children, they commonly accompanied their mothers on gathering expeditions; otherwise they were cared for by relatives or maternal friends with whom they were familiar: there was nothing analogous to day care. Because forager bands are small (20–50 members), children played in multi-age groups where older children had quasi-parental responsibilities and the younger learned from their elders. The nineteenth-century one-room schoolhouse was somewhat similar, but today's same-age elementary school classes contrast strikingly, with implications for development of self-esteem, competitiveness, caring, and responsibility.

Children grew older within a well-established and defined gender role structure, a pattern which persisted almost until the present. Whether the almost unlimited career choices now available for both boys and girls be viewed positively or negatively, they are undeniably novel relative to prior human psychological development.

In the Pleistocene that development was experienced in small groups. Bands of 15 to 50 individuals were typically components of tribes which numbered 200 to 1000. Almost everyone had friends and acquaintances in common: most were distant cousins. Social transactions in such settings are different from those in today's wider world. A person's reputation was widely known: both honourable and dishonourable actions became public knowledge. In such circumstances, group dynamics work against hereditary social stratification, at least in comparison with subsequent forms of societal organization. There was no parallel to today's media which provides unprecedented psychological interaction, exposing each of us to the entire world. We are subject to influence by media personalities made familiar by constant exposure. We can each compare ourselves, our mates, and our children to the most attractive, witty, athletically talented, and intelligent real or fictional competition. These novel influences have potential for good and ill; their relation to psychiatric problems like depression provides a rich opportunity for research.

CHRONIC DEGENERATIVE DISEASES

The palaeolithic lifestyle helped shape the current human genome; its continuing relevance is evident in the risk factors for common degenerative diseases. These influences were identified mainly with epidemiological methods; only later did it become apparent that they reiterated deviations from Pleistocene lifestyles which characterize life in affluent Western nations. A comparison of palaeolithic and current lifestyles helps clarify the pathophysiology of these illnesses and provides clues to their prevention. Disorders ranging from dental caries, high-frequency hearing loss, and myopia to cancer, atherosclerosis, and diabetes illustrate these principles. Insulin resistance, coronary heart disease, and age-related fractures are examples that show how changes from the ancestral human lifestyle promote illness and demonstrate the complexity of chronic disease causation. For each, several genetic and environmental factors interact to influence the initiation and progression of pathology.

Age-related fractures

Postcranial robusticity—the size, shape, and overall strength of the skeleton excluding the skull and mandible—has declined about 17 per cent since the Cro-Magnon period (Ruff *et al.* 1993). Bone strength is a function of both mass and structural geometry. Bone mass is influenced by overall size and by bone mineral density; persons with low bone mineral density (i.e. osteoporosis) are at increased risk of fracture. The significance of bone geometry is less appreciated but of equal or greater importance in resisting mechanical stresses. As humans grow, mature, and age, bones undergo continuous remodelling. During adulthood, the medullary cavity expands and overall bone diameter increases, compensating for reduced cortical thickness (another remodelling effect) so that overall bone strength is fairly well maintained. Increases in bone diameter are influenced by physical activity: in contemporary males, activity levels remain sufficient to produce this remodelling effect, but for most women in affluent nations, physical exertion is now too low for bone expansion to occur so that decreases in cortical thickness are no longer offset by increased diameter (Ruff 1992). The result is marked reduction in bone strength. Bone shape is also exercise related. The cross-sectional shape of both preagricultural and preindustrial human long bones is more oval than those of current affluent populations: oval bones are stronger.

These remodelling effects are compounded by the decreases in bone mineral density which normally occur with age, especially in menopausal women. During childhood, adolescence, and early adulthood, dietary calcium and habitual physical activity interact to increase bone mass. Despite their lack of dairy foods, the calcium intake of preagricultural humans was twice current levels due to the substantial calcium content of uncultivated vegetable foods (Eaton 1990). High dietary calcium together with obligatory physical exertion produced peak bone mass greater than that typical of affluent Western societies. The same two factors also slowed age-related bone loss. Bone mass invariably decreases with age, but for human ancestors this process began at a higher level and proceeded more slowly so that the critical level at which age-related fractures became common was rarely reached.

Age-related fracture rates have nearly doubled over the past generation (Kannus *et al.*

1995); deviations from palaeolithic standards in current exercise and diet help explain this trend and suggest how it might be reversed.

Insulin resistance

The ability of insulin to modulate carbohydrate metabolism varies widely among individuals. The lower range of this continuum is defined as insulin resistance: a subnormal capacity for insulin to effect glucose removal from the blood after eating carbohydrates. Increasing evidence suggests that insulin resistance and the compensatory mechanisms evoked in response, especially hyperinsulinaemia, lead to impaired glucose tolerance and non-insulin-dependent diabetes mellitus. They also appear related to abnormal blood lipid profiles, decreased fibrinolysis, hyperuricaemia, and hypertension.

The epidemiological relationship between obesity and insulin resistance is well established, and prevalence of non-insulin-dependent diabetes mellitus has increased with obesity rates during the past 40 years. Paradoxically, ingestion of dietary energy appears to have declined during the same period: a 30 per cent decrease for average Britons between 1956 and 1990 (Ministry of Agriculture, Fisheries, and Foods 1995) and 25 per cent for rural Japanese men between 1969 and 1983 (Shimamoto *et al.* 1989). Human energy utilization can be depicted as:

$$\frac{\text{energy}}{\text{intake}} = \frac{\text{energy}}{\text{expenditure}} + \frac{\text{energy}}{\text{storage}}$$

and, if energy storage (as adipose tissue) has increased while energy intake has decreased, then energy expenditure (of which physical exertion is the most quantitatively important discretionary component) must have decreased substantially. This analysis is plausible because the last 40 years have witnessed a proliferation of labour-saving equipment, motorized transportation, and sedentary recreation. Current estimates of dietary energy intake and/or availability for Americans (about 1865 kcal/day) matches British and Japanese (Shimamoto *et al.* 1989) data. Such figures are much below observed daily energy expenditure for hunter–

gatherers (2850 kcal/day) (Cordain *et al.* 1997) and traditional agriculturists (3250 kcal/day) (Heini *et al.* 1995) who are exceedingly lean (Eaton *et al.* 1988).

Thus increasing adiposity has been accompanied by decreases in skeletal muscle mass— sarcopenia. Average Westerners now have less muscle (and more adipose tissue) than comparable individuals of 50 years ago, or those living before agriculture. After a carbohydrate-containing meal, insulin clears excess glucose from the blood and a sizeable fraction enters skeletal muscle. Although adipocytes have insulin receptors, relatively little glucose can enter them. Skeletal muscle/adipose tissue disproportion mandates that, for any given insulin secretory pulse, too many insulin molecules interact with adipose tissue insulin receptors and too few with those of skeletal muscle. This retards uptake of glucose by muscle, impairs formation of muscle glycogen, and slows glucose clearance from the blood. As a result, more insulin secretion is required for a given carbohydrate load, and this relative hyperinsulinaemia can induce down-regulation of cellular insulin receptors, glucose transporters, and intracellular enzymatic sequences. The ultimate result can be development of the intrinsic insulin resistance known to characterize the muscle, adipose tissue, and liver of patients with non-insulin-dependent diabetes mellitus.

This experientially induced pathophysiology operates against the backdrop of various genetic (Groop *et al.* 1993), developmental (Chapter 21), and dietary factors (e.g. an abundance of fatty and high glycaemic index foods). Among these interacting factors, abnormal body composition resulting in a numerical imbalance between skeletal muscle and adipose tissue insulin receptors stands out as an increasingly important pathophysiological flaw— one temporally correlated with the increasing prevalence of non-insulin-dependent diabetes mellitus during the past generation. The biochemical processes relevant to carbohydrate metabolism were selected when adipose and muscle tissue proportions matched those of recently studied foragers. Maintenance or restoration of comparable body composition is a logical preventive and therapeutic measure for non-insulin-dependent diabetes mellitus.

Coronary heart disease

The multifactorial nature of chronic degenerative diseases is particularly well illustrated by coronary atherosclerosis, the initiation and progression of which are determined by the interaction of over 100 different genes (Sing *et al.* 1996, and Chapter 20) with numerous experiential factors. The latter include cigarette exposure, exercise, body composition, personality, and nutrition. These variables interact over an individual's life to determine whether or not atherosclerosis develops and, if so, with what time course and to what ultimate extent. The important physiological correlates include serum cholesterol and insulin levels, haemostasis, and blood pressure. Fetal growth appears very significant as a predictor of this chiefly adult disease, and tobacco abuse exacerbates several pathophysiological mechanisms.

Coronary heart disease is another example of how lifestyle influences that are new in evolutionary experience become risk factors; that this applies to nearly all the variables affecting this disease can hardly be coincidental. In almost every case (alcohol is a pseudoexception) pertinent differences between ancestral and present experience increase the risk of coronary heart disease.

No palaeolithic humans are available for physiological measurements, but recently studied hunter–gatherers can serve as acceptable surrogates. For such people, blood pressures are usually low (110/70 mmHg is typical), do not rise with age (as in many Westerners), and clinical hypertension is virtually nonexistent (Eaton *et al.* 1988). The serum cholesterol levels of five hunter–gatherer groups, from three continents, averaged 125 mg/dl (3.2 mmol/l) (Eaton *et al.* 1988), a value within the range (70–135 mg/dl; 1.8–3.5 mmol/l) observed in free-living non-human primates (Eaton 1992) and much below that in Western nations (typically 200–220 mg/dl; 5.2–5.7 mmol/l). Hunter–gatherers (Joffe *et al.* 1971; Merimee *et al.* 1972) and rudimentary horticulturalists

(Spielmann *et al.* 1982) exhibit high insulin sensitivity and low serum insulin levels. Fibrinolysis is greater (Gillman *et al.* 1957; Goldrick and Whyte 1959) and platelet aggregatability less (Dyerberg *et al.* 1975) for hunter–gatherers and other preliterate peoples than for most affluent Westerners.

The strong correlation between fetal growth and adult disease, including coronary heart disease (see Chapter 21), should greatly increase attention devoted to maternal nutrition during pregnancy. While changes that would reduce the risk of adult disease have not yet been determined, a reasonable hypothesis is that replicating the nutrition typically available to women during palaeolithic pregnancies will prove beneficial (Godfrey *et al.* 1996).

Current advice for the prevention of coronary heart disease generally parallels the palaeolithic lifestyle. Nearly all authorities advocate a diet low in cholesterol-raising saturated fatty acids and sodium, but rich in fibre and micronutrients. Tobacco avoidance, plentiful exercise, and maintenance of a muscle:fat ratio likely to forestall insulin resistance are other points of agreement between preventive cardiology and preagricultural existence.

However, in certain areas, medical authorities disagree with each other. The Mediterranean diet, with about 35 per cent fat that is chiefly monounsaturated, contrasts with the 10 per cent fat in the East Asian diet: each has its advocates. The importance of dietary cholesterol is disputed as is the proper proportion of carbohydrate. Some authorities suggest expending 1000 kcal/week in physical exercise, others recommend 2000 or even 3000 kcal/week. Some suggest a serum cholesterol of 5.2 mmol/l (200 mg/dl) is desirable, others contend the target should be substantially lower. In each case, comparison with palaeolithic experience can help select the more promising preventive approach.

CONCLUSION

This chapter is not an indictment of 'civilization' nor of technological and cultural advances.

Writing, communications, libraries, laboratories, computers, medical care, mathematics, and the scientific method have all greatly improved our quality of life. Still less should it be taken as glorifying the Palaeolithic. Existence then was not as nasty, brutish, and short as Hobbes depicted it: average life expectancy in the seventeenth century was similar to that for hunter–gatherers and execution by beheading —as by electrocution—was undeniably brutal. But Stone Age people would have eagerly accepted many advances now available. Medicines and supportive therapy can save children who would otherwise die of pneumonia or diarrhoea. Compound fractures, penetrating chest wounds, and placenta praevia were generally fatal before the last century. The extraordinary prevalence of degenerative arthritis among palaeolithic humans testifies to the physical hardship of their lives.

However, our ancestors were generally spared from the chronic diseases which cause most mortality and late life morbidity in Western nations. This appears to be because nutrition, exercise, reproduction, psychological development, and interpersonal relationships have all been vastly altered. While evolution has continued during the past 10000 years, selection has mainly acted on traits expressed in young, reproductive humans, affecting chronic degenerative disease pathophysiology indirectly or not at all: our genetic susceptibility to these illnesses is unlikely to have been altered. Burkitt and Eaton (1989) wrote '... modern Western man has, in a very short period of time by evolutionary standards, deviated greatly from the biological environment to which his body has been adapted. This is the best explanation for ... the high frequency of Western diseases within the communities that have deviated most from the lifestyle of their ancestors'.

The rapidity of cultural evolution, especially during the past century, has far outpaced any possible genetic response. An appreciation of this discordance between our genes and our lives addresses the ultimate nature of chronic illness as it provides vital clues to their proximate prevention.

SUMMARY

Evolutionary medicine is like a multifaceted gem whose surfaces provide distinctive views of an integrated central reality, one at the core of all biomedical science. The facet focused on chronic degenerative diseases illuminates the disparity between ancestral life experiences and those now common in affluent, Western-style nations. Its premises are that (i) the genetic component of current susceptibility to such multifactorial illnesses as atherosclerosis, cancer, diabetes, and osteoporosis was selected in the remote past; and (ii) differences between contemporary Western-style human existence and that during the remote past foster the 'afflictions of affluence'. To be sure, genetic evolution has continued during the Holocene, and even since the Industrial Revolution, but the effects of such 'recent' evolutionary experience must be quantitatively small when compared with that which occurred during the Pleistocene. That epoch lasted 200 times as long and its selective pressures established the adaptations which make us uniquely human— our height, brain size, body proportions, resting metabolic rate, and day range.

To an astonishing extent, risk factors for chronic degenerative diseases, elucidated by epidemiological, mechanistic, and clinical investigations, recapitulate experiential differences between preagricultural and current affluent living. Also, recommendations for health promotion and disease prevention, generated by organizations little concerned with evolutionary science, nevertheless increasingly resemble palaeoanthopological reconstructions of the ancestral human lifestyle.

At present the chief role of evolutionary medicine is to generate hypotheses which can be rigorously tested and, perhaps, to serve as a fundamental reference standard against which the sometimes discrepant findings of orthodox investigations can be assessed. In the future, its ultimate contribution may be to provide an overarching theoretical framework capable of unifying the pathophysiology, genetics, epidemiology, and prevention of chronic degenerative diseases.

23

TESTING EVOLUTIONARY HYPOTHESES ABOUT MENTAL DISORDERS

Randolph M. Nesse

Do mental disorders differ from other medical disorders because they are 'mental'? Not at all. The capacities for grief and anxiety were shaped by natural selection no less than the capacities for nausea and physical pain. The continuity between normal and abnormal anxiety is no different from normal and abnormal blood sugar. Mental and other medical disorders arise from the same kinds of vulnerabilities, and the evolutionary approach applies equally well to both. Because psychiatry is beset by conceptual difficulties even beyond those faced by the rest of medicine, it may benefit even more.

Most people find it hard to grasp the extent of morbidity and mortality caused by mental disorders. According to WHO data from 1990, a single psychiatric disease, unipolar depression, accounts for more disability-adjusted lost years (DALYs) in the world population than all but three other causes (pneumonia, diarrhoea, and perinatal causes) and several times more than HIV, war, and malnutrition (Murray and Lopez 1996: Fig. 23.1). By the year 2020 it

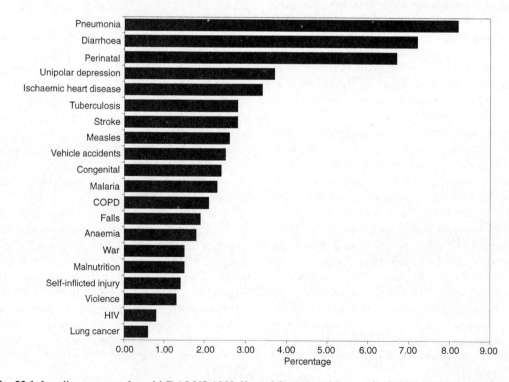

Fig. 23.1 Leading causes of world DALYS 1990 (from Murray and Lopez (WHO) 1996. *Science* **274**: 741.)

is predicted to be the second most significant health problem, after myocardial infarction. Current estimates for women of reproductive age in developed countries are astounding. In this group, unipolar depression accounts for 19 per cent of all DALYs, three times higher than the next most severe causes: schizophrenia, traffic accidents, bipolar disorder, obsessive compulsive disorder, alcoholism, osteoarthritis, chlamydia, and self-inflicted injuries. Mental disorders cause a high proportion of all human suffering, disability, and death, and should be a high priority for Darwinian medicine.

PSYCHIATRIC DISORDERS

This chapter shows how hypotheses can be formulated and tested about how natural selection can give rise to vulnerabilities that result in specific mental disorders. Several major psychiatric disorders are considered in terms of the eight kinds of vulnerabilities outlined in Chapter 2 (Table 23.1). This exercise is useful on two counts. First, it illustrates the main point of Chapter 2—that each kind of vulnerability requires substantially different kinds of hypotheses and tests. Second, it outlines what we do not know about the origins of mental disorder, a sort of 'encyclopaedia of ignorance'. It does not summarize the many recent contributions to evolutionary psychiatry reviewed elsewhere in articles, (McGuire *et al.*

Table 23.1 Categories of evolutionary explanation for vulnerability to disease

1. Defence—what we think is disease or defect is actually an adaptation
2. Infection and other coevolving aspects of the biological environment
3. Novel aspects of the physical environment
4. Genetic quirks that are harmful only in a novel environment
5. Design trade-offs at the level of the gene
6. Design trade-offs at the level of the trait
7. Path dependence
8. Random factors

1992; Nesse, in press) and books (Wenegrat 1990; Stevens and Price 1996; McGuire and Troisi 1997). Instead of providing another review, this chapter focuses on specific psychopathological syndromes, the evolutionary origins of our vulnerabilities to these disorders, and, for each kind of vulnerability, the kinds of hypotheses and studies that are especially likely to advance our knowledge.

Anxiety disorders

Anxiety, like cough, fever, and pain, is a defence elicited and useful in the presence of certain dangers. Anxiety disorders result from dysregulation of the normal system (Barlow 1988). The question of why so many people have anxiety that seems excessive can be addressed only after we know when it is normal, and we cannot know when it is normal until we understand its function and regulation.

Most people readily acknowledge the utility of some anxiety, so it might be thought unnecessary to conduct formal tests of the hypothesis. However, an attempt to define adequate tests reveals how little we really know and offers a model for studying other more problematic emotions like sadness, grief, guilt, suspicion, jealousy, and boredom. Tests are based on the same approaches used to study other possibly protective mechanisms like allergic responses, fever, nausea during pregnancy, menstrual bleeding, and blood iron depletion during infection.

The methods for testing the hypothesis that anxiety is an evolved defence are based on the nine kinds of predictions about defences listed in Chapter 2. The first is that individual differences in anxiety responses should influence the degree of protection against a threat. Some individuals have hypofunctional anxiety systems, whether from genetic, traumatic, or toxin-induced defects that prevent anxiety expression, from drugs that block anxiety, or simply from being at the low end of the anxiety distribution. They should show decreased ability to avoid and escape dangers and increased rates of death and harm. Conversely, individuals who readily experience anxiety should be relatively protected from dangers. The evidence for these predictions is remarkably skimpy (Wilson *et al.* 1994).

Second, the form of anxiety should match its function. Almost every aspect of the 'fight–flight response'—increased heart rate, glucose, clotting, sweating, and breathing—matches what is needed in the face of serious danger (Cannon 1929). These reactions are sometimes neglected by modern psychiatrists who study them as 'panic attacks'. One solid line of thought proposes that panic is designed to protect against suffocation. While I am not convinced that this is the primary function of panic, some evidence supports the idea (Klein 1996). Third, the characteristics of subtypes of anxiety should be different from global anxiety in ways that better meet certain kinds of threats. Fear of heights does seem to cause a response, freezing, that is probably adaptive in high places, but systematic data are lacking. Panic attacks occur initially in most patients where they would be helpful—away from home (Lelliott *et al.* 1989). The fear of blood is the only anxiety associated with fainting, and this may well be an adaptation (Marks 1988). The correspondence between the atheoretical categories of anxiety disorders in the *Diagnostic and statistical manual of mental disorders* of the American Psychiatric Association (1994) and the dangers people faced in the ancestral environment (Marks and Nesse 1994) is remarkable, but more work could be done to assess the match between the responses characteristic of each kind of anxiety, and the utility of those responses in the face of a particular danger.

Fourth, proximate mechanisms should match expectations concerning the function of the defence. If, for instance, the high heart rate in anxiety was an epiphenomenon resulting from skin vasodilation, this would argue against its functional significance. In fact, the high heart rate associated with anxiety results from co-ordinated actions of sympathetic innervation on the heart and vessels to increase blood flow to muscles. Adrenal cortical hormone secretion in the face of danger adjusts the body for energy expenditure—breaking down glycogen in the liver, increasing the availability of glucose and its ability to be utilized by muscles, and

decreasing its entry into cells involved in digestion and repair. Much has been made of high cortisol secretion in the face of psychological stress (Mason 1975). It remains unclear if this secretion is an adaptation, a mistake, or if, for our ancestors, such stress was routinely accompanied by the need for action (Sapolsky 1992). Considerable evidence suggests that cortisol is designed to protect against tissue damage caused by other aspects of the stress syndrome (Munck *et al.* 1984). Many components of the primary stress system should be harmful to tissues or expensive or disruptive of routine activity, otherwise they would be expressed all the time instead of packaged and tightly controlled.

The fifth and sixth predictions can be tested with comparative studies. A species should not show fear of heights if it is never endangered by heights, either because trees are its usual habitat like gibbons, or because they hardly ever encounter high places, like muskrats. Baboons, who often spend time on cliffs, might well benefit from experiencing more fear in high places compared with other primates. A comparative test of such predictions would be welcome.

The seventh and eighth predictions, about regulation of a defence, are the most important for understanding the relationship between normal and abnormal anxiety. Is anxiety expressed when it is useful? Do dangerous situations reliably arouse anxiety? How these responses are regulated remains unclear, as do the differences between normal people and people with anxiety disorders. People with anxiety disorders do not misjudge the likelihood of danger, they seem instead to overreact to cues of danger. Like normal people, they consistently underestimate the likelihood of common dangers and, also, like other people, they believe that they are less likely to suffer harm than the average person (Nesse and Klaas 1994).

Are fears of snakes and spiders 'innate' or 'learned'? Experiments show that the question is too simple. Vervet monkeys do not show fear on initial exposure to a snake, but after a single observation of another monkey's fearful response, an enduring fear is created, one that cannot be created to other novel stimuli like flowers (Mineka *et al.* 1984). Human studies show that fear responses are more readily conditioned to cues of objects that were dangerous in the ancestral environment, like snakes, than to objects dangerous only now, like guns (Öhman *et al.* 1985; Cook *et al.* 1986). There are trade-offs. Systems that express innate fear at first contact with the object protect an isolated individual on first exposure to a danger, but are of little help with novel dangers. Systems that require experience do not protect initially but offer better protection against new dangers, at the cost of learning to fear some stimuli that are not dangerous (Staddon 1983).

Why are people more often more anxious than seems useful? The encounter with any potentially dangerous stimulus poses a signal detection problem. For a given signal, will expressing the defence yield a net benefit or a net cost? If anxiety is inexpensive, and a lack of response is costly or fatal, natural selection will shape regulation mechanisms to a hair trigger with many resulting false alarms (Nesse and Williams 1995). On a more human level, social threats invisible to an observer, and perhaps not noticed consciously by the individual, can arouse profound subjective anxiety, making clinical psychiatry challenging indeed (Slavin and Kriegman 1992).

Finally, why are anxiety disorders so common? Each of the evolutionary origins of vulnerability contributes. Much anxiety that seems pathological is, in fact, a normal defence. There are also trade-offs: too little anxiety is as disadvantageous as too much. In a population distribution, some people will be at the maladapted extremes. Genetic variation is probably maintained because environments, especially social ones, differ in their dangerousness and the particular dangers they pose. In addition, novel dangers create new anxiety disorders, such as an unrealistic fear of AIDS in some people, and too little fear of AIDS in others. A full evolutionary explanation of anxiety disorders would address, for each specific kind of anxiety: normal functions, regulation, and how dysregulation can arise from novel aspects of

the environment; trade-offs at the level of the trait and the gene; and historical constraints and random factors. We are a long way from this goal.

Depression

Like anxiety, depression seems to be an exaggeration of a normal response. Unlike anxiety, the utility of ordinary sadness and grief is not obvious. Anxiety precedes a danger and induces action, so it is easy to see how it can prevent harm, while low mood usually follows a loss and induces lack of action, so its value is more difficult to see. None the less, ordinary sadness and grief are so reliably aroused by the same kinds of losses in almost everyone that they must have some utility. Low mood must somehow influence future behaviour to increase reproductive success. There are many possibilities. Low mood can motivate avoidance of a situation that may cause further losses, it can facilitate changes in strategy that are necessary after a loss, it can facilitate submissive behaviour after a loss of status. These hypotheses need testing. To understand depression, we will need to understand the evolutionary functions and regulation of normal mood. We are far from such an understanding.

One way to investigate the functions of mood would be to find people who lack a capacity for low mood and identify the disadvantages they experience. However, it is hard to distinguish those people who have little capacity for low mood from people who are simply fortunate. We know that tendencies to low mood are strongly influenced by genetic factors and, when the specific genes are identified, it will be much easier to address this question.

In the meanwhile, we should consider the possibility that depression is a growing worldwide epidemic created by some novel aspect of our environment. This is supported by the extraordinary frequency, morbidity, and mortality for depression (Kessler *et al.* 1994) and by cohort data that indicate rapid rises in prevalence (Klerman and Weissman 1989; Cross-National Collaborative Group 1992). To answer this question, of great public health significance, we urgently need epidemiological studies to determine the rates of depression in different kinds of cultures, especially primitive horticulturists and hunter–gatherers.

Manic depressive illness

Manic depressive illness appears to be a genetic disorder with heritability over 50 per cent. A review of seven studies found 69 per cent concordance rates for monozygotic twins reared together, 67 per cent for monozygotic twins reared apart, and 13 per cent for dizygotic twins (Rush et al. 1991). How can genes persist that cause an illness that severely interferes with function and is fatal in at least 20 per cent of cases (Goodwin and Jamison 1990)? The main possibilities are: (i) the genes are recent mutations, (ii) they are quirks that cause illness only in modern populations, and (iii) they somehow give an advantage that outweighs the disadvantage. The first and second possibilities are unlikely in view of the long recognition of this disorder in many societies and the strong selection against it (Goodwin and Jamison 1990). The third possibility has been considered in some detail because of the high rates of manic depressive illness in successful creative people (Jamison 1993). Mood disorders are more frequent both in accomplished writers and in their relatives (Andreasen 1987). Preliminary results also suggest that relatives of patients with manic depressive illness score high for creativity (Richards et al. 1988). If this is confirmed, the next step would be to determine the effects of creativity and manic depressive illness on fitness, preferably in premodern societies. Several routes to increased reproductive success seem possible. Creative accomplishment seems to increase sexual attractiveness, and it has been suggested that creativity is a product of sexual selection (Miller 1994). Another route to reproductive success could be via the social success of people with manic depressive illness or their relatives. When specific genes for manic depressive illness are identified, it will become much easier to study advantages that may counterbalance the high mortality rate.

If genes for manic depressive illness conferred a selective advantage that was not frequency dependent, we would all have the manic depressive illness phenotype. Would we still recognize it as a disease, or would we identify as diseased those rare people who lack wide mood fluctuations? This illustrates how natural selection could result in 'universal diseases'—syndromes like ageing that decrease everyone's health but none the less persist because they give an overall reproductive advantage. The 'young male syndrome'—risk taking and violent competition arising from sexual competition in men—may be an example (Wilson and Daly 1985).

Eating disorders

Eating disorders (bulimia and anorexia nervosa) are essentially unknown in traditional cultures, were rare in modern societies until this century, recently affected nearly 10 per cent of college females, and are now decreasing somewhat (Pate et al. 1992). They seem to be caused by novel features of our environment. What are these novel factors, and how do they interact with the evolved mechanisms that regulate eating? The selective forces that shaped mechanisms to limit excessive weight gain probably were feeble compared with those designed to prevent starvation, so it is not surprising that the ready availability, at any time with little effort, of a variety of tasty high-calorie foods has caused an increase in the frequency of obesity.

Serious eating disorders almost always begin with an attempt to lose weight by restricting food intake. The predictable adaptive response is aroused—overwhelming impulses to consume any available source of calories. At some point these impulses overcome will-power just as certainly as any attempt to hold one's breath ends with a deep inspiration. This bingeing causes a profound sense of loss of control that makes dieters even more fearful that their eating will make them obese. So they try even harder to diet, and an escalating feedback cycle ensues. Studies with volunteers show that a very low calorie diet itself induces profound psychological and behavioural changes (Keys 1950). If food is available, most dieters cannot resist their impulses and binge. However, a few people, by will-power, exercise, laxatives, vomiting, and a profoundly distorted self-image, can limit their food intake and stay thin. The required discipline absorbs much of their energy, and if their self-image is distorted enough, they believe they are obese even as they die of starvation.

This simplified perspective on eating disorders explains only why they arise in a modern environment, not why some people get them and others do not. Knowledge of the mechanisms that regulate eating is already impressive and growing larger; it includes information on the genetic, psychological, and social factors that make some people especially vulnerable to eating disorders. It should soon be possible to integrate these perspectives. Many aspects may fit together: the much higher frequency of these disorders in women, the onset of eating disorders at sexual maturity, their rarity after menopause, the finding that men may be designed to seek mates with a waist/hip ratio of 0.7 (Singh and Young 1995), and the increase of these disorders in cultures where women must attract their own mates and where media constantly display visual images of exaggerated female forms (Pate *et al.* 1992). We may soon find out what happens when new drugs make it possible to adjust weight to any level desired. Evolutionary analysis should help us assess the likelihood of two outcomes. Will the availability of such agents prevent the cycles of weight gain and dieting that result in eating disorders, or will people use them to achieve extraordinary thinness in an arms race that will be fatal for some? The frequency of plastic surgery and breast augmentation suggest that we should prepare for the latter.

Substance abuse

An evolutionary approach to substance abuse follows the same lines as for eating disorders, but substance abuse is more common and seems to be a product not just of the very recent modern environment, but of any environment where drugs or alcohol are available. Hunter–gatherer cultures use drugs and alcohol, and to excess, but supplies are erratic, impure, and limited in amount and methods of use. Only modern societies have the organizational and economic structures providing the steady supplies that foster alcoholism and other addictions.

Current research emphasizes brain mechanisms of addiction, and social and genetic differences that explain why some people abuse substances and others do not. However, all these factors may merely influence which individuals succumb to the effects of novel factors acting on brain mechanisms we all share, mechanisms that were not evolved to cope with such exogenous chemicals. Chemicals are involved in neurotransmission, and drugs can act directly on the mechanisms that regulate behaviour, so consuming them can easily come to dominate life (Pomerleau 1997). The mystery is why so few people become dependent on substances. This perspective undercuts the hope that drug use is a problem mainly for people with certain deficits or mainly in certain disadvantaged or stressful environments, and suggests instead that substance abuse is a universal problem for smart organisms with access to pure chemicals (Nesse 1994, Nesse and Berridge 1997).

Child abuse

Child protection workers have long laboured to discover why people would hurt their own children, but it took two ethologists to suggest that people might not be hurting their own children. Daly and Wilson checked the rates of fatal abuse for children living with two genetic parents, as compared with those living with at least one step-parent. Having a step-parent in the home increased the fatality rate for children from birth to the age of 2 years, not by a factor of two or three, but by at least 50 times (Daly and Wilson 1987; Gelles and Lancaster 1987). This does not explain why these children are killed, nor does it mean that such murders are an adaptation, but it does demonstrate the protection that comes from living with both biological parents and it illustrates the utility of an evolutionary approach.

Schizophrenia

Schizophrenia is strongly influenced by genetic factors (Kendler 1983), yet it substantially decreases fitness (MacSorley 1964). Evolutionary hypotheses must account for this mystery. The prevalence is relatively constant worldwide at about 1 per cent of most populations (Jablensky *et al.* 1992), making it unlikely that the frequency could be accounted for by recent mutations and drift unless many genes are involved. The genes responsible for this disease may also give a selective advantage (Crow 1995). Most hypotheses assume that the advantages are in the cognitive/emotional system and that the advantages accrue to people with the disorder. It seems more likely that any benefits accrue to people with the genes but no disorder; the benefits may have nothing to do with mental life—they might protect against some infection, for example. The test, of course, is to examine relatives of people with schizophrenia for traits that would have given a selective advantage in the palaeolithic. If the selection force no longer exists, that would remind us that some evolutionary questions may have answers that we cannot find. If the genes that cause schizophrenia are in mutation–selection balance, then the eighth source of vulnerability, random factors, would be the closest we can get to an explanation. At present, this final possibility seems the most likely explanation, but the alternative cannot be ruled out. The discovery of specific genes for schizophrenia would help us assess these hypotheses.

Normality

The psychiatric condition that may be most illuminated by an evolutionary perspective is normality. Long the subject of controversy, normality has proved an elusive concept. A leading psychiatric textbook previously included a chapter with a dozen different definitions, based on statistical norms, social expectations, or subjective experience. The WHO definition of health is optimistic: 'Health is a state of complete physical, mental and social wellbeing and not merely the absence of disease and infirmity'. A Darwinian perspective suggests that there is no one answer to this question. The effect of a condition on reproductive success gives important information on its likely evolutionary status, but we do not call normal many behaviours that increase reproductive success, and we do not call abnormal many that decrease reproductive success, such as birth control. So, a Darwinian approach does not define normality, but it helps to explain why the search for a definition has proved so difficult. Much suffering and conflict arises from normal operations of normal evolved systems, but medical or social intervention is none the less often justified. Far from oversimplifying, an evolutionary approach reveals the manifold complexities of mental disorders.

SUMMARY

From an evolutionary perspective, explaining mental disorders is no different from explaining other medical disorders, and the categories of causes, and kinds of hypotheses and tests, apply equally well to both kinds of disorders. Anxiety is a defence like pain, and dysregulation can result in disorders characterized by too little as well as too much anxiety, illustrating the trade-offs involved. Sadness and grief are almost certainly also defences, but until studies are conducted to help us better understand their functions, we will have a hard time understanding whether depression is a defence, an overshoot of a defence, or a trade-off. Manic depressive illness results from genetic variation that may offer fitness advantages as well as disadvantages, and the same might be true for schizophrenia, although it is more likely that mutation–selection balance is the correct explanation. Eating disorders are caused by novel factors in the modern environment interacting with nutrition regulation mechanisms that evolved to solve very different problems. Alcoholism and substance abuse are, at root, a product of our brains encountering environmental novelties. Explicitly stating such hypotheses, with attention to the different kinds of tests needed depending on the postulated origins of vulnerability, may facilitate studies that will help us better understand the origins of mental disorders and how to treat them.

24

THE EVOLUTION OF NON-INFECTIOUS AND DEGENERATIVE DISEASE

John Maynard Smith (rapporteur), David J.P. Barker, Caleb E. Finch,
Sharon L.R. Kardia, S. Boyd Eaton, Thomas B.L. Kirkwood, Ed LeGrand,
Randolph M. Nesse, George C. Williams, and Linda Partridge (chair)

Recent improvements in hygiene and nutrition have led to a dramatic decrease in the early mortality caused by infectious disease, and so to an increase in mean lifespan. Because more people survive into old age, more people encounter the diseases associated with old age. This simple effect is so strong that now most people requiring medical help in developed countries are suffering from diseases whose incidence increases with age.

Evolutionary medicine should, therefore, pay special attention to such diseases, yet at first sight they seem especially puzzling for a Darwinist. Senescence is the decline with age of precisely that property, adaptedness, that we expect natural selection to increase. The first section of this report, therefore, is devoted to the mechanisms causing senescence, and to describing the theories that have been proposed to explain their evolution.

Most participants felt that evolutionary considerations make it unreasonable, at least in the short run, to expect a large increase in human longevity, beyond those that are already emerging around the world, and that our efforts should be devoted to delaying the onset, or reducing the severity, of specific diseases, thus prolonging healthy life. However, animal studies have demonstrated a remarkable plasticity in life-history strategies, and this has led some to hope for more dramatic changes in human life expectancy. This is a controversy that we cannot resolve, but the arguments on both sides are briefly summarized below.

We discuss the causes of the differences between individuals, and the relative roles of environmental and genetic factors. Finally, we discuss three groups of diseases—the reproductive cancers of women, mental disorders (in particular, manic depression and unipolar depression), and coronary heart disease and some disorders associated with it. Our aims are to show how an evolutionary approach can aid our understanding, and to point to areas of ignorance requiring further research.

EVOLUTIONARY THEORIES OF SENESCENCE

Darwinian accounts of the evolution of senescence start from the observation that, in nature, most animals die before they reach old age. Consequently, genetic mutations (whether acting early or late) that have their effects on adaptation late in life are selected against only weakly, in comparison with mutations affecting fitness early in life. This leads to two main theories.

Trade-off theories

Imagine a mutation that increases survival or fecundity early in life, at the expense of reducing adaptedness late in life. Such a 'pleiotropic' mutation would be favoured by natural selection, provided that accidental death is frequent enough to ensure that few individuals survive to the later age. There are many reasons why such trade-offs between success early and late in life are to be expected. A well-studied example is the trade-off between fecundity and fertility in *Drosophila*. Almost any environmental treatment (including 5000 rad of X-rays) that reduces the rate at which a female lays eggs prolongs her life. Artificial selection for increased longevity reduces early fecundity, and vice versa. A more general phenomenon, relevant also to humans, is the existence of post-mitotic cells. A brain composed of non-dividing neurones probably works better, but its irreplaceable neurones leave its possessor at high risk of irreversible damage from vascular disease and Alzheimer's disease.

A different way of thinking about trade-offs is to identify stochastic cellular processes that lead to degeneration. Possible processes include changes in the nuclear genome, erroneous transcription and translation, oxidative damage, damage to mitochondria, and damage to non-replaced, or slowly replaced, proteins such as collagen. This is not an exhaustive list. Several of these processes are likely to be more severe in non-dividing cells, because in a dividing tissue, damaged molecules undergo a two-fold dilution at each cell division. But why are they relevant to a trade-off theory of ageing? The idea is that such processes can be slowed down, even if they cannot be wholly eliminated. For example, more accurate replication and translation, or more effective removal of oxidizing agents, will slow them down. If such protection is costly, a trade-off would exist, for energy diverted to maintenance would not be available for reproduction. It is plausible that such costs exist, but more evidence is needed.

Mutation–selection balance theories

Deleterious mutant genes exist in populations. Each is likely to be at low frequency, maintained by a balance between mutation generating new mutant alleles, and selection eliminating them. The selection against a gene whose effects on fitness are late in life will be relatively weak, and its frequency in the population correspondingly high, even if it has no counterbalancing beneficial effects earlier in life. Thus we expect late-acting deleterious alleles to accumulate, and to cause senescence.

There is evidence from animal experiments

that these processes occur, although their relative importance is hard to estimate. In humans, single mutations with deleterious effects late in life are relatively rare (although they do exist—e.g. Huntington's chorea, which emerges during middle age), compared with deleterious mutations acting early in life (e.g. phenylketonuria, and other inborn errors of metabolism). However, this observation does not rule out a mutation–selection balance theory of human ageing, because it may turn out that some common diseases of later life are caused by mutations at a number of different loci, either singly or in combination.

It is a striking fact that humans live, on average, some 40 times as long as mice, yet they show many of the same degenerative diseases of ageing, for example vascular changes and insulin resistance. This shows that the rate of degeneration can change genetically. But is it plausible that it could be changed by relatively simple means, either genetic or environmental? This is a question on which members of the group disagreed. One view is that it is unreasonable to suppose that the rate could be changed substantially by altering one or a few genes. An obvious fact about senescence is that it affects many systems, in many different ways. Although there is no reason, in principle, why a human-like animal should not live for, say, 200 years, the number of genetic changes needed to bring this about would be very large. Further, if the trade-off theory is correct, there is a risk that prolongation of life would be accompanied by deleterious changes earlier in life.

However, an alternative view is possible. The evolutionary flexibility of animal life histories suggests that a substantial prolongation of human lifespan may not be impossible.

One observation that may bear on this controversy is that there has been a world-wide decrease in the rate of change in mortality rates with age at advanced ages. Whereas mortality rate doubles every 8 years from 15 to 65 years of age, this acceleration slows down, and may reach a constant maximum value of 0.3–0.5 of mortality risk per year by 105–110 years of age. This mortality risk is about 10 000 times higher than at age 15, but it is interesting that it seems no longer to be increasing with age. Why should this be so?

There are, in principle, two kinds of explanation. One is that the population is heterogeneous: there is a subpopulation of unusual individuals with a lower mortality rate in old age, for either genetic or environmental reasons. The other supposes a homogeneous population. For example, old people may change their behaviour so as to reduce potential mortality factors (the first author of this report is much more careful crossing roads than he used to be), or they may be better looked after. If intrinsic changes in life-history strategy are occurring, they too could produce the observed change in mortality rates in a homogeneous population. It is hard to decide between these possibilities, but it is relevant that a similar phenomenon has been observed in some animal experiments, including genetically uniform populations of *Drosophila* (where the explanation could not be genetic heterogeneity). Whatever the explanation, the consensus was that the observation should not lead us to hope for a dramatic extension of human lifespan: the best we can hope for is a modest extension of healthy life.

THE CAUSES OF INDIVIDUAL DIFFERENCES

Everyone dies of something. Those certified as dying of old age, if such a certification could ever be correct, died of everything at once, or at least of many interacting causes. Today, most people die of a disease whose incidence increases with age. It is sensible to ask why some people die of stroke, some of coronary heart disease, some of breast cancer, and so on, and why they do so at different ages. A general answer is not possible, but three general points are worth making.

1. We should not expect often to find single mutant genes causing conditions that are both common and which cause substantial lowering of Darwinian fitness. If we did find such a gene (and sickle cell is an example), we would have

to seek for some reason why it was common: possible explanations are that it gives a powerful selective advantage in the heterozygote, or that it is favoured by selection at the gene level (e.g. by segregation distortion), or that it has some feature causing an exceptionally high mutation rate. Although these phenomena are possible, and should be looked for, they are unlikely to be typical and should be relatively easy to spot. In general, common diseases, such as those discussed below, are unlikely to be caused by mutations at single loci.

2. Genes often have different effects in different contexts—that is, in different environments, or in different genetic backgrounds. We should not expect to find one 'normal' or 'wild-type' genotype that is superior in all environments. This is particularly important in the context of senescence, or diseases of old age, because evolution theory leads us to expect trade-offs between effects at different ages. Because the point is important, we give two examples. Myopia is common in literate societies, and has a genetic component. It is absent among Eskimos, but develops in a proportion of them if they are taught to read. Here we have a genetic difference that is expressed only in some environments (see Box 10.1, Chapter 10). A more complex but immediately relevant example concerns different Apo E genotypes. In a random sample of female Caucasians, smoking was associated with an 8 mg/dl increase in cholesterol levels, relative to non-smokers, for the $\varepsilon33$ genotype, whereas smoking was associated with a 30 mg/dl increase in cholesterol in the $\varepsilon32$ genotype.

3. Despite these complications, there is convincing evidence of genetic involvement in many non-infectious and chronic diseases. This conclusion is based on estimates of 'heritability' (defined as the proportion of variance caused by genes: note, therefore, that the heritability of a trait is not universal to a species, but specific to a particular population in a particular range of environments). Genetic involvement is deduced from twin and adoption studies and from the identification of candidate genes. However, it is typical that several to

many genes are involved, and much work will be needed to discover what these genes do, and how they interact with each other and with the environment.

WHY ARE PARTICULAR DISEASES COMMON?

The following diseases are, we think, typical: they are certainly important. None are simple, and the way evolutionary considerations are relevant varies.

Women's reproductive cancers

The reproductive experience of women in current affluent societies differs in many ways from that of recently studied foragers, whose lives are the best available approximation to our evolutionary past (see Chapters 8 and 22). Typical members of affluent societies experience earlier menarche, later first birth (so that the interval between menarche and first birth is three to four times as great), less nursing, lower parity, and later menopause. The net effect is to increase the exposure of reproductive tissues to oestrogenic hormones, which in turn increases cell proliferation (e.g. breast duct cell turnover rates are up to 20 times higher between menarche and first birth than after the first full-term pregnancy). Cells that are dividing frequently are more likely to develop clinical malignancy. Information about women's cancer rates in foraging societies, let alone in preagricultural humans, is unknown, but it may well be that women in affluent societies have a risk that is from 10- to 100-fold greater.

Neither increased parity nor earlier first pregnancy can responsibly be advocated as preventive measures for breast, ovarian, and endometrial malignancies, but interventional endocrinology—menarche delay, early pseudopregnancy, and/or oestrogen-lowering oral contraception—could perhaps recreate an ancestral hormonal milieu. Improved diagnosis and treatment of breast cancer have failed to reduce age-corrected mortality—the rate in 1990 was precisely that in 1930—so new approaches to actual *prevention* of women's cancers, based on evolutionary medical principles, deserve investigation and social discussion.

Manic depressive illness

Manic depressive illness is fairly common, affecting nearly 1 per cent of adults. It substantially lowers fitness: suicide rates are high (20 per cent), and comorbid substance abuse also causes many premature deaths. The heritability is high (0.5), but the number of genes involved is unknown. These features suggest that we should seek selective advantages that could counterbalance the powerful disadvantages. There are hints that people of high 'creativity' have a high frequency of manic depressive illness. The alternative, and more powerful, test of looking for high creativity in the relatives of those suffering from manic depressive illness is promising. But there remains the obvious difficulty that high creativity is not the same as high Darwinian fitness.

How else, other than by a compensating selective advantage, could the high frequency of manic depressive illness be explained? The idea that there are many mutations, each able to cause the behaviour, and maintained by mutation–selection balance, seems implausible. So there are good reasons for seeking a compensating advantage.

Unipolar depression

This is an important disease, with a lifetime rate greater than 10 per cent in most Western populations. It is, by a large margin, the major cause of illness in women of reproductive age in the United States. It has a heritability of 0.3–0.4. Its frequency varies greatly between populations: for example, there is a greater than fivefold variation in frequency between European countries. Its effect in reducing fitness, although real, is less than in the case of manic depressive illness. (It is interesting to contrast it with schizophrenia, which has a similar heritability, a more obvious adverse effect on fitness, but does not vary substantially between populations.)

Perhaps the most fruitful way of thinking

about depression is to see it as an extreme, and maladaptive, manifestation of variation in 'mood'. People can be happy or sad. The variation is universal, and elicited by similar circumstances (e.g. sadness is elicited by the death of a relative) in different societies. It is easy to see how it could be adaptive. For example, to feel sad if a child dies is likely to lead one to avoid performing, or continuing, actions tending to cause the same result in future, to reconsider the actions that led to the death, and so on. To oversimplify, sadness can be seen as a stimulus leading to long-term changes in behaviour, much as pain can lead to short-term changes.

However, it is hard to see clinical depression as adaptive. We seem to be faced with an adaptive response which, in extreme circumstances, and particularly in a species living in a changed environment, can overshoot, resulting in non-adaptive behaviour. Such an explanation would fit with the observation that acute depression is characteristic of individuals who find themselves in situations from which no escape seems possible.

The idea that an adaptive response may, in some circumstances, result in maladaptive behaviour may apply more generally. For example, Daly and Wilson have found that infanticide is some 50 times more frequent (allowing for the relative frequencies of the relationships) in step-parents than in biological parents. There is no claim being made that, in Canada where the study was carried out, infanticide of stepchildren increases fitness, still less that it is inevitable, or morally right. But it can be seen as an exaggerated and maladaptive expression—an 'overshoot'—of an evolved adaptation to invest more heavily in one's own children.

One possible explanation for between-individual variation in mood is 'frequency-dependent selection'. In animals, there is both theoretical and observational support for the idea that populations are variable in the degree of aggression or submission, because if most other individuals are aggressive, it pays to be submissive, and vice versa. The possibility that an analogous process is involved in the evolution of human variability is worth investigating.

Coronary heart disease

At first sight, coronary heart disease seems rather straightforward. Its prevalence has increased dramatically in Western societies during the last 50 years, although it is now flattening out, and even decreasing. Three environmental factors exist that could explain the increase—smoking, excess fat in the diet, and lack of exercise. This explanation must be part of the truth, but there are snags (see Chapter 21). For example, coronary heart disease is increasing rapidly in India, although in many areas women do not smoke, and a dietary explanation is implausible (coronary heart disease is commoner in lower socio-economic groups, and men with this disease are often thin).

An alternative approach is to see coronary heart disease as one outcome of a wider complex of 'metabolic diseases', including diabetes, hypertension, and obesity. A large body of evidence shows that many genes influence the risk of coronary heart disease. The genes identified so far come from multiple metabolic systems (lipid metabolism, blood pressure regulation, carbohydrate metabolism, and haemostasis). Since environmental factors affect these metabolic systems, the influence of gene–environment and gene–gene interactions is important in determining risk. As these candidate genes are concerned with the normal homeostasis of the metabolic system, they should not be regarded as 'disease genes'. The nature of variation in these genes, and their interaction with environmental factors, remain important research issues.

A further complication is that events during fetal life have been shown to be important in causing coronary heart disease. The essential notion is that the fetus makes responses to poor nutrition which, although adaptive in preserving brain growth, can lead to coronary heart disease in adult life. The key point here is that an adaptive response in the fetus can have deleterious effects later in life (coronary heart disease, diabetes)—precisely the sort of trade-off predicted by evolutionary biology.

CONCLUSIONS

1. There has been a recent, rapid increase in the relative importance of age-related diseases.

2. The evolutionary theory of senescence leads us to expect trade-offs between deleterious effects late in life and beneficial effects earlier in life. It is important to identify these trade-offs, and to be alert to the selective advantages that may counterbalance the more obvious disadvantages of genes causing degenerative changes. At this time, few examples of trade-offs have been established.

3. A pathological condition may be an extreme, and maladaptive, expression of a response which evolved because it is adaptive in most circumstances. Such overshoots are likely to be common in a species living in an environment different from that in which it evolved.

4. Common non-infectious and degenerative diseases tend to have a complex causation, involving many genetic and environmental factors. This is to be expected, both because it will be unusual for single gene mutations with deleterious effects on fitness to be common, and because of the trade-offs we expect to be associated with life-history traits. However, it is useful to identify single gene mutations associated with such diseases, for they will make it easier to identify these trade-offs, so

that, when aiming to alleviate disease, we do not do more harm than good.

5. It is important to do further research on the fundamental causes of animal and human senescence. If there are several independent degenerative processes synchronized only by natural selection, which is weakest on the systems that last longest, then a substantial further prolongation of human lifespan will be very difficult. In contrast, if there is a unitary cause of the many symptoms of senescence, then hopes for a dramatic increase in human lifespan are not wholly unreasonable.

KEY REFERENCES

Barker, D.J.P. 1994. *Mothers, babies and disease in later life*. BMJ Publishing Group, London.

Eaton, S.B., Shostack, M., and Korner, M. 1988. *The palaeolithic prescription*. Harper & Row, New York.

Ricklefs, R. and Finch, C.E. 1995. *Aging, a natural history*. W.H. Freeman (Scientific American Library), New York.

Nesse, R. and Williams, G.C. 1994. *Why we get sick: the new science of Darwinian medicine*. Times Books, New York.

Stearns, S.C. 1992. *The evolution of life histories*. Oxford University Press.

Wachter, K. and Finch, C.E. (eds.) 1997. *The biodemography of aging*. NAS Press, Washington.

Weiss, K.M. 1995. Genetic variation and human disease. Principles and evolutionary approaches. Cambridge University Press.

REFERENCES

Aaby, P. 1988. Malnutrition and overcrowding-exposure in severe measles infection. A review of community studies. *Rev. Infect. Dis.* **10**: 478–91.

Aaby, P., Bukh, J., Hoff, G., *et al.* 1986. High measles mortality in infancy related to intensity of exposure. *J. Pediatr.* **109**: 40–4.

Aaby, P., Samb, B., Simondon, F., *et al.* 1995. Non-specific beneficial effect of measles immunization: analysis of mortality studies from developing countries. *Br. Med. J.* **311**: 481–5.

Aaby, P., Samb, B., Andersen, M., and Simondon, F. 1996. No long-term excess mortality after measles infection: a community study from Senegal. *Am. J. Epidemiol.* **143**: 1035–41.

Abel, L. and Demenais, F. 1988. Detection of major genes for susceptibility to leprosy and its subtypes in a Caribbean island: Desirade island. *Am. J. Hum. Genet.* **42**: 256–66.

Abrams, P.A. and Ludwig, D. 1995. Optimality theory, Gompertz law and the disposable soma theory of senescence. *Evolution* **49**, 1055–66.

Abrams, R.H. 1957. Double pregnancy. Report of a case with thirty-five days between deliveries. *Obstet. Gynecol.* **9**: 435–8.

Addington, W. 1979. Patient compliance: the most serious problem in control of tuberculosis in the United States. *Chest* **76** (Suppl.): 741–3.

Agnew, P. and Koella, J.C. 1997. Virulence, parasite mode of transmission, and host fluctuating asymmetry. *Proc. Roy. Soc. London B* **264**: 9–15.

Aiello, L.C. and Dean, C. 1990. *An introduction to human evolutionary anatomy.* Academic Press, London.

Ainsworth, M.D., Blehar, M.C., Waters, E., and Wall, S. 1978. *Patterns of attachment: a psychological study of the strange situation.* Erlbaum, Hillsdale, New Jersey.

Aklillu, E., Persson, I., Bertilsson, L., *et al.* 1996. Frequent distribution of ultrarapid metabolizers of debrisoquine in an Ethiopian population carrying duplicated and multiduplicated functional CYP2D6 alleles. *J. Pharmacol. Exp. Ther.* **278**: 441–6.

Alcolado, J. and Alcolado, R. 1991. Importance of maternal history of non-insulin dependent diabetes patients. *Br. Med. J.* **302**: 1178–80.

Alexander, G. 1962. Energy metabolism in the starved new-born lamb. *Aust. J. Agric. Res.* **13**: 144–64.

Alexander, R.D. 1979. *Darwinism and human affairs.* University of Washington Press, Seattle.

Alexander, R.D. 1987. *The biology of moral systems.* Aldine de Gruyter, New York.

Alexander, R.M. 1997. *Optima for animals*, 2nd edn. Arnold, London.

Allison, A.C. 1964. Polymorphism and natural selection in human populations. *Cold Spring Harbor Symp. Quant. Biol.* **29**: 137–49.

Ambrosone, C.B., Freudenheim, J.L., Graham, S., *et al.* 1996. Cigarette smoking, *N*-acetyltransferase 2 genetic polymorphisms, and breast cancer risk. *JAMA* **13**: 1494–501.

American Heart Association. 1994. *1994 Heart and stroke facts: 1994 statistical supplement.* American Heart Association National Center, Dallas.

American Psychiatric Association. 1994. *Diagnostic and statistical manual of mental disorders*, 4th edn (DSM-IV). APA Press, Washington DC.

American Thoracic Society. 1992. Control of tuberculosis in the United States. *Am. Rev. Resp. Dis.* **146**: 1623–33.

Ammerman, A.J. and Cavalli-Sforza, L.L. 1984. *The neolithic transition and the genetics of populations in Europe.* Princeton University Press.

Anderson, A.B.M., Laurence, K.M., and Turnbull, A.C. 1969. The relationship in anencephaly between the size of the adrenal cortex and the length of gestation. *J. Obstet. Gynaecol. Br. Commonw.* **76**: 196–9.

Anderson, R.M. and May, R.M. 1982. Coevolution of hosts and parasites. *Parasitology* **85**: 411–26.

Anderson, R.A. and May, R.M. 1991. *Infectious diseases of humans: dynamics and control.* Oxford University Press.

Anderson, R.M. and Garnett, G.P. 1996. Low-efficacy HIV vaccines: potential for community-based intervention programmes. *Lancet* **348**: 1010–13.

Anderson, R.M., Gupta, S., and May, R.M. 1991. Potential of community-wide chemotherapy or immunotherapy to control the spread of HIV-1. *Nature* **350**: 356–9.

Anderson, R.M., Swinton, J.S., and Garnett, G.P. 1995. Potential impact of low efficacy HIV-1 vaccines in populations with high rates of infection. *Proc. Roy. Soc. London B* **261**: 147–51.

Andersson, M. 1994. *Sexual selection*. Princeton University Press.

Andreasen, N.C. 1987. Creativity and mental illness: prevalence rates in writers and their first-degree relatives. *Am. J. Psychiatry* **144**: 1288–92.

Andrews, F.W. 1922. Studies on group agglutination. *J. Pathol. Bacteriol.* **25**: 515–21.

Ankel-Simons, F. and Cummins, J.M. 1996. Misconceptions about mitochondria and mammalian fertilization: implications for theories on human evolution. *Proc. Natl. Acad. Sci. USA* **93**: 13859–63.

Anon. 1997. Human monkeypox in Kasai Oriental, Zaire (1996–1997). *Weekly Epidemiol. Rec.* **72**: 101–4.

Antia, R. and the EcLF. 1997. The epidemiology of drug resistance in acute infections (in prep.)

Antia, R., Levin, B.R., and May, R.M. 1994. Within-host population dynamics and the evolution and maintenance of microparasite virulence. *Am. Nat.* **144**: 457–72.

Antonovics, J. 1971. The effects of a heterogeneous environment on the genetics of natural populations. *Am. Sci.* **59**: 593–9.

Apple, R.J., Erlich, H.A., Klitz, W., *et al.* 1994. HLA DR-DQ associations with cervical carcinoma show papillomavirus-type specificity. *Nature Genet.* **6**: 157–62.

Apte, S.V. and Iyengar, L. 1972. Composition of the human foetus. *Br. J. Nutr.* **27**: 305–12.

Apter, D. and Vihko, R. 1982. Early menarche, a risk factor for breast cancer, indicates early onset of ovulatory cycles. *J. Clin. Endocrin. Metab.* **57**: 82–6.

Arbuckle, T.E., Wilkins, R., and Sherman, G.J. 1992. Birth weight percentiles by gestational age in Canada. *Obstet. Gynecol.* **81**: 39–48.

Arcavi, L. and Benowitz, N.L. 1993. Clinical significance of genetic influences on cardiovascular drug metabolism. *Cardiovasc. Drug Ther.* **7**: 311–24.

Arnold, F. 1993. Sex preference and its demographic and health implications. *Int. Fam. Plan. Persp.* **18**: 93–7.

Arnould, J.P.Y., Boyd, I.L., and Socha, D.G. 1996. Milk consumption and growth efficiency in Antarctic fur seal (*Arctocephalus gazella*) pups. *Can. J. Zool.* **74**: 254–66.

Arranz, M.J., Collier, D., Munro, I., *et al.* 1996. Probing therapeutic targets in psychosis by association studies. *Br. J. Clin. Pharmacol.* **42**: 545–50.

Arthur, M. and Courvalin, P. 1993. Genetics and mechanisms of glycopeptide resistance in enterococci. *Antimicrob. Agents Chemother.* **37**: 1563–71.

Åstrand, P.-O. 1988. Whole body metabolism. In E. Horton and R. Terjung (eds) *Exercise, nutrition, and energy metabolism*. MacMillan, New York, pp. 1–8.

Atkinson, J. 1993. Infant vision screening: prediction and prevention of strabismus and amblyopia from refractive screening in the Cambridge photo refraction program. In K. Simons (ed.) *Early visual development, normal and abnormal*. Oxford University Press, New York, pp. 335–48.

Ausman, L.M., Powell, E.M., Mercado, D.L., *et al.* 1982. Growth and developmental body composition of the cebus monkey (*Cebus albifrons*). *Am. J. Primatol.* **3**: 211–27.

Austad, S.N. 1994. Menopause: an evolutionary perspective. *Exp. Gerontol.* **29**: 255–63.

Austin, D.J. and Anderson, R.M. 1996. Immunodominance, competition and evolution in immunological responses to helminth parasite antigens. *Parasitology* **113**: 157–72.

Austin, D.J. and Anderson, R.M. 1998. Mathematical models of the impact of different drug treatment regimes on the dynamics of parasite population growth and the evolution of drug resistant strains. *J. Antimicrob. Ther.* (submitted).

Austin, D.J., Bonten, M.J.M., Slaughter, S., *et al.* 1998*a*. Vancomycin-resistant enterococci (VRE) in intensive care hospital settings: transmission dynamics, persistence and the impact of infection control programs. *Proc. Nat. Acad. Sci.* (in press).

Austin, D.J., Kakehashi, M., and Anderson, R.M. 1997. The transmission dynamics of antibiotic-resistant bacteria: the relationship between resistance in commensal organisms and antibiotic consumption. *Proc. Roy. Soc. London B* **264**: 1629–38.

Austin, D.J., Kristinsson, K.G. and Anderson, R.M. 1998*b*. The relationship between the volume of antimicrobial consumption in human communities and the frequency of resistance. *Proc. Roy. Soc. Ser. B.* (in press).

Austin, D.J., White, N., and Anderson R.M. 1998c. The dynamics of drug action on the within-host population growth of infectious agents: melding pharmacokinetics with pathogen population dynamics. *J. Theor. Biol.* (in press).

Aviram, M. 1996. Interaction of oxidized low density lipoprotein with macrophages in atherosclerosis, and the antiatherogenicity of antioxidants. *Eur. J. Clin. Chem. Clin. Biochem.* **34**: 599–608.

Axelrod, R. and Hamilton, W.D. 1981. The evolution of cooperation. *Science* **211**: 1390–6.

Ayala, F.J., Escalante, A., O'Huigin, C., and Klein, J. 1994. Molecular genetics of speciation and human origins. *Proc. Natl. Acad. Sci. USA* **91**: 6787–94.

Ayala, F.J., Escalante, A., O'Huigin, C., and Klein, J. 1995. Molecular genetics of speciation and human origins. In W.M. Fitch and F.J. Ayala (eds) *Tempo and mode in evolution: genetics and paleontology 50 years after Simpson.* National Academy Press, Washington DC, pp. 187–211.

Bachman, D.L., Wolf, P.A., Linn, R., *et al.* 1992. Prevalence of dementia and probable senile dementia of the Alzheimer type in the Framingham Study. *Neurology* **42**: 115–19.

Badimon, J.J., Fuster, V., Chesebro, J.H., and Badimon, L. 1993. Coronary atherosclerosis a multifactorial disease. *Circulation* **87**: II-3–15.

Baird, D.T. and Michie, E.A. (eds). 1985. *Mechanism of menstrual bleeding.* Raven Press, New York.

Baker, T.G. 1963. A quantitative and cytological study of germ cells in human ovaries. *Proc. Roy. Soc. London B* **158**: 417–33.

Bankier, A.T. *et al.* 1991. The DNA sequence of the human cytomegalovirus genome. *DNA Seq.* **2**: 1–12.

Baquero, F. and Reig, M. 1992. Resistance of anaerobic bacteria to antimicrobial agents in Spain. *Eur. J. Clin. Microbiol. Infect. Dis.* **11**: 1016–30.

Barbour, M. L., Mayon-Whilte, R. T., Coles, C., *et al.* 1995. The impact of conjugate vaccine on carriage of *Haemophilus influenzae* type b. *J. Infect. Dis.* **171**: 93–8.

Barbujani, G. and Pilastro, A. 1993. Genetic evidence on origin and dispersal of human populations speaking languages of the Nostratic macrofamily. *Proc. Natl. Acad. Sci. USA* **90**: 4670–3.

Barbujani, G. and Sokal, R.R. 1991. Zones of sharp genetic change in Europe are also linguistic boundaries. *Proc. Natl. Acad. Sci. USA* **87**: 1816–19.

Barbujani, G., Nasidze, I.N., and Whitehead, G.N. 1994a. Genetic diversity in the Caucasus. *Hum. Biol.* **66**: 639–68.

Barbujani, G., Pilastro, A., De Domenico, S., and Renfrew, C. 1994b. Genetic variation in North Africa and Eurasia: Neolithic demic diffusion versus Paleolithic colonisation. *Am. J. Phys. Anthropol.* **95**: 137–54.

Barbujani, G., Oden, N.L., and Sokal, R.R. 1995. Indo-European origins: A computer-simulation test of five hypotheses. *Am. J. Phys. Anthropol.* **96**: 109–32.

Barbujani, G., Magagni, A., Minch, E., and Cavalli-Sforza, L.L. 1997. An apportionment of human DNA diversity. *Proc. Natl. Acad. Sci. USA* **94**: 4516–19.

Barker, D.J.P. 1989. The rise and fall of western diseases. *Nature* **338**: 371–2.

Barker, D.J.P. 1995. Fetal origins of coronary heart disease. *Br. Med. J.* **311**: 171–4.

Barker, D.J.P, Osmond, C., Winter, P.D., *et al.* 1989. Weight in infancy and death from ischaemic heart disease. *Lancet* **ii**: 577–80.

Barker, D.J.P., Osmond, C., Simmonds, S.J., and Wield, G.A. 1993. The relation of small head circumference and thinness at birth to death from cardiovascular disease in adult life. *Br. Med. J.* **306**: 422–6.

Barker, D.J.P., Martyn, C.N., Osmond, C., and Wield, G.A.. 1995. Abnormal liver growth *in utero* and death from coronary heart disease. *Br. Med. J.* **310**: 703–4.

Barkow, L., Cosmides, L., and Tooby, J. (eds). 1992. *The adapted mind.* Oxford University Press.

Barlow, D.H. 1988. *Anxiety and its disorders.* Guilford, New York.

Bartlett, M. S. 1957. Measles periodicity and community size. *J. Roy. Stat. Soc. A* **120**: 40–70.

Baudoin, M. 1975. Host castration as a parasitic strategy. *Evolution* **29**: 335–52.

Baumberg, S., Young, J.P.W., Wellington, E.M.H., and Saunders, J.R. (eds) 1995. *Population genetics of bacteria.* Cambridge University Press.

Bell, J.I. and Lathrop, G.M. 1996. Multiple loci for multiple sclerosis. *Nature Genet.* **13**: 377–8.

Bellamy, R., Ruwende, C., Corrah, T. *et al.* 1998. Variations in the NRAMP1 gene and susceptibility to tuberculosis in West Africans. *New Engl. J. Med.* **338**: 640–4.

Benjamin, J., Li, L., Patterson, C., *et al.* 1996. Population and familial association between the D4 dopamine receptor gene and measures of novelty seeking. *Nature Genet.* **12**: 81–4.

Bereczkei, T. and Dunbar, R.I.M. 1997. Female-biased reproductive strategies in a Hungarian Gypsy population. *Proc. Roy. Soc. London B* **264**: 17–22.

Berenson, G.S. 1985. Epidemiologic investigations of cardiovascular risk factor variables in childhood —an overview. *Acta Paediatr. Scand. Suppl.* **318**: 7–9.

Berg, P.H., Voit, E.O., and White, R.L. 1996. A pharmacodynamic model for the action of the antibiotic imipenem on *Pseudomonas aeruginosa* populations *in vitro. Bull. Math. Biol.* **58**: 923–38

Berkowitz, F.E. 1995. Antibiotic resistance in bacteria. *Southern Med. J.* **88**: 797–804.

Bertilsson, L., Alm, C., de Las Carreras, C., *et al.* 1989. Debrisoquine hydroxylation polymorphism and personality. *Lancet* I: 555.

Bertilsson, L., Dahl, M.L., Sjoqvist, F., *et al.* 1993. Molecular basis for rational megaprescribing in ultrarapid hydroxylators of debrisoquine. *Lancet* **341**: 63.

Bettinger, R.L. 1991. *Hunter-gatherers. Archaeological and evolutionary theory.* Plenum Press, New York, pp. 83–111.

Betzig, L. 1997. *Human nature: A critical reader.* Oxford University Press.

Betzig, L., Borgerhof Mulder, M., and Turke, P. (eds) 1989. *Human reproductive behaviour,* Cambridge University Press.

Beutler, E. 1994. G6PD deficiency. *Blood* **84**: 3613–36.

Beutler, E. 1996. G6PD: population genetics and clinical manifestations. *Blood Rev.* **10**: 45–52.

Beutler, E., Kuhl, W., Vives-Corrons, J.L., and Prchal, J.T. 1989. Molecular heterogeneity of glucose-6-phosphate dehydrogenase A. *Blood* **74**: 2550–5.

Biraben, J. 1968. Certain demographic characteristics of the plague epidemic in France, 1720–1722. *Daedalus* **97**: 536–45.

Biraben, J.N. 1979. Essai sur l'évolution du nombre des hommes. *Population* **1**: 13–25.

Birkhead, T.R. and Møller, A.P. 1992. Sperm competition in birds: evolutionary causes and consequences. Academic Press, London.

Birkhead, T.R. and Møller, A.P. 1993. Female control of paternity. *TREE* **8**: 100–4.

Birkhead, T.R., Møller, A.P., and Sutherland, W.J. 1993. Why do females make it so difficult for males to fertilize their eggs? *J. Theor. Biol.* **161**: 51–60.

Bisdee, J.T., James, W.P.T., and Shaw, M.A. 1989.

Changes in energy expenditure during the menstrual cycle. *Br. J. Nutr.* **61**: 187–99.

Bishop, J.D.D. 1996 Female control of paternity in the internally fertilizing compound ascidian *Diplosoma listerianum.* I. Autoradiographic investigation of sperm movement in the female reproductive tract. *Proc. Roy. Soc. London B* **263**: 369–76.

Bishop, J.D.D., Jones, C.S., and Noble, L.R. 1996 Female control of paternity in the internally fertilizing compound ascidian *Diplosoma listerianum.* II. Investigation of male mating success using RAPD markers. *Proc. Roy. Soc. London B* **263**: 401–7.

Black, F.L. 1975. Infectious diseases in primitive societies. *Science* **187**: 515–18.

Black, F.L. 1980. Modern isolated pre-agricultural populations as a source of information on prehistoric epidemic patterns. In N.F. Stanley and R.A. Joske (eds) *Changing disease patterns and human behavior.* Academic Press, London, pp. 37–54.

Black, F.L. 1992. Why did they die? *Science* **258**: 1739–41.

Blackwell, C.C. 1989. The role of ABO blood groups and secretor status in host defences. *FEMS Microbiol. Immunol.* **1**: 341–9.

Blass, J.P. 1985. Alzheimer's disease. *Disease a Month* **31**: 1–69.

Block, G., Patterson, B., and Subar, A. 1992. Fruit, vegetables, and cancer prevention: a review of the epidemiological evidence. *Nutr. Cancer* **18**: 1–29.

Bloom, B.R. 1995. A perspective on AIDS vaccines. *Science* **272**: 1888–90.

Bloom, B.R. and Murray, C.J.L. 1992. Tuberculosis —commentary on a reemerging killer. *Science* **257**: 1055–64.

Blower, S.M. and McLean, A.R. 1994. Prophylactic vaccines, risk behaviour change, and the probability of eradicating HIV in San Francisco. *Science* **265**: 1451–4.

Blower, S.M, Small, P.M., and Hopewell, P.C. 1996. Control strategies for tuberculosis epidemics: new models for old problems. *Science* **273**: 497–500.

Bodmer, J., Cambon-Thomsen, A., Hors, J. *et al.* (1997). Anthropology report. Introduction. In D.Charron (ed.) *Genetic diversity of HLA : Functional and medical implications,* vol. II. EDK, Sèvres.

Bogin, B. 1988. *Patterns of human growth.* Cambridge University Press.

Bogin, B. 1994. Adolescence in evolutionary perspective. *Acta Pæd. Suppl.* **406**: 29–35.

Bogin, B. 1996. Human growth and development from an evolutionary perspective. In C.J.K. Henry and S.J. Ulijaszek (eds) *Long-term consequences of early environment: growth, development and the lifespan developmental perspective.* Cambridge University Press, pp. 7–24.

Bonduelle, M., Legein, J., Buysse, A., *et al.* 1996. Prospective follow-up study of 423 children born after intracytoplasmic sperm injection. *Hum. Reprod.* **11**: 1558–64.

Bongaarts, J. 1980. Does malnutrition affect fecundity? A summary of evidence. *Science* **208**: 564–9.

Bongaarts, J. and Potter, R. 1983. Fertility, biology and behaviour: an analysis of the proximate determinants. Academic Press, New York.

Bonhoeffer, S. and Nowak, M.A. 1994. Mutation and the evolution of virulence. *Proc. Roy. Soc. London B* **258**: 133–40.

Bonhoeffer, S., Lenski, R.E., and Ebert, D. 1996. The curse of the pharaoh: the evolution of virulence in pathogens with long living propagules. *Proc. Roy. Soc. London B* **263**: 715–21.

Bonhoeffer, S.M., Lipsitch, M., and Levin, B.R. 1997. Evaluating treatment protocols to prevent antibiotic resistance. *Proc. Natl. Acad. Sci. USA* (in press).

Boone, J.L. 1989. Parental investment, social subordination and population processes among the 15th and 16th Century Portuguese nobility. In Betzig *et al.* (1989), pp 201–20.

Boren, T., Falk, P., Roth, K.A. *et al.* 1993. Attachment of *Helicobacter pylori* to human gastric epithelium mediated by blood group antigens. *Science* **262**: 1892–5.

Borgerhoff Mulder, M. 1987. On cultural and reproductive success, with an example from the Kipsigis. *Am. Anthropol.* **89**: 617–34.

Borgerhoff Mulder, M. 1991. Human behavioral ecology. In Krebs and Davies (3rd edn, 1991), pp. 69–98.

Borman, A.M., Paulous, S., and Clavel, A. 1996. Resistance of human immunodeficiency virus type 1 to protease inhibitors: selection of resistance mutations in the presence and absence of the drug. *J. Gen. Virol.* **77**: 419–26

Borst, P. and Greaves, D.R.. 1987. Programmed gene rearrangements altering gene expression. *Science* **235**: 658–67.

Borst, P. and Rudenko, G. 1994. Antigenic variation in African typanosomes. *Science* **264**: 1872–3.

Bothwell, T.H., Charlton, R.W., and Motulsky, A.G. 1995. Hemochromatosis. In Scriver *et al.* (1995), Ch. 69.

Bowcock, A.M., Kidd, J.K., Mountain, J.L., *et al.* 1991. Drift, admixture and selection in human evolution: A study with DNA polymorphisms. *Proc. Natl. Acad. Sci. USA* **88**: 839–43.

Bowcock, A.M., Ruiz-Linares, A., Tomfohrde, J., *et al.* 1994. High resolution evolutionary trees with polymorphic microsatellites. *Nature* **368**: 455–7.

Bowlby, J. 1969. Attachment theory, separation anxiety, and mourning, vol. 6. Basic Books, New York.

Boyce, A.J. and Mascie-Taylor, C.J.N. (eds) 1996. *Molecular biology and human diversity.* Cambridge University Press.

Boyd, R. and Richerson, P.J. 1985. *Culture and the evolutionary process.* University of Chicago Press.

Bräuer, G., Yokoyama, Y., Falguères, C., and Mbua, E. 1997. Modern human origins backdated. *Nature* **386**: 337–8.

Brauer, M., Bartlett, K., Regaldo-Pineda, J., and Perez-Padilla, R. 1996. Assessment of particulate concentrations from domestic biomass combustion in rural Mexico. *Environ. Sci. Tech.* **30**: 104–9.

Bray, G.A. 1992. Obesity. In King *et al.* (1992), pp. 507–28.

Bremermann, H.J. and Pickering, J. 1983. A game-theoretical model of parasite virulence. *J. Theor. Biol.* **100**: 411–26.

Breslow, J.L. 1988. Apolipoprotein genetic variation and human disease. *Physiol. Rev.* **68**: 85–132.

Bridges, B.A. 1997. Hypermutation under stress. *Nature* **387**: 557–8.

Bridges, P.S. 1994. Vertebral arthritis and physical activities in the prehistoric southeastern United States. *Am. J. Phys. Anthropol.* **93**: 83–93.

Brinton, L.A., Scharier, C., Hoover, R.N., and Fraumeni, J.F. 1988. Menstrual factors and risk of breast cancer. *Cancer Invest.* **6**: 245–54.

Britto, M.R., McKean, H.E., Bruckner, G.G., and Wedlund, P.J. 1991. Polymorphisms in oxidative drug metabolism: relationship to food preference. *Br. J. Clin. Pharm.* **32**: 235–7.

Brockmoller, J., Cascorbi, I., Kerb, R., and Roots, I. 1996. Combined analysis of inherited polymorphisms in arylamine *N*-acetyltransferase 2, glutathione *S*-transferases M1 and T1, microsomal epoxide hydrolase, and cytochrome P450 enzymes as modulators of bladder cancer risk. *Cancer Res.* **56**: 3915–25.

Brooks, A., Lithgow, G.J., and Johnson, T.E. 1994. Mortality rates in a genetically heterogeneous population of *Caenorhabditis elegans*. *Science* **263**: 668–71.

Brosen, K., de Morais, S.M.F., Meyer, U.A., and Goldstein, J.A. 1995. A multifamily study on the relationship between CYP2C19 genotype and S-mephenytoin oxidation phenotype. *Pharmacogenetics* **5**: 312–17.

Brown, A.H.D., Feldman, M.W. and Nevo, E. 1980. Multilocus structure of populations of *Hordeum spontaneum*. *Genetics* **96**: 523–6.

Brown, J.L. 1997. A theory of mate choice based on heterozygosity. *Behav. Ecol.* **8**: 60–5.

Brown, J.L. and Eklund, A. 1994. Kin recognition and the major histocompatibility complex: an integrative review. *Am. Nat.* **143**: 435–61.

Brown, M.S. and Goldstein, J.L. 1983. Lipoprotein metabolism in the macrophage: implications for cholesterol deposition in atherosclerosis. *Annu. Rev. Biochem.* **52**: 223–61.

Bucher, E. and Woosley, R.L. 1992. Clinical implications of variable antiarrhythmic drug metabolism. *Pharmacogenetics* **2**: 2–11.

Bugos, P.E. and McCarthy, L.M. 1984. Aroyeo infanticide: a case study. In G. Hausfater and S.B. Hrdy (eds) *Infanticide: comparative and evolutionary perspectives*. Aldine, New York, pp. 503–20.

Bulatao, R.A. and Stephens, P.W. 1992. Global estimates and projections of mortality by cause 1970–2015. Pre-working paper 1007. Population, Health and Nutrition Department, World Bank, Washington DC.

Bull, J. J. 1994. Virulence. *Evolution* **48**: 1423–37.

Bull, J. J. and Molineux, I.J. 1992. Molecular genetics of adaptation in an experimental model of cooperation. *Evolution* **46**: 882–95.

Bull, J. J., Molineux, I.J., and Rice. W.R. 1991. Selection of benevolence in a host–parasite system. *Evolution* **45**: 875–82.

Bulmer, M. 1994. *Theoretical evolutionary ecology*. Sinauer, Sunderland, Massachusetts.

Burke, D.S., Nisalak, A., Johnson, D.E., and Scott, R.M. 1988. A prospective study of dengue infections in Bangkok. *Am. J. Trop. Med. Hyg.* **38**: 172–80.

Burke, J.R., Wingfield, M.S., Lewis, K.E., *et al.* 1994. The Haw River syndrome: Dentatorubralpallido-luysian atrophy (DRPLA) in a African-American family. *Nature Genet.* **7**: 521–4.

Burke, W. and Motulsky, A.G. 1992. Hypertension. In King *et al.* (1992), pp. 170–91.

Burkitt, D.P. and Eaton, S.B. 1989. Putting the wrong fuel in the tank. *Nutrition* **5**: 189–91.

Burkitt, D.P., Walker, A.R.P., and Palmer, N.S. 1972. Effect of dietary fibre on stools and transit times, and its role in the causation of disease. *Lancet* II: 1408–12.

Butzer, K.W. 1977. Environment, culture, and human evolution. *Am. Sci.* **65**: 572–84.

Cabana, T., Jolicoeur, P., and Michaud, J. 1993. Prenatal and postnatal growth and allometry of stature, head circumference, and brain weight in Québec children. *Am. J. Hum. Biol.* **5**: 93–9.

Cabrera, M., Shaw, M.A., Sharples, C., *et al.* 1995. Polymorphism in tumor necrosis factor genes associated with mucocutaneous leishmaniasis. *J. Exp. Med.* **182**: 1259–64.

Cairns, J., Overbaugh, J., and Miller, S. 1988. The origin of mutants. *Nature* **335**: 142–5.

Calloway, D.H. and Kurzer, M.S. 1982. Menstrual cycle and protein requirements of women. *J. Nutr.* **112**: 356–66.

Campbell, K.L. and Wood, J.W. 1989. Fertility in traditional societies. In P. Diggory, M. Potts, and S. Teper (eds) *Natural human fertility: Social and biological mechanisms*. Macmillan Press, London.

Cann, R.L., Stoneking, M., and Wilson, A.C. 1987. Mitochondrial DNA and human evolution. *Nature* **325**: 31–6.

Cannon, W.B. 1929. Bodily changes in pain, hunger, fear, and rage. Researches into the function of emotional excitement. Harper & Row, New York.

Carey, J.R., Liedo, P., Orzoco, D., and Vaupel, J.W. 1992. Slowing of mortality rates at older ages in large medfly cohorts. *Science* **258**: 457–61.

Carman, W.F., Zanetti, A.R., Karayiannis, P., *et al.* 1990. Vaccine-induced escape mutant of hepatitis B virus. *Lancet* **336**: 325–9.

Castelloe, J. and Templeton, A.R. 1994. Root probabilities for intraspecific gene trees under neutral coalescent theory. *Mol. Phylogenet. Evol.* **3**: 102–13.

Catalano, P.M., Tysbir, E.D., Allen, S.R., *et al.* 1992. Evaluation of fetal growth by estimation of neonatal body composition. *Obstet. Gynecol.* **79**: 46–50.

Caugant, D.A., Mocca, L.F., Frasch, C.E., *et al.* 1987. Genetic structure of *Neisseria meningitidis* populations in relation to serogroup, serotype, and outer membrane protein pattern. *J. Bacteriol.* **168**: 2781–92.

Cavalli-Sforza, L.L. and Bodmer, W. 1971. *The genetics of human populations*. Freeman, San Francisco.

Cavalli-Sforza, L.L., Piazza, A., Menozzi, P., and Mountain, J. 1988. Reconstruction of human evolution: bringing together genetic, archaeologic and linguistic data. *Proc. Natl. Acad. Sci. USA* **85**: 6002–6.

Cavalli-Sforza, L.L., Wilson, A.C., Cantor, C.R., *et al*. 1991. Call for a worldwide survey of human genetic diversity: a vanishing opportunity for the Human Genome Project. *Genomics* **11**: 490–1.

Cavalli-Sforza, L.L., Menozzi, P., and Piazza, A. 1994. *The history and geography of human genes*. Princeton University Press.

Chadwick, D. and Cardew, G. (eds). 1996. *Variation in the human genome*. John Wiley, New York.

Chagnon, N.A. 1988. Life histories, blood revenge, and warfare in a tribal population. *Science* **239**: 985–92.

Chapman, T., Liddle, L.F., Kalb, J.M., *et al*. 1995. Cost of mating in *Drosophila melanogaster* females is mediated by male accessory gland products. *Nature* **373**: 241–4.

Charbonneau, H., Desjardins, B., Guillemette, A., *et al*. 1993. *The first French Canadians. Pioneers in the St Lawrence Valley*. University of Delaware Press, Newark.

Chard, T. 1993. Placental radar. *J. Endocrinol.* **138**: 177–9.

Charlesworth, B. 1980. (2nd edn, 1994) *Evolution in age-structured populations*. Cambridge University Press.

Charlesworth, B. 1996. Evolution of senescence: Alzheimer's disease and evolution. *Curr. Biol.* **6**: 20–2.

Charlesworth, B. and Hughes, K.A. 1996. Age-specific inbreeding depression and components of genetic variance in relation to the evolution of senescence. *Proc. Natl. Acad. Sci. USA* **93**: 6140–5.

Charnov, E.L. 1982. *The theory of sex allocation*. Princeton University Press.

Chen, E.Y., Liao, Y.-C., Smith, D.H., *et al*. 1989. The human growth hormone locus: nucleotide sequence, biology, and evolution. *Genomics* **4**: 479–97.

Chen, S., Chou, W.-H., Blouin, R.A., *et al*. 1996. The cytochrome P450 2D6 (CYP2D6) enzyme polymorphism: screening costs and influence on clinical outcomes in psychiatry. *Clin. Pharmacol. Ther.* **60**: 522–34.

Chen, Y.-S., Torroni, A., Excoffier, L., *et al*. 1995. Analysis of mtDNA variation in African populations reveals the most ancient of all human continent-specific haplogroups. *Am. J. Hum. Genet.* **57**: 133–49.

Cheverud, J.M. and Routman, E.J. 1995. Epistasis and its contribution to genetic variance components. *Genetics* **139**: 1455–61.

Chippindale, A.K., Hoang, D.T., Service, P.M., and Rose, M.R.. 1994. The evolution of development in *Drosophila melanogaster* selected for postponed senescence. *Evolution* **48**: 1880–99.

Chippindale, A.K., Chu, T.J.F., and Rose, M.R. 1996. Complex trade-offs and the evolution of starvation resistance in *Drosophila melanogaster*. *Evolution* **50**: 753–66.

Clapp, J.F. and Capeless, E.L. 1990. Neonatal morphometrics after endurance exercise during pregnancy. *Am. J. Obstet. Gynecol.* **163**: 1805–11.

Clarke, J. 1994. The meaning of menstruation in the elimination of abnormal embryos. *Hum. Reprod.* **9**: 1204–7.

Cleland, J. 1995. Obstacles to fertility decline in developing countries. In R.I.M. Dunbar (ed.) *Human reproductive decisions*. MacMillan, London, pp. 207–29.

Clements, C.J., Strassburg, M., Cutts, F.T., and Torel, C. 1992. The epidemiology of measles. *World Health Stat. Q.* **45**: 285–91.

Clutton-Brock, T.H. 1982. Sons and daughters. *Nature* **298**: 11–13.

Clutton-Brock, T.H. and Parker, G.A. 1992. Potential reproductive rates and the operation of sexual selection. *Q. Rev. Biol.* **67**: 437–56.

Cohan, F.M. 1994. Genetic exchange and evolutionary divergence in prokaryotes. *TREE* **9**: 175–80.

Cohen, J. 1997. Developing prescriptions with a personal touch. *Science* **275**: 776.

Cohen, M.L. 1992. Epidemiology of drug resistance: implications for a post-antimicrobial era. *Science* **257**: 1050–5.

Cohen, M.L. 1994. Antimicrobial resistance: prognosis for public health. *Trends Microbiol.* **2**: 422–55.

Cohen, M.N. 1977. *The food crisis in prehistory*. Yale University Press, New Haven.

Cohen, M.N. 1989. *Health and the rise of civilization*. Yale University Press, New Haven.

Cohen, M.N. and Armelagos, G.J. (eds) 1984. *Paleopathology at the origins of agriculture*. Academic Press, New York.

Cohen, M.N. and Armelagos, G.J. 1984. Editors' summation. In Cohen and Armelagos (1984), pp. 585–601.

Collings D.A., Sithole, S.D., and Martin, K.S. 1990. Indoor woodsmoke pollution causing lower respiratory disease in children. *Trop. Doctor* **20**: 151–55.

Collins, F. and Galas, D. 1993. A new five-year plan for the U.S. Human Genome Project. *Science* **262**: 43–6.

Comfort, A. 1979. *The biology of senescence*, 3rd edn. Churchill Livingstone, Edinburgh.

Comstock, G.W. 1978. Tuberculosis in twins: a re-analysis of the Prophit survey. *Am. Rev. Resp. Dis.* **117**: 621–4.

Comuzzie, A.G., Hixson, J.E., Almasy, L., *et al.* 1997 A major quantitative trait locus determining serum leptin levels and fat mass is located on human chromosome 2. *Nature Genet.* **15**: 273–7.

Contreras, V.T., Araque, W., and Delgado, V.S. 1994. *Trypanosoma cruzi*: metacyclogenesis *in vitro*—I. Changes in the properties of metacyclic trypomastigotes maintained in the laboratory by different methods. *Mem. Inst. Oswaldo Cruz, Rio de Janeiro* **89**: 253–9.

Conway, D.J., Holland, M.J., Bailey, R.L., *et al.* 1997. Scarring trachoma is associated with poly-morphism in the tumor necrosis alpha gene promoter and with elevated TNF levels in tear fluid. *Infect. Immun.* **65**: 1003–6.

Cook, E.W., Hodes, R.L., and Lang, P.J. 1986. Preparedness and phobia: effects of stimulus con-tent on human visceral conditioning. *J. Abnorm. Psychol.* **95**: 195–207.

Cooper, D.M. 1994. Evidence for and mechanisms of exercise modulation of growth. *Med. Sci. Sports Exerc.* **26**: 733–40.

Cordain. L., Gotshall, R.W., and Eaton, S.B. 1997. Evolutionary aspects of exercise. *World Rev. Nutr. Diet* **81**: 49–60.

Coronado, V.G., Beck-Sague, C.M.,.Hutton, M.D., *et al.* 1993. Transmission of multidrug-resistant *Mycobacterium* tuberculosis among persons with human immunodeficiency virus infection in an urban hospital: epidemiologic and restriction fragment length polymorphism analysis. *J. Infect. Dis.* **168**: 1052–5.

Cronk, L. 1991. Human behavioral ecology. *Annu. Rev. Anthropol.* **20**: 25–53.

Cross, G.A.M. 1996. Antigenic variation in trypano-somes: secrets surface slowly. *Bioessays* **18**: 283–91.

Cross-National Collaborative Group 1992. The changing rate of major depression. Cross-national comparisons. *JAMA* **268**: 3098–105.

Crouse, J.R. 1984. Progress in coronary artery disease risk-factors research: What remains to be done? *Clin. Chem.* **30**: 1125–7.

Crow, J.F. and Kimura, M. 1971. *An introduction to population genetics theory*. Harper & Row, New York.

Crow, T.J. 1995. A Darwinian approach to the origins of psychosis. *Br. J. Psych.* **167**: 12–25.

Cumming, A.M. and Robertson, F. 1983. Poly-morphism at the apo E locus in relation to risk of coronary disease. *Clin. Genet.* **25**: 310–13.

Curtsinger, J.W., Fukui, H., Townsend, D., and Vaupel, J. 1992. Failure of the limited-lifespan paradigm in genetically homogeneous popu-lations of *Drosophila melanogaster*. *Science* **258**: 461–3.

Cuthbert, A.W., Halstead, J., Ratcliff, R., *et al.* 1995. The genetic advantage hypothesis in cystic fibrosis heterozygotes: a murine study. *J. Physiol. London* **482**: 449–54.

Dalvit, S.P. 1981. The effect of the menstrual cycle on patterns of food intake. *Am. J. Clin. Nutr.* **34**: 1811–15.

Daly, A.K., Brockmöller, J., Broly, F., *et al.* 1996. Nomenclature for human CYP2D6 alleles. *Pharmacogenetics* **6**: 193–201.

Daly, M. and Wilson, M. 1985. Competitiveness, risk taking, and violence: the young male syndrome. *Ethol. Sociobiol.* **6**: 59–73.

Daly, M. and Wilson, M. 1987. Children as homicide victims. In R.J. Gelles and J.B. Lancaster (eds) *Child abuse and neglect: biosocial dimensions.* Aldine DeGruyter, New York, pp. 201–14.

Daly, M. and Wilson, M. 1988. *Homicide*. Aldine de Gruyter, New York.

Daniels, S.E., Bhattacharrya, S., James, A., *et al.* 1996. A genome-wide search for quantitative trait loci underlying asthma. *Nature* **383**: 247–50.

Dard, P., Schreiber, Y., Excoffier, L., *et al.* 1992. Polymorphisme des loci de classe I: HLA-A, -B, et -C, dans la population Mandenka du Sénégal Oriental. *Comptes-Rendu de l'Académie des Sciences Paris* **314**: 573–8.

Dausset, J., Cann, H., Cohen, D., *et al.* 1990. Centre d'Etude du Polymorphisme Humain (CEPH): collaborative genetic mapping of the human genome. *Genomics* **6**: 575–7.

Davignon, J. 1993. Apolipoprotein E polymorphism and atherosclerosis. In G. Born and C. Schwartz (eds) *New horizons in coronary heart disease.* Current Science, New York. pp. 5.1.–5.22.

Davignon, J., Gregg, R.E., and Sing, C.F. 1988. Apolipoprotein E polymorphism and athero-sclerosis. *Arteriosclerosis* **8**: 1–21.

Davis, D.H.S. *et al.* 1975. Plague. In W.T. Hubbert (ed.) *Diseases transmitted from animals to man*, 6th edn. C.C. Thomas, Springfield, pp. 147–73.

Davis, P.V. and Bradley, J.G. 1996. The meaning of normal. *Perspect. Biol. Med.* **40**: 68–77.

Dawkins, R. 1976. *The selfish gene.* Oxford University Press.

Dawkins, R. 1986. *The blind watchmaker.* Longman, Harlow.

de Andrade, M., Thandi, I., Brown, S., *et al.* 1995. Relationship of the apolipoprotein E polymorphism with carotid artery atherosclerosis. *Am. J. Hum. Genet.* **56**: 1379–90.

de Bruin, N.C., van Velthoven, K.A.M., de Ridder, M., *et al.* 1996. Standards for total body fat and fat-free mass in infants. *Arch. Dis. Child.* **74**: 386–99.

de la Chapelle, A. 1993. Disease gene mapping in isolated human populations: the example of Finland. *J. Med. Genet.* **30**: 857–65.

de Morais, S.M.F., Wilkinson, G.R., Blaisdell, J., *et al.* 1994*a*. Identification of a new genetic defect responsible for the polymorphism of *S*-mephenytoin metabolism in Japanese. *Mol. Pharmacol.* **46**: 594–8.

de Morais, S.M.F., Wilkinson, G.R., Blaisdell, J., *et al.* 1994*b*. The major defect responsible for the polymorphism of *S*-mephenytoin metabolism in humans. *J. Biol. Chem.* **269**: 15419–22.

de Vries, R.R., Fat, R.F., Nijenhuis, L.E., and van-Rood, J.J. 1976. HLA-linked genetic control of host response to *Mycobacterium leprae. Lancet* **ii**: 1328–30.

Dean, M., Carrington, M., Winkler, C., *et al.* 1996. Genetic restriction of HIV-1 infection and progression to AIDS by a deletion allele of the CKR5 structural gene. *Science* **273**: 1856–62.

Dearsly, A.L., Sinden, R.E., and Self, I.A. 1990. Sexual development in malarial parasites: gametocyte production, fertility and infectivity to the mosquito vector. *Parasitology* **100**: 359–68.

Deeb, S., Failor, A., Brown, B.G., *et al.* 1986. Molecular genetics of apolipoproteins and coronary heart disease. *Cold Spring Harbor Symp. Quant. Biol.* **51**: 403–9.

Deitsch, K.W., Moxon, E.R., and Wellems, T.E. 1997. Shared themes of antigenic variation and virulence in bacterial, protozoal and fungal infections. *Microbiol. Mol. Biol. Rev.* **61**: 281–93.

Deka, R., Majumder, P.P., Shriver, M.D., *et al.* 1996. Distribution and evolution of CTG repeats at the myotonin protein kinase gene in human populations. *Genome Res.* **6**: 142–54.

Dekaban, A.S. and Sadowsky, D. 1978. Changes in brain weights during the span of human life: relation of brain weights to body heights and body weights. *Ann. Neurol.* **4**: 345–56.

Dekaban, G.A., Digilio, L., and Franchini, G. 1995. The natural history and evolution of human and simian T cell leukemia/lymphotropic viruses. *Curr. Opin. Genet. Dev.* **5**: 807–13.

Devoto, M. 1991. Origin and diffusion of the major CF mutation in Europe. The European Working Group on CF Genetics (EWGCFG). *Adv. Exp. Med. Biol.* **290**: 63–71.

Di Rienzo, A. and Wilson, A.C. 1991. Branching pattern in the evolutionary tree for human mitochondrial DNA. *Proc. Natl. Acad. Sci. USA* **88**: 1597–601.

Diamond, J.M. 1987. Causes of death before birth. *Nature* **329**: 487–8.

Dickemann, M. 1979. Female infanticide, reproductive strategies and social stratification: a preliminary model. In W. Irons and N. Chagnon (eds) *Evolutionary biology and human social behaviour.* Duxbury, North Scituate, Massachusetts, pp. 321–68.

Dickemann, M. 1981. Paternal confidence and dowry competition: a biocultural analysis of purdah. In R. Alexander and D. Tinkle (eds) *Natural selection and social behavior.* Chiron Press, New York.

Dietz, W.H. and Gortmaker, S.L. 1985. Do we fatten our children at the television set? Obesity and television viewing in children and adolescents. *Pediatrics* **75**: 807–12.

Dockery, D.W., Pope, C.A., Xu, X., *et al.* 1993. An association between air pollution and mortality in six U.S. cities. *New Engl. J. Med.* **329**: 1753–9.

Doherty, P.C. and Zinkernagel, R.M. 1975. Enhanced immunological surveillance in mice heterozygous at the H-2 gene complex. *Nature* **256**: 50–2.

Donnelly, P., Tavaré, S., Balding, D.J., and Griffiths, R.C. 1996. Estimating the age of the common ancestor of men from the ZFY intron. *Science* **272**: 1357–9.

Dorgan, L.T. and Clarke, P.E. 1956. Uterus didelphys with double pregnancy. *Am. J. Obstet. Gynecol.* **72**: 663–6.

Dorit, R. L., Akashi, H., and Gilbert, W. 1995. Absence of polymorphism at the ZFY locus on the human Y chromosome. *Science* **268**: 1183–5.

Dorman, C.J. and Higgins, C.F. 1987. Fimbrial phase variation in *Escherichia coli*: dependence on integration host factor and homologies with other site specific recombinases. *J. Bacteriol.* **169**: 3840–3.

Dorman, J.B., Albinder, B., Shroyer, T., and Kenyon, C. 1995. The age-1 and daf-2 genes function in a common pathway to control the lifespan of *Caenorhabditis elegans*. *Genetics* **141**: 1399–406

Dorman, J.S., Laporte, R.E., Stone, R.A., and Trucco, M. 1990. Worldwide differences in the incidence of type 1 diabetes are associated with amino acid variation at position 57 of the HLA DQB chain. *Proc. Natl. Acad. Sci. USA* **87**: 7370.

Dowson, C.G., Hutchison, A., Brannigan, J.A., *et al.* 1989. Horizontal transfer of penicillin-binding genes in penicillin resistant clinical isolates of *Streptococcus pneumoniae*. *Proc. Natl. Acad. Sci. USA* **86**: 8842–6.

Drake, J.W. 1991. A constant rate of spontaneous mutation in DNA-based microbes. *Proc. Natl. Acad. Sci. USA* **88**: 7160–4.

Drake, J.W. 1993. Rates of spontaneous mutation among RNA viruses. *Proc. Natl. Acad. Sci. USA* **90**: 4171–5.

Du Bose, R.F., Dykhuizen, D.E., and Hartl, D. L. 1988. Genetic exchange between natural isolates of bacteria: recombination within the phoA gene of *Escherichia coli*. *Proc. Natl. Acad. Sci. USA* **85**: 7036–40.

Dugatkin, L.A. 1992. Tendency to inspect predators predicts risk in the guppy (*Poecilia reticulata*). *Behav. Ecol.* **3**: 124–7.

Dunbar, R.I.M. 1991. On sociobiological theory and the Cheyenne case. *Curr. Anthropol.* **32**: 169–73.

Dunbar, R.I.M. and Spoor, M. 1995. Social networks, support cliques and kinship. *Hum. Nature* **6**: 273–90.

Dunbar, R.I.M., Clarke, A., and Hurst, N. 1995. Cooperation and conflict among the Vikings. *Ethol. Sociobiol.* **16**: 233–46.

Durham, W. 1991. *Coevolution: genes, culture and human diversity*. Stanford University Press.

Duvekot, J.J., Cheriex, E.C., Pieters, F.A.A., *et al.* 1993. Early pregnancy changes in hemodynamics and volume homeostasis are consecutive adjustments triggered by a primary fall in systemic vascular tone. *Am. J. Obstet. Gynecol.* **169**: 1382–92.

Dye, C. and Williams, G.G. 1997. Multigenic drug resistance among inbred malaria parasites. *Proc. Roy. Soc. London B* **264**: 61–7.

Dyer, C. 1989. *Standards of living in the later Middle Ages*. Cambridge University Press.

Dyerberg, J., Bang, H.O., and Hjorne, N. 1975. Fatty acid composition of the plasma lipids in Greenland Eskimos. *Am. J. Clin. Nutr.* **28**: 958–66.

Dykhuizen, D.E. and Green, L. 1986. DNA sequence variation, DNA phylogeny and recombination. *Genetics* **113**: 171.

Dykhuizen, D.E. and Green, L. 1991. Recombination in *Escherichia coli* and the definition of biological species. *J. Bacteriol.* **173**: 7257–68.

Dykhuizen, D.E., Polin, D.S., Dunn, J.J., *et al.* 1993. *Borrelia burgdorferi* is clonal: implications for taxonomy and vaccine development. *Proc. Natl. Acad. Sci. USA* **90**: 10163–7.

Eaton, S.B. 1990. Fibre intake in prehistoric times. In A.R. Leeds (ed.) *Dietary fibre perspectives: reviews and bibliography 2*. John Libbey, London, pp. 27–40.

Eaton, S.B. 1992. Humans, lipids, and evolution. *Lipids* **27**: 814–20.

Eaton, S.B. and Cordain, L. 1997. Evolutionary aspects of diet: old genes, new fuels. *World Rev. Nutr. Diet* **81**: 26–37.

Eaton, S.B. and Konner, M. 1985. Paleolithic nutrition. A consideration of its nature and current implications. *New Engl. J. Med.* **312**: 283–9.

Eaton, S.B. and Nelson, D.A. 1991. Calcium in evolutionary perspective. *Am. J. Clin. Nutr.* **54**: 281S–287S.

Eaton, S.B., Konner, M., and Shostak, M. 1988. Stone Agers in the fast lane: chronic degenerative diseases in evolutionary perspective. *Am. J. Med.* **84**: 739–49.

Eaton, S.B., Pike, M.C., Short, R.V., *et al.* 1994. Women's reproductive cancers in evolutionary context. *Q. Rev. Biol.* **69**: 353–67.

Eaton, S.B., Eaton, S.B. III, Konner, M.J., and Shostak, M. 1996. An evolutionary perspective enhances understanding of human nutritional requirements. *J. Nutr.* **126**: 1732–40.

Eaton, S.B., Eaton, S.B. III, and Konner, M.J. 1997. Paleolithic nutrition revisited: a twelve-year retrospective on its nature and implications. *Eur. J. Clin. Nutr.* **51**: 207–16.

Eberhard, W.G. 1996. *Female control: sexual selection by cryptic female choice*. Princeton University Press.

Ebert, D. 1994. Virulence and local adaptation of a horizontally transmitted parasite. *Science* **265**: 1084–6.

Ebert, D. 1995. The ecological interactions between a microsporidian parasite and its host *Daphnia magna*. *J. Anim. Ecol.* **64**: 361–9.

Ebert, D. and Hamilton, W.D. 1996. Sex against virulence: the coevolution of parasitic diseases. *TREE* **11**: 79–82.

Ebert, D. and Herre, E.A. 1996. The evolution of parasitic diseases. *Parasitol. Today* **12**: 96–100.

Ebert, D. and Mangin, K.L. 1997. The influence of host demography on the evolution of virulence of a microsporidian gut parasite. *Evolution* **51**: 1828–37.

Ebert, D. and Weisser, W.W. 1997. Optimal killing for obligate killers: the evolution of life histories and virulence of semelparous parasites. *Proc. Roy. Soc. London B* **264**: 985–91.

Ebert, D., Rainey, P., Embley, T.M., and Scholz, D. 1996. Development, life cycle, ultrastructure and phylogenetic position of *Pasteuria ramosa* Metchnikoff 1888: rediscovery of an obligate endoparasite of *Daphnia magna* Straus. *Phil. Trans. Roy. Soc. London B* **351**: 1689–701.

Ebstein, R.P., Novick, O., Umansky, R., *et al.* 1996. Dopamine D4 receptor (D4DR) exon III polymorphism associated with the human personality trait of novelty seeking. *Nature Genet.* **12**: 78–80.

Edeki, T.I., Goldstein, J.A., de Morais, S.M.F., *et al.* 1996. Genetic polymorphism of *S*-mephenytoin 4'-hydroxylation in African-Americans. *Pharmacogenetics* **6**: 357–60.

Edmond, M.B., Ober, J.F., Dawson, J.D., *et al.* 1996. Vancomycin resistant enterococcal bacteremia: natural history and attribute mortality. *Clin. Infect. Dis.* **23**: 1234–9.

Edwards, C.R.W., Benediktsson, R., Lindsay, R.S., and Seckl, J.R. 1993. Dysfunction of placental glucocorticoid barrier: link between fetal environment and adult hypertension? *Lancet* **341**: 355–7.

Eggen, D.A. and Solberg, L.A. 1968. Variation of atherosclerosis with age. *Lab. Invest.* **18**: 571–9.

Egid, K. and Brown, J.L. 1989. The major histocompatibility complex and female mating preferences in mice. *Anim. Behav.* **38**: 548–50.

Ehret, C. 1973. Patterns of Bantu and Central Sudanic settlement in central and southern Africa. *Trans. J. Hist.* **3**: 1–71.

Eichelbaum, M. and Gross, A.S. 1990. The genetic polymorphism of debrisoquine/sparteine metabolism—clinical aspects. *Pharmacol. Ther.* **46**: 377–94.

Eichner, J.E., Kuller, L.H., Orchard, T.J., *et al.* 1993. The relationship of apolipoprotein E phenotype to myocardial infarction and mortality from coronary artery disease. *Am. J. Cardiol.* **71**: 160–5.

Elena, S.F., Cooper, V.S., and Lenski, R.E. 1996. Punctuated evolution caused by selection of rare beneficial mutations. *Science* **272**: 1802–4.

Ellish, N.J., Saboda, K., O'Connor, J., *et al.* 1996. A prospective study of early pregnancy loss. *Hum. Reprod.* **11**: 406–12.

Ellison, P.T., Panter-Brick, C., Lipson, S.F., and O'Rourke, M.T. 1993. The ecological context of human ovarian function. *Hum. Reprod.* **8**: 2248–58.

Epstein, C.J., Martin, G.M., Schultz, A.L., and Motulsky, A.G. 1966. Werner's syndrome: a review of its sympotmatology, natural history, pathologic features, genetics and relationship to the natural aging process. *Medicine* **45**: 177–221.

Eschenbach, D. 1976. Acute pelvic inflammatory disease: etiology, risk factors and pathogenesis. *Clin. Obst. Gynecol.* **19**: 147–69.

Eshel, I. and Feldman, M.W. 1991. The handicap principle in parent–offspring conflict: comparison of optimality and population genetic analyses. *Am. Nat.* **137**: 167–85.

Essunger, P. and Perelson, A.S. 1994. Modeling HIV infection of CD4+ T-cell supopulations. *J. Theor. Biol.* **170**: 367–91.

Evans, D.A., Funkenstein, H.H., Albert, M.S., *et al.* 1989. Prevalence of Alzheimer's disease in a community population of older persons. Higher than previously reported. *JAMA* **264**: 2551–6.

Evans, D.A.P. 1993. *Genetic factors in drug therapy. Clinical and molecular pharmacogenetics.* Cambridge University Press.

Evans, D.A.P., Krahn, P., and Narayanan, N. 1995. The mephenytoin (cytochrome P450 2C 19) and dextromethorphan (cytochrome P450 2D6) polymorphisms in Saudi Arabians and Filipinos. *Pharmacogenetics* **5**: 64–71.

Ewald, P.W. 1980. Evolutionary biology and the treatment of signs and symptoms of infectious disease. *J. Theor. Biol.* **86**: 169–76.

Ewald, P.W. 1983. Host–parasite relations, vectors, and the evolution of disease severity. *Annu. Rev. Ecol. Syst.* **14**: 465–85.

Ewald, P.W. 1991. Waterborne transmission and the evolution of virulence among gastrointestinal bacteria. *Epidemiol. Infect.* **106**: 83–119.

Ewald, P.W. 1994. *Evolution of infectious diseases.* Oxford University Press.

Ewald, P.W. 1996. Vaccines as evolutionary tools: the virulence-antigen strategy. In S.H.E. Kaufmann (ed.) *Concepts in vaccine development.* de Gruyter and Co., Berlin, pp. 1–25.

Ewbank, J.J., Barnes, T.M., Lakowski, B., *et al.* 1997. Structural and functional conservation of the *Caenorhabditis elegans* timing gene clk-1. *Science* **275**: 980–3.

Excoffier, L. and Langaney, A. 1989. Origin and differentiation of human mitochondrial DNA. *Am. J. Hum. Genet.* **44**: 73–85.

Excoffier, L., Pellegrini, B., Sanchez-Mazas, A., *et al*. 1987. Genetics and history of sub-Saharan Africa. *Yearb. Phys. Anthropol*. **30**: 151–94.

Excoffier, L., Harding R.M., Sokal, R.R., *et al*. 1991. Spatial differentiation of RH and GM haplotype frequencies in sub-Saharan Africa and its relation to linguistic affinities. *Hum. Biol*. **63**: 273–307.

Excoffier, L., Poloni, E.S., Santachiara-Benerecetti, A.S., *et al*. 1996. The molecular diversity of the Niokholo Mandenkalu from Eastern Senegal: an insight into West Africa genetic history. In Boyce and Mascie-Taylor (1996), pp. 141–55.

Faddy, M.J., Gosden, R.G., Gougeon, A., *et al*. 1992. Accelerated disappearance of ovarian follicles in mid-life—implications for forecasting menopause. *Hum. Reprod*. **7**: 1342–6.

Falkner, F. and Tanner, J.M. (eds) 1978. *Human growth. Volume 2. Postnatal growth*. Plenum Press, New York.

Falkow, S. 1975. *Infectious multiple drug resistance*. Pion Press, London.

Fan, W., Kasahara, I., Gutknecht, J., *et al*. 1989. Shared class II MHC polymorphisms between humans and chimpanzees. *Hum. Immun*. **26**: 107–21.

Feder, J.N., Gnirke, A., Thomas, W., *et al*. 1996. A novel MHC class I-like gene is mutated in patients with hereditary haemochromatosis. *Nature Genet*. **13**: 399–408.

Felger, I., Tavul, L., Kabintik, S., *et al*. 1994. *Plasmodium falciparum*: extensive polymorphism in merozoite surface antigen 2 alleles in an area with endemic malaria in Papua New Guinea. *Exp. Parasitol*. **79**: 106–16.

Fenner, F. 1970. The effects of changing social organisation on the infectious diseases of man. In S.V. Boyden (ed.) *The impact of civilization on the biology of man*. University of Toronto Press, pp. 48–76.

Fenner, F. and Ratcliffe, F.N. 1965. *Myxomatosis*. Cambridge University Press.

Ferin, M., Jewelewicz, R., and Warren, M. 1993. *The menstrual cycle: physiology, reproductive disorders, infertility*. Oxford University Press, New York.

Fields, S.J., Livshits, G., Sirotta, L., and Merlob, P. 1996. Path analysis of risk factors leading to premature birth. *Am. J. Hum. Biol*. **8**: 433–43.

Figlerowicz, M., Nagy, P.D., and Bujarski, J.J. 1997. A mutation in the putative RNA polymerase gene inhibits nonhomologous, but not homologous, genetic recombination in an RNA virus. *Proc. Natl. Acad. Sci. USA* **94**: 2073–8.

Figueroa, F., Günther, E., and Klein, J. 1988. MHC polymorphism predating speciation. *Nature* **335**: 265–7.

Finch, C.E. 1990. *Longevity, senescence and the genome*. University of Chicago Press.

Finch, R. and Phillips, I. (eds). 1986. The carrier state. *J. Antimicrobial Chemother*. **18**, Suppl. A.

Fine, P.E.M. 1975. Vectors and vertical transmission: an epidemiologic perspective. *Ann. NY Acad. Sci*. **266**: 173–94.

Finn, C.A. 1987. Why do women and some other primates menstruate? *Perspect. Biol. Med*. **30**: 566–74.

Finn, C.A. 1994. The meaning of menstruation. *Hum. Reprod*. **9**: 1202–3.

Finn, C.A. 1996. Why do women menstruate? Historical and evolutionary review. *Eur. J. Obstet. Gynecol. Reprod. Biol*. **70**: 3–8.

Fischer, C.S. 1982. *To dwell among friends: personal networks in town and city*. University of Chicago Press.

Fishel, R., Lescoe, M.K., Rao, M.R.S., *et al*. 1993. The human mutator gene homolog MSH2 and its association with hereditary non-polyposis: cancer. *Cell* **75**: 1027–38.

Fisher, R.A. 1930. *The genetical theory of natural selection*. Clarendon Press, Oxford. (2nd edn, 1958, Dover.)

Fitch, W.M. 1997. Networks and viral evolution. *J. Mol. Evol*. **44**: S65–S75.

Flannery, K.V. 1973. The origins of agriculture. *Annu. Rev. Anthropol*. **2**: 271–310.

Flatz, G. 1992. Lactase deficiency: biologic and medical aspects of the adult human lactase polymorphism. In King *et al*. (1992), pp. 305–25.

Fleischmann, R.D., Adams M.D., White O., *et al*. 1995. Whole-genome random sequencing and assembly of *Haemophilus influenzae* Rd. *Science* **269**: 496–512.

Flinn, M. and England, B. 1995. Childhood stress and family environment. *Curr. Anthropol*. **36**: 854–66.

Flint, J., Harding, R.M., Boyce, A.J., and Clegg, J.B. 1993*a*. The population genetics of the haemoglobinopathies. *Baillières Clin. Haematol*. **6**: 215–62.

Flint, J., Harding, R.M., Clegg, J.B., and Boyce, A.J. 1993*b*. Why are some genetic diseases common?: Distinguishing selection from other processes by molecular analysis of globin gene variants. *Hum. Genet*. **91**: 91–117.

Flood, D.G. and Coleman, P.D. 1986. Failed compensatory dendritic growth as a pathophysiological process in Alzheimer's disease. *Can. J. Neurol. Sci.* **13**: 475–9.

Foley, R. 1995/1996. The adaptive legacy of human evolution: a search for the environment of evolutionary adaptedness. *Evol. Anthropol.* **4**: 194–203.

Foley, R.A. and Lahr, M.M. 1992. Beyond 'out of Africa.' *J. Hum. Evol.* **22**: 523–9.

Folstad, I. and Karter, A.K. 1992. Parasites, bright males and the immunocompetence handicap. *Am. Nat.* **139**: 603–22.

Fomon, S.J., Haschke, F., Ziegler, E.E., and Nelson, S.E. 1982. Body composition of reference children from birth to age 10 years. *Am. J. Clin. Nutr.* **35**: 1169–75.

Forbes, G.B. 1987. *Human body composition.* Springer-Verlag, New York.

Fornage, M., Amos, C.I., Kardia, S., *et al.* 1997. Variation in the region of the angiotensin-converting enzyme gene influences interindividual differences in blood pressure levels in young Causcasian males. *Circulation* (in press).

Forsen, T., Eriksson, J.G., Tuomilehto, J., *et al.* 1997. Mother's weight in pregnancy and coronary heart disease in a cohort of Finnish men: follow up study. *Br. Med J.* **315**: 837–40.

Forster, P., Harding, R., Torroni A., and Bandelt, H.-J. 1996. Origin and evolution of native American mtDNA variation: a reappraisal. *Am. J. Hum. Genet.* **59**: 935–45.

Foster, P.L. 1993. Adaptive mutation: the uses of adversity. *Annu. Rev. Microbiol.* **47**: 467–504.

Foster, P.L. and Trimarchi, J.M. 1994. Adaptive reversion of a frame-shift mutation in *Escherichia coli* by simple based deletions in homopolymeric runs. *Science* **265**: 407–9.

Frank, S.A. 1996. Models of parasite virulence. *Q. Rev. Biol.* **71**: 37–78.

Franklin-Tong, N.V.E. and Franklin, C.F.C.H. 1993. Gametophytic self-incompatibility: contrasting mechanisms for Nicotiana and Papaver. *Trends Cell Biol.* **3**: 340–5.

Freedman, A.M. 1995. The biopsychosocial paradigm and the future of psychiatry. *Comp. Psychiat.* **36**: 397–406.

Friedman, D.B. and Johnson, T.E. 1988*a*. A mutation in the age-1 gene in *Caenorhabditis elegans* lengthens life and reduces hermaphrodite fertility. *Genetics* **118**: 75–86.

Friedman, D.B. and Johnson, T.E. 1988*b*. Three mutants that extend both mean and maximum lifespan of the nematode *Caenorhabditis elegans* define the age-1 gene. *J. Gerontol.* **43**: 102–9.

Frost, S.D. and McLean, A.R. 1994. Germinal center destruction as a major pathway of HIV pathogenesis. *J. AIDS* **7**: 236–44.

Fu, Y.-X. and Li, W.-H. 1996. Estimating the age of the common ancestor of men from the ZFY intron. *Science* **272**: 1356–7.

Fukuchi, K., Martin, G.M., and Monnat, R.J. Jr. 1989. Mutator phenotype of Werner syndrome is characterized by extensive deletions. *Proc. Natl. Acad. Sci. USA* **86**: 5893–7.

Fukuchi, K., Tananka, K., Kumahara, Y., *et al.* 1990. Increased frequency of 6-thioguanine-resistant peripheral blood lymphocytes in Werner syndrome patients. *Hum. Genet.* **84**: 249–52.

Fullerton, S. M. 1996. Allelic sequence diversity at the human β-globin locus. In Boyce and Mascie-Taylor (1996), pp. 225–41.

Fullerton, S.M., Harding, R.M., Boyce, A.J., and Clegg, J.B. 1994. Molecular and population genetic analysis of allelic sequence diversity at the human β-globin locus. *Proc. Natl. Acad. Sci. USA* **91**: 1805–9.

Futuyma, D. J. 1995. The uses of evolutionary biology. *Science* **267**: 41–2.

Gabriel, S.E., Brigman, K.N., Koller, B.H., *et al.* 1994. Cystic fibrosis heterozygote resistance to cholera toxin in the cystic fibrosis mouse model. *Science* **266**: 107–9.

Gabunia, L. and Vekua, A. 1995. A Plio-Pleistocene hominid from Dmanisi, East Georgia, Caucasus. *Nature* **373**: 509–12.

Gadalla, S., McCarthy, J., and Campbell, O. 1985. How the number of sons influences contraceptive use in Menoufia Governorate, Egypt. *Stud. Fam. Planning* **16**: 164–9.

Galton, F. 1875. The history of twins as a criterion of the relative powers of nature and nurture. *J. Br. Anthropol. Inst.* **5**: 391.

Gao, F., Yue, L., Robertson, D.L., *et al.* 1994. Genetic diversity of human immunodeficiency virus type 2: evidence for distinct sequence subtypes with differences in virus biology. *J. Virol.* **68**: 7433–47.

Garenne, M. and Aaby, P. 1990. Pattern of exposure and measels mortality in Senegal. *J. Infect. Dis.* **161**: 1088–94.

Garrett, E.R. 1978. Kinetics of antimicrobial action. *Scand. J. Infect. Dis. Suppl.* **14**: 54–85.

Garrett, E.R. and Heman-Ackmah, S. 1973. Microbial kinetics and dependencies of individual and combined antibiotic inhibitors of protein synthesis. *Antimicrob. Agents Chemother.* **4**: 574–84.

Garrett, L. 1994. *The coming plague.* Farrar Straus and Giroux, USA.

Gaspar, C., Lopes-Cendes, I., DeStefano, A.L., *et al.*

1996. Linkage disequilibrium analysis in Machado-Joseph disease patients of different ethnic origins. *Hum. Genet.* **98**: 620–4.

Gaulin, S.J.C. and Robbins, C.J. 1991. Trivers-Willard effect in contemporary North American society. *Am. J. Phys. Anthropol.* **85**: 61–9.

Gavan, J.A. 1953. Growth and development of the chimpanzee; a longitudinal and comparative study. *Hum. Biol.* **25**: 93–143.

Gaynes, R. 1995. Surveillance of antibiotic resistance; learning to live with bias. *Infect. Control Hosp. Epidemiol.* **18**: 623–8.

Gelles, R.J. and Lancaster, J.B. (eds) 1987. *Child abuse and neglect.* Aldine, New York.

Gemmill, A.W., Viney, M. E., and Read, A.F. 1997. Host immune status determines sexuality in a parasitic nematode. *Evolution* **51**: 393–401.

Genco, C.A. and Desai, P.J. 1996. Iron acquisition in the pathogenic bacteria. *TIB* **4**: 179–84.

Gentry, L.O. 1991. Bacterial resistance. *Orthop. Clin. North Am.* **22**: 379–88.

Gerber, A.U., Craig, W.A., Brugger, H.P., *et al.* 1983. Impact of dosing intervals of gentamicin and ticarcillin against *Pseudomonas aeniginosa* in granulocytopic mice. *J. Infect. Dis.* **147**: 910–17

Gerdes, L.U., Klausen, I.C., Sihm, I., and Faergeman, O. 1992. Apolipoprotein E polymorphism in a Danish population compared to findings in 45 other study populations around the world. *Genet. Epidemiol.* **9**: 155–67.

Gibbs, M.J. 1995. The Luteovirus supergroup: rampant recombination and persistent partnerships. In A. Gibbs, C.H. Calisher, and F. García Arenal (eds) *Molecular basis of virus evolution.* Cambridge University Press, pp. 351–68.

Gilbert, A.N., Yamazaki, K., Beauchamp, G.K., and Thomas, L. 1986. Olfactory discrimination of mouse strains (*Mus musculus*) and major histocompatibility types by humans (*Homo sapiens*). *J. Comp. Psychol.* **100**: 262–5.

Gilbert, S.C., Plebanski, M., Gupta, S. *et al.* 1998. Association of malaria parasite population structure, HLA, and immunological antagonism. *Science* **279**: 1173–7.

Gill, P. and Evett, I. 1995. Population genetics of short tandem repeat (STR) loci. *Genetica* **96**: 69–87.

Gillespie, J.H. 1975. Natural selection for resistance to epidemics. *Ecology* **56**: 493–5.

Gillman, T., Naidoo, S.S., and Hathorn, M. 1957. Fat, fibrinolysis and atherosclerosis in Africans. *Lancet* **ii**: 696–7.

Gimbutas, M. 1979. Three waves of Kurgan people into old Europe, 4200–2500 B.C. *Arch. Suisses Anthropol. Gen.* **43**: 113–37.

Godfray, H.C.J. 1995. Evolutionary theory of parent–offspring conflict. *Nature* **376**: 133–8.

Godfrey, K., Robinson, S., Barker, D.J.P., Osmond, C., and Cox, V. 1996. Maternal nutrition in early and late pregnancy in relation to placental and fetal growth. *Br. Med. J.* **312**: 410–4.

Goldman, A., Krause, A., Ramsay, M., and Jenkins, T. 1996. Founder effect and the prevalence of myotonic dystrophy in South Africans: molecular studies. *Am. J. Hum. Genet.* **59**: 445–52.

Goldrick, R.B. and Whyte, H.M. 1959. A study of blood clotting and serum lipids in natives of New Guinea and Australians. *Aust. Ann. Med.* **8**: 238–44.

Goldsmith, T.H. 1990. Optimization, constraint, and history in the evolution of eyes. *Q. Rev. Biol.* **65**: 281–322.

Goldstein, D.B., Zhivotovsky, L.A., Nayar, K., *et al.* 1996. Statistical properties of the variation at linked microsatellite loci: implications for the history of human Y chromosomes. *Mol. Biol. Evol.* **13**: 1213–18.

Goldstein, J., Hobbs, H., and Brown, M. 1995. Familial hypercholesterolemia. In C.R. Scriver, A.L. Bendet, and W.S. Sly (eds) *The metabolic and molecular bases of inherited disease*, 7th edn. McGraw-Hill, New York, pp. 1981–2030.

Goldstein, J.A. and de Morais, S.M.F. 1994. Biochemistry and molecular biology of the human CYP2C subfamily. *Pharmacogenetics* **4**: 285–99.

Gompertz, B. 1825. On the nature and function expressive of the law of human mortality and on a new mode of determining life contingencies. *Phil. Trans. Roy. Soc. London* **115**: 513–85.

Gonzalez, F.J. and Idle, J.R. 1994. Pharmacogenetic phenotyping and genotyping. Present status and future potential. *Clin. Pharmacokinet.* **26**: 59–70.

Gonzalez, F.J. and Nebert, D.W. 1990. Evolution of the P450 gene superfamily. *Trends Genet.* **6**: 182–6.

Goodman, A., Martin, D., Armelagos, G., and Clark, G. 1984. Indications of stress from bones and teeth. In Cohen and Armegalos (1984a), pp. 13–39.

Goodwin, F.K. and Jamison, K.R. 1990. *Manic-depressive illness.* Oxford University Press.

Gosden, R.G. 1985. *Biology of menopause: the causes and consequences of ovarian aging.* Academic Press, London, pp. 188.

Gosden, R.G. and Faddy, M.J. 1994. Ovarian aging, follicular depletion and steroidogenesis. *Exp. Geront.* **29**: 265–74.

Gosden, R.G. and Telfer, E. 1987. Numbers of follicles in mammalian ovaries and their allometric relationships. *J. Zool.* **211**: 169–75.

Goss, D.A. 1994. Effect of spectacle correction on the progression of myopia in children—a literature review. *J. Am. Optom. Assoc.* **65**: 117–28.

Grafen, A. 1990. Biological signals as handicaps. *J. Theor. Biol.* **144**: 517–46.

Grahn, M., Langefors, A., and von Schantz, T. 199?. The importance of mate choice in improving viability in captive populations. In T. Caro (ed.) *Behavioral ecology and conservation biology.* Oxford University Press.

Grant, P. 1986. *Ecology and evolution of Darwin's finches.* Princeton University Press.

Graven, L., Passarino, G., Semino, O., *et al.* 1995. Evolutionary correlation between control region sequence and RFLP diversity pattern in the mitochondrial genome of a Senegalese sample. *Mol. Biol. Evol.* **12**: 334–45.

Gray, G.E., Pike, M.C., and Henderson, B.E. 1979. Breast-cancer incidence and mortality rates in different countries in relation to known risk factors and dietary practices. *Br. J. Cancer* **39**: 1–7.

Greenberg, J.H. 1963. *The languages of Africa.* Mouton, The Hague.

Greene, M.H. 1997. Genetics of breast cancer. *Mayo Clinic Proc.* **72**: 54–65.

Groop, L.C., Kankuri, M., Sehalin-Jantti, C., *et al.* 1993. Association between polymorphism of the glycogen synthase gene and non-insulin-dependent diabetes mellitus. *N. Engl. J. Med.* **328**: 10–14.

Grossman, C.J. 1985. Interactions between the gonadal steroids and the immune system. *Science* **227**: 257–61.

Groube, L. 1996. The impact of diseases on the emergence of agriculture. In D.R. Harris (ed.) *The origins and spread of agriculture and pastoralism in Eurasia.* Smithsonian Institution Press, Washington DC, pp. 101–29.

Groves, C.P. and Lahr, M.M. 1994. A bush not a ladder: speciation and replacement in human evolution. *Perspect. Hum. Biol.* **4**: 1–11.

Grundschober, C., Sanchez-Mazas, A., Excoffier, L., *et al.* 1994. HLA-DPB1 DNA polymorphism in the Swiss population: linkage disequilibrium with other HLA loci and population genetic affinities. *Eur. J. Immunogenet.* **21**: 143–57.

Guastello, S.J. Alcohol and drug use and involvement in automobile accidents. *J. Psychol.* **121**: 335–40.

Gupta, S., Ferguson, N.M., and Anderson, R.M. 1997. Vaccination and the population structure of antigenically diverse pathogens that exchange genetic material. *Proc. Roy. Soc. Lond. B* **264**: 1435–43.

Gupta, S. and Hill, A.V.S. 1995. Dynamic interactions in malaria: host heterogeneity meets parasite polymorphism. *Proc. Roy. Soc. London B.* **261**: 271–7.

Gupta, S., Hill, A.V.S., Kwiatkowski, D., *et al.* 1994*a*. Parasite virulence and disease patterns in *P. falciparum* malaria. *Proc. Natl. Acad. Sci. USA* **91**: 3715–19.

Gupta, S., Swinton, J., and Anderson, R.M. 1994*b*. Theoretical studies of the effects of heterogeneity in the parasite population on the transmission dynamics of malaria. *Proc. Roy. Soc. London B* **256**: 231.

Gupta, S., Trenholme, K., Anderson, R.M., and Day, K.P. 1994*c*. Antigenic diversity and the transmission dynamics of *Plasmodium falciparum*. *Science* **263**: 961–3.

Gupta, S., Maiden, M., Feavers, I., *et al.* 1996. The maintenance of strain structure in populations of recombining infectious agents. *Nature Medicine* **2**: 437–42.

Gusella, J.F. and MacDonald, M.E. 1996. Trinucleotide instability: a repeating theme in human inherited disorders. *Ann. Rev. Med.* **47**: 201–9.

Gustaffson, L., Qvarnström, A., and Sheldon, B.C. 1995. Trade-offs between life-history traits and a secondary sexual character in male collared flycatchers. *Nature* **375**: 311–13.

Gutierrez-Lopez, M.D., Bertera, S., Chantres, M.T., *et al.* 1992. Susceptibility to type 1 (insulin-dependent) diabetes mellitus in Spanish patients correlates quantitatively with expression of HLA-DQ alpha Arg 52 and HLA-DQ beta non-Asp 57 alleles. *Diabetologia* **35**: 583–5.

Guttman, D.S. and Dykhuizen, D.E.. 1994. Clonal divergence in *Escherichia coli* as a result of recombination, not mutation. *Science* **266**: 1380–3.

Hafner, M.S., Sudman, P.S., Villablanca, F.X., Spradling, F.A., Demastes, J.W., and Nadler, S.A. 1994. Disparate rates of molecular evolution in cospeciating hosts and parasites. *Science* **265**: 1087–90.

Hagelberg, E., Quevedo, W., Turbon, D., and Clegg, J.B. 1994. DNA from ancient Easter Islanders. *Nature* **369**: 25–6.

Haig, D. 1987. Kin conflict in seed plants. *TREE* **2**: 337–40.

Haig, D. 1990. Brood reduction and optimal parental investment when offspring differ in quality. *Am. Nat.* **136**: 550–6.

Haig, D. 1993. Genetic conflicts in human pregnancy. *Q. Rev. Biol.* **68**: 495–532.

Haig, D. 1996*a*. Gestational drive and the green-bearded placenta. *Proc. Natl. Acad. Sci. USA* **93**: 6547–51.

Haig, D. 1996*b*. Placental hormones, genomic imprinting, and maternal–fetal communication. *J. Evol. Biol.* **9**: 357–80.

Haig, D. 1996*c*. Altercation of generations: genetic conflicts of pregnancy. *Am. J. Reprod. Immunol.* **35**: 226–32.

Haig, D. and Graham, C. 1991. Genomic imprinting and the strange case of the insulin-like growth factor-II receptor. *Cell* **64**: 1045–6.

Haig, D. and Westoby, M. 1989. Parent-specific gene expression and the triploid endosperm. *Am. Nat.* **134**: 147–55.

Haldane, J.B.S. 1932*a*. The causes of evolution. Longmans, Green, & Co., London.

Haldane, J.B.S. 1932*b*. The time of action of genes, and its bearing on some evolutionary problems. *Am. Nat.* **66**: 5–24.

Hales, C.N. and Barker, D.J.P. 1992. Type 2 (non-insulin-dependent) diabetes mellitus: the thrifty phenotype hypothesis. *Diabetologia* **35**: 595–601.

Hales, C.N., Barker, D.J.P., Clark, P.M.S., *et al.* 1991. Fetal and infant growth and impaired glucose tolerance at age 64. *Br. Med. J.* **303**: 1019–22.

Hales, C.N., Desai, M., Ozanne, S.E., and Crowther, N.J. 1996. Fishing in the stream of diabetes: from measuring insulin to the control of fetal organogenesis. *Biochem. Soc. Trans.* **24**: 341–50.

Hallman, D.M., Boerwinkle, E., Saha, N., *et al.* 1991. The apolipoprotein E polymorphism: a comparison of allele frequencies and effects in nine populations. *Am. J. Hum. Genet.* **49**: 338–49.

Halstead, S.B. 1988. Pathogenesis of dengue: challenges to molecular biology. *Science* **239**: 476–81.

Hamilton, W.D. 1964. The genetical evolution of social behavior I and II. *J. Theor. Biol.* **7**: 1–52.

Hamilton, W.D. 1966. The moulding of senescence by natural selection. *J. Theor. Biol.* **12**: 12–45.

Hamilton, W.D. and Zuk, M. 1982. Heritable true fitness and bright birds: a role for parasites? *Science* **218**: 384–7.

Hamilton, W.D., Axelrod, R., and Tanese, R. 1990. Sexual reproduction as an adaption to resist parasites (a review). *Proc. Natl. Acad. Sci. USA* **87**: 3566–73.

Hammer, M., Spurdle, A., Karafet, T., *et al.* 1997. The geographic distribution of human Y chromosome variation. *Genetics* **145**: 787–805.

Hammer, M.F. 1995. A recent common ancestry for human Y chromosomes. *Nature* **378**: 376–8.

Handyside, A.H., Kontogianni, E.H., Hardy, K., and Winston, R.M.L. 1990. Pregnancies from biopsied human preimplantation embryos sexed by Y-specific DNA amplification. *Nature* **344**: 768–70.

Hanis, C.L., Hewett-Emmett, D., Douglas, T.C., *et al.* 1991. Effects of the apolipoprotein E polymorphism on levels of lipids, lipoproteins and apolipoprotein among Mexican-Americans in Starr Country, Texas. *Arterioscler. Thromb.* **11**: 362–70.

Haraguchi, Y. and Sasaki, A. 1996. Host–parasite arms race in mutation modifications: indefinite escalation despite a heavy load. *J. Theor. Biol.* **183**: 121–37.

Harder, J.D., Stonerook, M.J., and Pondy, J. 1993. Gestation and placentation in two New World opossums: *Didelphis virginiana* and *Monodelphis domestica*. *J. Exp. Zool.* **266**: 463–79.

Hardin, G. 1968. The tragedy of the commons. *Science* **162**: 1243–8.

Harding, R., Fullerton, S., Griffiths, R., *et al.* 1997. Archaic African and Asian lineages in the genetic ancestry of modern humans. *Am. J. Hum. Genet.* **60**: 772–89.

Hardy, J. 1996*a*. Molecular genetics of Alzheimer's disease. *Acta Neurol. Scand.* **93**: 13–17.

Hardy, J. 1996*b*. New insights into the genetics of Alzheimer's disease. *Ann. Med.* **28**: 255–8.

Harlan, J.R. 1971. Agricultural origins: centers and non-centers. *Science* **174**: 468–74.

Harpending, H., Sherry, S.T., Rogers, A., and Stoneking, M. 1993. The genetic structure of ancient human populations. *Curr. Anthropol.* **34**: 483–96.

Harrison, D.E. and Archer, J.R. 1988. Natural selection for extended longevity from food restriction. *Growth Dev. Aging* **52**: 65.

Harsanyi, J.C. 1953. Cardinal utility in welfare economics and in the theory of risk-taking. *J. Polit. Econ.* **61**: 434–5.

Hartung, J. 1982. Polygyny and the inheritance of wealth. *Curr. Anthropol.* **23**: 1–12.

Harvey, P.H. and Pagel, M.D. 1991. *The comparative method in evolutionary biology.* Oxford University Press.

Harvey, P.H., Leigh Brown, A.J., Maynard Smith, J., and Nee, S. (eds). 1996. *New uses for new phylogenies.* Oxford University Press.

Harvey, P.H., Holmes, E.C., Mooers, A.Ø., and Nee, S. 1994. Inferring evolutionary processes from molecular phylogenies. In R.W. Scotland,

D.J. Siebert, and D.M. Williams (eds) *Models in phylogeny reconstruction. Systematics Association Special Volume Series* **52**: 313–33.

Hassold, T., Hunt, P.A., and Sherman, S. 1993. Trisomy in humans: incidence, origin and etiology. *Curr. Opin. Genet. Dev.* **3**: 398–403.

Hassold, T.J. 1986. Chromosome abnormalities in human reproductive wastage. *Trends Genet.* **2**: 105–10.

Hastbacka, J., de la Chapelle, A., Kaitila, I., *et al.* 1992. Linkage disequilibrium mapping in isolated founder populations: diastrophic dysplasia in Finland. *Nature Genet.* **2**: 204–11.

Hastings, I.M. 1997. A model for the origin and spread of drug-resistant malaria. *Parasitology* **115**: 133–44.

Hastings, I.M. and Wedgwood-Oppenheim, B. 1997. Sex, strains and virulence. *Parasitol. Today* **13**: 375–83.

Haubold, B. and Rainey, P.B. 1996. Genetic and ecotypic structure of a fluorescent pseudomonas population. *Mol. Ecol.* **5**: 747–61.

Haviland, M.B., Kessling, A.M., Davignon, J., and Sing, C.F. 1995. Cladistic analysis of the apolipoprotein AI-CIII-AIV gene cluster using a health French Canadian sample. I. Haploid analysis. *Ann. Hum. Genet.* **59**: 211–31.

Haviland, M.B., Ferrell, R.E., and Sing, C.F. 1997. Association between common alleles of the low-density lipoprotein receptor gene region and interindividual variation in plasma lipid and apolipoprotein levels in a population-based sample from Rochester, Minnesota. *Hum. Genet.* **99**: 108–14.

Hawkes, K., O'Connell, J.F., and Rogers, L. 1997. The behavioral ecology of modern hunter–gatherers, and human evolution. *TREE* **12**: 29–32.

Hazegawa, M., DiRienzo, A., Kocher, T.D., and Wilson, A.C. 1993. Toward a more accurate time scale for the human mitochondrial DNA tree. *J. Mol. Evol.* **37**: 347–54.

Hedrick, P.W. and Thomson, G. 1983. Evidence for balancing selection at HLA. *Genetics* **104**: 449–56.

Heim, M.H. and Meyer, U.A. 1992. Evolution of a highly polymorphic human cytochrome P450 gene cluster: CYP2D6. *Genomics* **14**: 49–58.

Heini, A.F., Mingholli, G., Diaz, E., *et al.* 1995. Free-living energy expenditure assessed by two different methods in lean rural Gambian farmers. *Am. J. Clin. Nutr.* **61**: 893 (abstract).

Henderson, B.E., Ross, R.K., Judd, H.L., *et al.* 1985. Do regular ovulatory cycles increase breast cancer risk? *Cancer* **56**: 1206–8.

Henderson, B.E., Ross, R.K., and Pike, M.C. 1993. Hormonal chemoprevention of cancer in women. *Science* **259**: 633–8.

Henry, L. 1961. Some data on natural fertility. *Eugen. Q.* **8**: 81–91.

Herre, E.A. 1993. Population structure and the evolution of virulence in nematode parasites of fig wasps. *Science* **259**: 1442–5.

Herre, E.A. 1995. Factors affecting the evolution of virulence: nematode parasites of fig wasps as a case study. *Parasitology* **111**: S179–S191.

Hetzel, C. and Anderson, R.M. 1996. The within-host cellular dynamics of bloodstage malaria: theoretical and experimental studies. *Parasitology* **113**: 25–38.

Hewlett, E.L. 1990. Bordetella species. In G.L. Mandell, R.G. Douglas, and J.E. Bennett (eds) *Principles and practice of infectious disease.* Wiley, New York, pp. 1757–62.

Higgins, C.F. and Dorman, C.J. 1990. DNA super-coiling: a role in the regulation of gene expression and bacterial virulence. In R. Rappuoli *et al.* (eds) *Bacterial protein toxins.* Gustav Fisher, New York, pp. 293–301.

Higgins, L.G. 1954. Prolonged pregnancy (partus serotinus). *Lancet* **267**: 1154–6.

Higgins, M.W. and Luepker, R.V. 1989. Trends and determinants of coronary heart disease mortality: internation comparisons. *Int. J. Epidemiol.* **18** (Suppl.1): 3.

High, N.J., Deadman, M.E., and Moxon, E.R. 1993. The role of a repetitive DNA motif ('5-CAAT-3') in the variable expression of the *Haemophilus influenzae* lipopolysaccharide epitope Gal (1–4) Gal. *Mol. Microbiol.* **9**: 1275–82.

High, N.J., Jennings M.P., and Moxon E.R. 1996. Tandem repeats of the tetramer 5′-CAAT-3′ present in lic2A are required for phase-variation but not lipopolysaccharide biosynthesis in *Haemophilus influenzae. Mol. Microbiol.* **20**: 165–74.

Higuchi, S., Matsushita, S., Murayama, M., *et al.* 1995. Alcohol and aldehyde dehydrogenase polymorphisms and the risk of alcoholism. *Am. J. Psychiatry* **152**: 1219–21.

Hill, A.V.S. 1992. Malaria resistance genes: a natural selection. *Trans. Roy. Soc. Trop. Med. Hyg.* **86**: 225–6.

Hill, A.V.S. 1996a. Genetic susceptibility to malaria and other infectious diseases: from the MHC to the whole genome. *Parasitology* **112** (Suppl.): S75–84.

Hill, A.V.S. 1996b. Genetics of infectious disease resistance. *Curr. Opin. Genet. Dev.* **6**: 348–53.

Hill, A.V.S. 1996c. HIV and HLA: confusion or complexity? *Nature Med.* **2**: 395–6.

Hill, A.V.S. 1997. MHC polymorphism and susceptibility to intracellular pathogens in humans. In Kaufmann, S.E. (ed.) *Host response to intracellular pathogens.* R.G. Landes, Austin, pp. 47–59.

Hill, A.V.S. 1998. Immunogenetics of human infectious diseases. *Annu. Rev. Immun.* **16**: 593–617.

Hill, A.V.S., Allsopp, C.E.M., Kwiatowski, D., *et al.* 1991. Common West African HLA antigens are associated with protection from severe malaria. *Nature* **352**: 595–600.

Hill, A.V.S., Elvin, J., Willis, A.C., *et al.* 1992. Molecular analysis of the association of HLA-B53 and resistance to severe malaria. *Nature* **360**: 434–9.

Hill, A.V.S., O'Shaughnessy, D.F., and Clegg, J.B. 1989. Haemoglobin and globin gene variants in the Pacific. In A.V.S Hill and S.W. Serjeantson (eds) *The colonization of the Pacific: a genetic trail.* Clarendon, Oxford. pp. 246–85.

Hill, K. and Hurtado, A.M. 1991. The evolution of premature reproductive senescence and menopause in human females, an evaluation of the 'grandmother hypothesis'. *Hum. Nature* **2**: 313–50.

Hill, K. and Hurtado, M. 1996. Ache life history: the ecology and demography of a foraging people. Aldine de Gruyter, New York.

Hill, K. and Kaplan, H. 1988. Tradeoffs in male and female reproductive strategies among the Ache: part 2. In Betzig *et al.* (1989), pp. 291–306.

Hillis, D.M. 1996. Inferring complex phylogenies. *Nature* **383**: 130–1.

Hirayama, K., Matsushita, S., Kikuchi, I., Iuchi, M., Ohta, N., and Sasazuki, T. 1987. HLA-DQ is epistatic to HLA-DR in controlling the immune response to schistomal antigen in humans. *Nature* **327**: 426–30.

Hirsch, V.M., Dapolito, G., Goeken, R., and Campbell, B.J. 1995. Phylogeny and natural history of the primate lentiviruses, SIV and HIV. *Curr. Opin. Genet. Dev.* **5**: 798–806.

Ho, D.D., Neuman, A.U., Perelson, A.S., *et al.* 1995. Rapid turnover of plasma virions and CD4 lymphocytes in HIV-1 infection. *Nature* **373**: 123–6.

Hobbs, H.H., Russell, D.W., Brown, M.S., and Goldstein, J.L. 1990. The LDL receptor locus in familial hypercholesterolemia: mutational analysis of a membrane protein. *Annu. Rev. Genet.* **24**: 133–70.

Hobcraft, J., McDonald, J.W., and Rutstein, S. 1983. Child-spacing effects on infant and early child mortality. *Pop. Index* **49**: 585–618.

Hochberg, M.E. 1991. Intra-host interactions between a braconid endoparasitoid, *Apanteles glomeratus,* and a baculovirus for larvae of *Pieris brassicae. J. Anim. Ecol.* **60**: 51–63.

Hoehn, H., Bryant, E.M., Au, K., *et al.* 1975. Variegated translocation mosaicism in human skin fibroblast cultures. *Cytogenet. Cell Genet.* **15**: 282–98.

Holguin, G. and Bashan, Y. 1992. Increased aggressiveness of *Alternaria macropora,* a causal agent of leaf-blight in cotton monoculture. *Can. J. Bot.* **70**: 1878–84.

Holliday, M.A. 1978. Body composition and energy needs during growth. In Falkner and Tanner (1978), pp. 117–39.

Holliday, R. 1989. Food, reproduction and longevity: is the extended life span of calorie-restricted animals an evolutionary adaptation? *BioEssays* **10**: 125–7.

Holliday, R. 1995. *Understanding ageing.* Cambridge University Press.

Holmes, E.C., Zhang, L.Q., Simmonds, P., *et al.* 1992. Convergent and divergent sequence evolution in the surface envelope glycoprotein of human immunodeficiency virus type 1 within a single infected patient. *Proc. Natl. Acad. Sci. USA* **89**: 4835–9.

Holmes, E.C., Gould, E.A., and Zanotto, P.M. de A. 1996. An RNA virus tree of life? In D.M. Roberts, P. Sharp, G. Alderson, and M.A. Collins (eds) *Evolution of microbial life. Soc. Gen. Microbiol. Symp.* **54**, Cambridge University Press, pp. 127–44.

Holmes, E.C., Nee, S., Rambaut, A., *et al.* 1995a. Revealing the history of infectious disease epidemics using phylogenetic trees. *Phil. Trans. Roy. Soc. London B* **349**: 33–40.

Holmes, E.C., Zhang, L.Q., Robertson, P., *et al.* 1995b. The molecular epidemiology of human immunodeficiency virus type 1 in Edinburgh, Scotland. *J. Infec. Dis.* **171**: 45–53.

Holmes, J.C. 1982. Impact of infectious disease agents on the population growth and geographical distribution of animals. In R.M. Anderson and R.M. May (eds) *Population biology of infectious diseases.* Springer, New York, pp. 37–51.

Holmes, J.C. and Bethel, W.M. 1972. Modification of intermediate host behaviour by parasites. *Zool. J. Linn. Soc.* Suppl. 1: 123–49.

Hood, D.W., Deadman, M.E., Jennings, M.P. *et al.* 1996a. DNA repeats identify novel virulence genes in *Haemophilus influenzae*. *Proc. Natl. Acad. Sci. USA* **93**: 11121–5.

Hood, D.W., Deadman, M.E., Allen, T., *et al.* 1996*b*. Use of the complete genome sequence information of *Haemophilus influenzae* strain Rd to investigate lipopolysaccharide biosynthesis. *Mol. Microbiol.* **22**: 951–65.

Hook-Nikanne, J., Sistonen, P., and Kosunen, T.U. 1990. Effect of ABO blood group and secretor status on the frequency of *Helicobacter pylori* antibodies. *Scand. J. Gastroenterol.* **25**: 815–18.

Horai, S., Hayasaka, K., Kondo, R., *et al.* 1995. Recent African origin of modern humans revealed by complete sequences of hominoid mitochondrial DNAs. *Proc. Natl. Acad. Sci. USA* **92**: 532–6.

Horuk, R., Chitnis, C.E., Darbonne, W.C., *et al.* 1993. A receptor for the malarial parasite *Plasmodium vivax*: the erythrocyte chemokine receptor. *Science* **261**: 1182–4.

Howard, R.S. and Lively, C.M. 1994. Parasitism, mutation accumulation and the maintenance of sex. *Nature* **367**: 554–7.

Howell, N., Kubacha, I., and Mackey, D.A. 1996. How rapidly does the human mitochondrial genome evolve? *Am. J. Hum. Genet.* **59**: 501–9.

Howells, W.W. 1992. The dispersion of modern humans. In S. Jones, R. Martin, and D. Pilbeam (eds) *The Cambridge encyclopaedia of human evolution.* Cambridge University Press, pp. 389–401.

Hrdy, S.B. 1979. Infanticide among animals: a review, classification and examination of the implications for the reproductive strategies of female. *Ethol. Sociobio.* **1**: 13–40.

Hrdy, S.B. 1998. The contingent nature of maternal love and its implications for 'adorable' babies. In S.B. Hrdy (ed.) *Mother nature.* Pantheon, New York.

Huang, Y., Paxton W.A., Wolinsky, S.M., *et al.* 1996. The role of a mutant CCR5 allele in HIV-1 transmission and disease progression. *Nature Med.* **2**: 1240–3.

Hudson, R.R. 1990. Gene genealogies and the coalescent process. *Oxford Surv. Evol. Biol.* **7**: 1–44.

Huelsenback, J.P. and Rannala, B. 1997. Phylogenetic methods come of age: testing hypotheses in an evolutionary context. *Science* **276**: 227–32.

Huggett, A.St G. and Widdas, W.F. 1951. The relationship between mammalian foetal weight and conception age. *J. Physiol.* **114**: 306–17.

Hughes, A.L. and Hughes, M.K. 1995. Natural selection on the peptide-binding regions of major histocompatibility complex molecules. *Immunogenetics* **42**: 233–43.

Hughes, A.L. and Nei, M. 1988. Pattern of nucleotide substitution at major histocompatibility complex class I loci reveals overdominant selection. *Nature* **335**: 167–70.

Hughes, A.L. and Nei, M. 1989. Nucleotide substitution at major histocompatibility complex class II loci: evidence for overdominant selection. *Proc. Natl. Acad. Sci. USA* **86**: 958–62.

Hughes, K.A. and Charlesworth, B. 1994. A genetic analysis of senescence in *Drosophila*. *Nature* **367**: 64–6.

Hung, L.F., Crawford, M.L.J., and Smith, E.L. 1995. Spectacle lenses alter eye growth and the refractive status of young monkeys. *Nature Med.* **1**: 761–5.

Hurst, L.D. 1992. Intragenomic conflict as an evolutionary force. *Proc. Roy. Soc. London B* **248**: 135–40.

Hurtado, A.M., Hawkes, K., Hill, K., and Kaplan, H. 1985. Female subsistence strategies among Aché hunter-gatherers of Eastern Paraguay. *Hum. Ecol.* **13**: 1–28.

Inoue, I., Nakajima, T., Williams, C.S., *et al.* 1997. A nucleotide substitution in the promoter of human angiotensinogen is associated with essential hypertension and effects of basal transcription *in vitro*. *J. Clin. Invest.* **99**: 1786–97.

Irons, W.I. 1979. Cultural and biological success. In N.A. Chagnon and W. Irons (eds) *Evolutionary biology and human social behavior: an anthropological perspective.* Duxbury Press, North Scituate, Massachusetts, pp. 257–72.

Irving, E.L., Sivak, J.G. and Callender, M.G. 1992. Refractive plasticity of the developing chick eye. *Ophthal. Physiol. Opt.* **12**: 448–56.

Irwin, C. 1989. The sociocultural biology of Netsilingmiut female infanticide. In Rasa *et al.* (1989), pp. 234–64.

Ison, C.A. 1988. Immunology of gonorrhoea. In D.J.M. Wright (ed.) *Immunology of sexually transmitted disease.* Kluwer Academic, London, pp. 95–116.

Istock, C.A., Duncan, K.E., Ferguson, N., and Zhou, X. 1992. Sexuality in a natural population of bacteria—*Bacillus subtilis* challenges the clonal paradigm. *Mol. Ecol.* **1**: 95–103.

Iverson, S.J., Bowen, W.D., Boness, D.J., and Oftedal, O.T. 1993. The effect of maternal size and milk energy output on pup growth in grey seals (*Halichoerus grypus*). *Phys. Zool.* **66**: 61–88.

Iwamoto, S., Li, J., Sugimoto, N., *et al.* 1996. Characterization of the Duffy gene promoter: evidence for tissue-specific abolishment of expression in Fy(a-b-) of black individuals. *Biochem. Biophys. Res. Commun.* **222**: 852–9.

Jablensky, Satorius, N., and Ernberg, G. 1992. Schizophrenia: manifestations, incidence and course in different cultures. A World Health Organization ten country study. *Psychol. Med. Suppl.* **20**: 1–97.

Jacobs, W.R., Barletta, R.G., Udani, R., *et al.* 1993. Rapid assessment of drug susceptibilities of *Mycobacterium tuberculosis* by means of luciferase reporter phages. *Science* **260**: 812–22. (See comments.)

Jaenike, J. 1978. An hypothesis to account for the maintainance of sex within populations. *Evol. Theory* **3**: 191–4.

Jamison, K.R. 1993. Touched with fire: manic-depressive illness and the artistic temperament. Free Press, New York.

Jazwinska, E.C., Cullen L.M., Busfield, F., *et al.* 1996. Haemochromatosis and HLA-H. *Nature Genet.* **14**: 249–51.

Jepson, A., Banya, W., Sisy-Joof, F., *et al.* 1997a. Quantification of the relative contribution of major histocompatibility complex (MHC) and non-MHC genes to human immune responses to foreign antigens. *Infect. Immun.* **65**: 872–6.

Jepson, A., Banya, W., Sisay-Joof, F. *et al.* 1997b. Genetic linkage of mild malaria to the major histocompatibility complex in Gambian children: study of affected sibling pairs. *Brit. Med. J.* **315**: 96–7.

Joffe, B.I., Jackson, W.P.U., Thomas, M.E., *et al.* 1971. Metabolic response to oral glucose in the Kalahari Bushmen. *Br. Med. J.* **4**: 206–8.

Johansson, I., Lundqvist, E., Bertilsson, L., *et al.* 1993. Inherited amplification of an active gene in the cytochrome P450 CYP2D locus as a cause of ultrarapid metabolism of debrisoquine. *Proc. Natl. Acad. Sci. USA* **90**: 11825–9.

Johnson, K.M. 1975. Yellow fever. In W.T. Hubbert (ed.) *Diseases transmitted from animals to man*, 6th edn. C.C. Thomas, Springfield, pp. 929–38.

Johnson, M.H. and Everitt, B.J. 1988. *Essential reproduction*, 3rd edn. Blackwell Scientific, Oxford.

Johnson, S.R., Petzold, C.R., and Galask, R.P. 1985. Qualitative and quantitative changes of the vaginal microbial flora during the menstrual cycle. *Am. J. Reprod. Immun. Microbiol.* **9**: 1–5.

Johnstone, R.A. 1995. Sexual selection, honest advertisment and the handicap principle: reviewing the evidence. *Biol. Rev.* **70**: 1–65.

Johnstone, R.A. 1997. The evolution of animal signals. In Krebs and Davies (1997), pp. 155–78.

Jones, J.S. and Rouhani, S. 1986. How small was the bottleneck? *Nature* **319**: 449–50.

Jorge, L.F., Arias, T.D., Griese, U., *et al.* 1993. Evolutionary pharmacogenetics of *CYP2D6* in Ngawbe Guaymi of Panama: allele-specific PCR detection of the *CYP2D6B* allele and RFLP analysis. *Pharmacogenetics* **3**: 231–8.

Judson, H. 1979. *The eighth day of creation: the makers of the revolution in biology.* Simon & Schuster, New York.

Jurmain, R., Nelson, H., Kilgore, L., and Trevathon, W. 1997. *Introduction to physical anthropology*, 7th edn. Wadsworth Publishing, Belmont.

Kaatz, G.W., Seo, S.M., Barriere, S.L., *et al.* 1991. Development of resistance to fleroxacin during therapy of experimental methicillin-susceptible *Staphylococcus aureus* endocarditis. *Antimicrob. Agents Chemother.* **35**: 1547–50.

Kaiserman-Abramof, I.R. and Padykula, H.A. 1989. Angiogenesis in the postovulatory primate endometrium: the coiled arteriolar system. *Anat. Rec.* **224**: 479–89.

Kakehashi, M. and Yoshinaga, F. 1992. Evolution of airborne infectious diseases according to changes in characteristics of the host population. *Ecol. Res.* **7**: 235–43.

Kalow, W. (ed.) 1992. *Pharmacogenetics of drug metabolism.* Pergamon Press, New York.

Kalow, W. and Bertilsson, L. 1994. Interethnic factors affecting drug response. In U.A.M. Testa (ed.) *Advances in drug research.* Academic Press, New York, pp. 2–53.

Kannel, W.B. and Abbott, R.D. 1984. Incidence and prognosis of unrecognized myocardial infarction. An update on the Framingham Study. *N. Engl. J. Med.* **311**: 1144–7.

Kannel, W.B. and Schatzin, A. 1985. Sudden death: lessons for subsets in population studies. *J. Am. Coll. Cardiol.* **5** (Suppl. 6): 141B–149B.

Kannus, P., Parkkari, J., and Niemi, S. 1995. Age-adjusted incidence of hip fractures. *Lancet* **346**: 50–1.

Kaplan, H. 1995. Does observed fertility maximise fitness among New Mexican men? A test of an optimality model and a new theory of parental investment in the embodied capital of offspring. *Hum. Nature* **6**: 325–60.

Kaprio, J., Ferrell, R.E., Kottke, B.A,. *et al.* 1991. Effects of polymorphisms in apolipoproteins E, A-IV, and H on quantitative traits related to risk for cardiovascular disease. *Arterioscler. Thromb.* **11**: 1330–48.

Kapur, V., Li, L.L., Hamrick, M.R., *et al.* 1995. Rapid *Mycobacterium* species assignment and unambiguous identification of mutations associated with antimicrobial resistance in *Myco-bacterium tuberculosis* by automated DNA sequencing. *Arch. Path. Lab. Med.* **119**: 131–8.

Kardia, S.L.R., Haviland, M.B., Ferrell, R.E., and Sing, C.F. 1996. A search for functional mutations in the lipoprotein lipase (LPL) gene region that influence quantitative intermediate risk factors for coronary artery disease. *Am. J. Hum. Genet.* **59**: A29.

Katzel, L.I., Fleg, J.L., Paidi, M., *et al.* 1993. Apo E4 polymorphism increases the risk for exercise-induced silent myocardial ischemia in older men. *Arterioscler. Thromb.* **13**: 1495–500.

Kehoe, M.A., Kapur, V., Whatmore, A.M., and Musser, J.M. 1996. Horizontal gene transfer among group A streptococci: implications for pathogenesis and epidemiology. *TIM* **4**: 436–43.

Kellam, P., Boucher, C.A.B, Tijnagel, J.M.G.H., and Larder, B.A. 1994. Zidovudine treatment results in the selection of human immuno-deficiency virus type 1 variants whose genotypes confer increasing levels of drug resistance. *J. Gen. Virol.* **75**: 341–51.

Kelly, R.J., Rouquier, S., Giorgi, D., *et al.* 1995. Sequence and expression of a candidate for the human secretor blood group alpha(1,2)fucosyl-transferase gene (FUT2). Homozygosity for an enzyme-inactivating nonsense mutation commonly correlates with the non-secretor phenotype. *J. Biol. Chem.* **270**: 4640–9.

Keltner, D. and Busswell, B. 1996. Evidence for the distinctness of embarassment, shame, and guilt: a study of recalled antecedents and facial ex-pressions of emotion. *Cognit. Emot.* **10**: 155–72.

Kempkes, M., Golka, K., Reich, S.S., *et al.* 1996. Glutathione S transferase GSTM1 and GSTT1 null genotypes as potential risk factors for urothelial cancer of the bladder. *Arch. Toxicol.* **71**: 123–6.

Kendler, K.S. 1983. Overview: a current perspective on twin studies of schizophrenia. *Am. J. Psych.* **140**: 1413–25.

Kennedy, J.L., Petronis, A., Gao, J., *et al.* 1994. Genetic studies of DRD4 and clinical response to neuroleptic medications. *Am. J. Hum. Genet.* **55** (Suppl. A): 190.

Kenyon, C. 1996. Ponce d'elegans: genetic quest for the fountain of youth. *Cell* **84**: 501–4.

Kessler, R.C., McGonagle, K.A., Zhao, S., *et al.* 1994. Lifetime and 12-month prevlece of DSM-III-R psychiatric disorders in the United States: results from the national comorbidity survey. *Arch. Gen. Psychiatry* **51**: 8–19.

Ketterer, B., Harris, J.-M., Talaska, G., *et al.* 1992. The human glutathione S-transferase supergene family, its polymorphism, and its effects on susceptibility to lung cancer. *Environ. Health Perspect.* **98**: 87–94.

Keys, A. 1950. *The biology of human starvation.* University of Minnesota Press, Minneapolis.

Kimura, K.D., Tissenbaum, H.A., Liu, Y., and Ruvkun, G. 1997. daf-2, An insulin receptor-like gene that regulates longevity and diapause in *Caenorhabditis elegans. Science* **277**: 942–6.

Kimura, M. 1967. On the evolutionary adjustment of spontaneous mutation rates. *Genet. Res.* **9**: 23–34.

Kimura, M. and Crow, J.F. 1964. The number of alleles that can be maintained in a finite popula-tion. *Genetics* **49**: 725–38.

King, D.G., Soller, M., and Yechezkel, K. 1997. Evolutionary tuning knobs. *Endeavour* **21**: 36.

King, R.A., Rotter, J.I., and Motulsky, A.G. (eds). 1992. *Genetic basis of common diseases.* Oxford University Press.

Kirkwood, T.B.L. 1977. Evolution of ageing. *Nature* **270**: 301–4.

Kirkwood, T.B.L. 1981. Evolution of repair: survival versus reproduction. In C.R. Townsend and P. Calow (eds) *Physiological ecology: an evolution-ary approach to resource use.* Blackwell Scientific, Oxford, pp. 165–89.

Kirkwood, T.B.L. 1985. Comparative and evolution-ary aspects of longevity. In C.E. Finch and E.L. Schneider (eds) *Handbook of the biology of aging,* 3rd edn. Van Nostrand Reinhold, New York, pp. 27–44.

Kirkwood, T.B.L. 1996. Human senescence. *BioEssays* **18**: 1009–16.

Kirkwood, T.B.L. and Cremer, T. 1982. Cyto-gerontology since 1881: a reappraisal of August Weismann and a review of modern progress. *Hum. Genet.* **60**: 101–21.

Kirkwood, T.B.L. and Franceschi, C. 1992. Is aging as complex as it would appear? *Ann. NY Acad. Sci.* **663**: 412–17.

Kirkwood, T.B.L. and Holliday, R. 1979. Evolution of ageing and longevity. *Proc. Roy. Soc. London B* **205**: 531–46.

Kirkwood, T.B.L. and Holliday, R. 1986. Ageing as a consequence of natural selection. In A.J. Collins and A.H. Bittles (eds) *The biology of human ageing.* Cambridge University Press, pp. 1–16.

Kirkwood, T.B.L. and Rose, M.R. 1991. Evolution of senescence: late survival sacrificed for repro-duction. *Phil. Trans. Roy. Soc. London B* **332**: 15–24.

Klass, M.R. and Hirsh, D. 1976. Non-ageing developmental variant of *Caenorhabditis elegans*. *Nature* **260**: 523–5.

Klein, D. 1996. Panic disorder and agoraphobia: hypothesis hothouse. *J. Clin. Psych.* **57** (Suppl. 6): 21–7.

Klein, J. 1986. Natural history of the major histocompatibility complex. John Wiley, New York.

Klein, J. 1987. Origin of major histocompatibility complex polymorphism. The trans-species hypothesis. *Hum. Immunol.* **19**: 155–62.

Klein, J. and Klein, D. 1991. Molecular evolution of the major histocompatibility complex. Springer-Verlag, Berlin.

Klein, J., Satta, Y., O'Huigin, C., and Takahata, N. 1993. The molecular descent of the major histocompatibility complex. *Annu. Rev. Immunol.* **11**: 269–95.

Klein, R.G. 1989. *The human career: human biological and cultural origins.* University of Chicago Press.

Klein, R.G. 1992. The archaeology of modern human origins. *Evol. Anthropol.* **1**: 5–14.

Klerman, G. and Weissman, M. 1989. Increasing rates of depression. *JAMA* **261**: 2229–35.

Klitz, W., Thomson, G., and Baur, M.P. 1986. Contrasting evolutionary histories among tightly linked HLA loci. *Am. J. Hum. Genet.* **39**: 340–9.

Knittle, J.L. 1978. Adipose tissue development in man. In Falkner and Tanner (1978), pp. 295–315.

Kolman, C.J., Sambuughin, N., and Bermingham, E. 1996. Mitochondrial DNA analysis of Mongolian populations and implications for the origin of New World founders. *Genetics* **142**: 1321–34.

Konner, M.J. 1988. The natural child. In S.B. Eaton, M. Shostak, and M.J. Konner (eds) *The paleolithic prescription. A program of diet and exercise and a design for living.* New York, Harper & Row, pp. 200–28.

Koonin, E.V. 1991. The phylogeny of RNA-dependent RNA polymerases of positive-strand RNA viruses. *J. Gen. Virol.* **72**: 2197–206.

Koonin, E.V. and Mushegian, A.R. 1996. Complete genome sequences of cellular life forms: glimpses of theoretical evolutionary genomics. *Curr. Opin. Genet. Dev.* **6**: 757–62.

Korona, R., Korona, B., and Levin, B.R. 1993. Sensitivity of naturally occurring coliphages to type I and type II restriction and modification. *J. Gen. Microbiol.* **139**: 1283–90.

Kowald, A. and Kirkwood, T.B.L. 1996. A network theory of ageing: the interactions of defective mitochondria, aberrant proteins, free radicals and scavengers in the ageing process. *Mutat. Res.* **316**: 209–36.

Kozlowski, J. and Stearns, S.C. 1989. Hypotheses for the production of excess zygotes: models of risk-aversion and progeny choice. *Evolution* **43**: 1369–77.

Krebs, J.R. and Davies, N.B. 1997 *Behavioural ecology. An evolutionary approach,* 4th edn. Blackwell Scientific, Oxford.

Kreiswirth, B., Kornblum, J., Arbeit, R.D., *et al.* 1993. Evidence for a clonal origin of methicillin resistance in *Staphylococcus aureus. Science* **259**: 227–30.

Kreitman, M. and Hudson, R. 1991. Inferring the evolutionary histories of the Adh and Adh-dup loci in *Drosophila melanogaster* from patterns of polymorphism and divergence. *Genetics* **127**: 565–82.

Krieger, M., Acton, S., Ashkenas, J., Pearson, A., Penman, M., and Resnick, D. 1993. Molecular flypaper, host defense, and atherosclerosis. Structure, binding properties, and functions of macrophage scavenger receptors. *J. Biol. Chem.* **268**: 4569–72.

Kristensen, K.G., Hjamarsdottir, M.A., and Steingrimsson, O. 1992. Increasing penicillin resistance in pneumococci in Iceland. *Lancet* **339**: 1606–7.

Kritshenwski, I.L. and Rubenstein, P.L. 1932. Über die Medikamenten-festigkeit der Esseger von Vogelmalaria (*Plasmodium praecox*). *Mitt. Z. Immun. Exp. Therap.* **76**: 506–14

Kroemer, H.K. and Eichelbaum, M. 1995. 'It's the genes, stupid'. Molecular bases and clinical consequences of genetic cytochrome P450 2D6 polymorphism. *Life Sci.* **56**: 2285–98.

Kroner, B.L., Goedert, J.J., Blattner, W.A., *et al.* 1995 Concordance of human leukocyte antigen haplotype-sharing, CD4 decline and AIDS in hemophilic siblings. Multicenter Hemophilia Cohort and Hemophilia Growth and Development Studies. *AIDS* **9**: 275–80.

Kuiken, C.L., Zwart, G., Baan, E., *et al.* 1993. Increasing antigenic and genetic diversity of the V3 variable domain of the human immunodeficiency virus envelope protein in the course of the AIDS epidemic. *Proc. Natl. Acad. Sci. USA* **90**: 9061–5.

Kumar, A., Kumar, A., Sinha, S., *et al.* 1995. Hind III genomic polymorphism of the C3b receptor (CR1) in patients with SLE: low erythrocyte CR1 expression is an acquired phenomenon. *Immunol. Cell. Biol.* **73**: 457–62.

Kumar, S. and Miller, L.K. 1987. Effects of serial

passage of Autographa californica nuclear poly-hedrosis virus in cell culture. *Virus Res.* **7**: 335–49.

Kunst, C.B., Zerylnick, C., Karickhoff, L., et al. 1996. FMR1 in global populations. *Am. J. Hum. Genet.* **58**: 513–22.

Kusano, K., Naito, T., Handa, N., and Kobayashi, I. 1995. Restriction-modification systems as genomic parasites in competition for specific sequences. *Proc. Natl. Acad. Sci. USA* **92**: 11095–9.

Kushi, L.H., Lew, R.A., Stare, F.J., et al. 1985. Diet and 20-year mortality from coronary heart disease. The Ireland–Boston diet–heart study. *N. Engl. J. Med.* **312**: 811–8.

Kuusi, T., Nieminen, M.S., Ehnholm, C., et al. 1989. Apolipoprotein E polymorphism and coronary artery disease: increased prevalence of apolipoprotein E-4 in angiographically verified coronary patients. *Arteriosclerosis* **9**: 237–41.

Laakso, M., Lehto, S., Penttila, I., and Pyorala, K. 1993. Lipids and lipoproteins predicting coronary heart disease mortality and morbidity in patients with non-insulin-dependent diabetes. *Circulation* **88**: 1421–30.

Lack, D. 1947. The significance of clutch size. *Ibis* **89**: 302–52.

Lahr, M.M. and Foley, R. 1994. Multiple dispersal and modern human origins. *Evol. Anthropol.* **3**: 48–60.

Laird, A.K. 1967. Evolution of the human growth curve. *Growth* **31**: 345–55.

Lakowski, B. and Hekimi, S. 1996. Determination of life-span in *Caenorhabditis elegans* by four clock genes. *Science* **272**: 1010–13.

Lande, R. 1981. Models of speciation by sexual selection on polygenic traits. *Proc. Natl. Acad. Sci. USA* **78**: 3721–5.

Langaney, A. and Sanchez-Mazas, A. 1987. The human genetic network and its history. *Ann. Démograph. Historique* **49**: 48–52.

Larsen, P.L. Albert, P.S., and Riddle, D.L. 1995. Genes that regulate development and longevity in *Caenorhabditis elegans*. *Genetics* **139**: 1567–83.

Latter, B.D.H. 1980. Genetic differences within and between populations of the major human subgroups. *Am. Nat.* **116**: 220–37.

Laurie, C.C., Bridgham, J.T., and Choudray, M. 1991. Associations between DNA sequence variation and variation in expression of the Adh gene in natural populations of *Drosophila melanogaster*. *Genetics* **129**: 489–99.

Law, C.M. and Shiell, A.W. 1996. Is blood pressure inversely related to birth weight? The strength of evidence from a systematic review of the literature. *J. Hypertens.* **14**: 935–41.

Lawlor, B.J., Read, A.F., Keymer, A.E., et al. 1990. Non-random mating in a parasitic worm: mate choice by males? *Anim. Behav.* **40**: 870–6.

Lawlor, D.A., Ward, F.E., Ennis, P.D., et al. 1988. HLA-A, B polymorphisms predate the divergence of humans and chimpanzees. *Nature* **335**: 268–71.

Lawrence, J.G. and Roth, J.R. 1996. Selfish operons: horizontal transfer may drive the evolution of gene clusters. *Genetics* **143**: 1843–60.

Layde, P.M., Webster, L.A., Baughman, A.L., et al. 1989. The independent associations of parity, age at first full term pregnancy, and duration of breast feeding with the risk of breast cancer. *J. Clin. Epidem.* **42**: 963–73.

Leakey, M.D. and Hay, R.L. 1979. Pliocene foot-prints in Laetoli beds at Laetoli, Northern Tanzania. *Nature* **278**: 317–23.

LeClerc, J.E., Li, B., Payne, W.L., and Cebula, T.A. 1996. High mutation frequencies among *Escherichia coli* and *Salmonella* pathogens. *Science* **274**: 1208–11.

Lederberg, J. and Iino, T. 1956. Phase variation in *Salmonella. Genetics* **41**: 744–57.

Lee, R.B. 1968. What hunters do for a living, or, how to make out on scarce resources. In R.B. Lee and I. DeVore (eds) *Man the hunter*. Aldine, Chicago, pp. 30–48.

Lee, R.B. 1979. The !Kung San: Men, women and work in a foraging society. Cambridge University Press.

Lehrman, S. 1996. Diversity Project: Cavalli-Sforza answers his critics. *Nature* **381**: 14.

Lehtimäki, T., Porkka, K., Viikari, J., et al. 1994. Apolipoprotein E phenotypes and serum lipids in newborns and 3-year-old children: the cardio-vascular risk in young Finns study. *Pediatrics* **94**: 489–93.

Leigh, E.G. 1973. The evolution of mutation rates. *Genetics* **73**: 1–18.

Leitner, T., Escanilla, D., Franzen, C., et al. 1996. Accurate reconstruction of a known HIV-1 transmission history. *Proc. Natl. Acad. Sci. USA* **93**: 10864–9.

Lelliott, P., Marks, I., McNamee, G., and Tobena, A. 1989. Onset of panic disorder with agoraphobia: toward an integrated model. *Arch. Gen. Psychiatry* **46**: 1000–4.

Lenski, R.E. 1993. Addressing the genetic structure of microbial populations. *Proc. Natl. Acad. Sci. USA* **90**: 4334–6.

Lenski, R.E. and Hattingh, S.E. 1986. Coexistence

of two competitors on one resource and one inhibitor: a chemostate model based on bacteria and antibiotics. *J. Theor. Biol.* **122**: 83–93.

Lenski, R.E. and May, R.M. 1994. The evolution of virulence in parasites and pathogens: reconciliation between two competing hypotheses. *J. Theor. Biol.* **169**: 253–66.

Lenski, R.E. and Mittler, J.E. 1993. The directed mutation controversy and neo-Darwinsm. *Science* **259**: 188–94.

Lenski, R.E. and Travisano, M. 1994. Dynamics of adaptation and diversification: a 10 000-generation experiment with bacterial populations. *Proc. Natl. Acad. Sci. USA* **91**: 6808–14.

Lenzen, H.J., Assmann, G., Buchwalsky, R., and Schulte, H. 1986. Association of apolipoprotein E polymorphism, low density lipoprotein cholesterol, and coronary artery disease. *Clin. Chem.* **32**: 778–81.

Leroi, A.M., Chen, W.R., and Rose, M.R. 1994*a*. Long-term laboratory evolution of a genetic life-history trade-off in *Drosophila melanogaster*. 2. Stability of genetic correlations. *Evolution* **48**: 1258–68.

Leroi, A.M., Chippindale, A.K., and Rose, M.R. 1994*b*. Long-term laboratory evolution of a genetic life-history trade-off in *Drosophila melanogaster*. 1. The role of genotype-by-environment interaction. *Evolution* **48**: 1244–57.

Leutenegger, W. 1974. Functional aspects of pelvic morphology in simian primates. *J. Hum. Evol.* **3**: 207–22.

Levin, B. 1980. Conditions for the existence of R-plasmids in bacterial populations. In S. Misuhashi, L. Rosival, and V. Krcmery (eds) *Fourth International Symposium on Antibiotic Resistance*, pp. 197–202. Springer-Verlag, New York.

Levin, B.R. 1996. The evolution and maintenance of virulence in microparasites. *Emerg. Infect. Dis.* **2**: 93–192.

Levin, B.R. and Bull, J.J. 1994. Short-sighted evolution and the virulence of pathogenic microorganisms. *TIB* **2**: 76–81.

Levin, B.R. and Lenski, R.E. 1983. Coevolution of bacteria and their viruses and plasmids. In D.J. Futuyama and M. Slatkin (eds) *Coevolution*. Sinauer, Sunderland, Massachusetts, pp. 99–127.

Levin, B.R., Lipsitch, M., Perrot, V., *et al.* 1997. The population genetics of antibiotic resistance. *Clin. Infect. Dis.* **24** (Suppl.1): S9–316.

Levin, B.R. and Svanborg-Edén, C. 1990. Selection and evolution of virulence in bacteria: an ecumenical excursion and modest suggestion. *Parasitology* **100**: S103–S115.

Levin, S. and Pimentel, D. 1981. Selection of intermediate rates of increase in parasite–host systems. *Am. Nat.* **117**: 308–15.

Levinson, G. and Gutman, G.A. 1987. Slipped-strand mispairing: a major mechanism for DNA sequence evolution. *Mol. Biol. Evol.* **4**: 203–21.

Levy, S.B. 1992. *The antibiotic paradox: how miracle drugs are destroying the miracle*. Plenum Press, New York.

Levy, S.B. 1994. Balancing the drug-resistance equation. *Trends Microbiol.* **2**: 341–42.

Lewin, R. 1993. *Human evolution*. Blackwell Scientific, Boston.

Lewin Group. 1997. Cost of illness for *Staphycloccus aureus* infections in New York City. Public Health Research Institute (PHRI), BARG/IDC program, The Rockefeller University and the Bacterial Antibiotic Resistance Group (BARG).

Lewontin, R.C. 1972. The apportionment of human diversity. *Evol. Biol.* **6**: 381–98.

Li, W.-H. and Sadler, L.A. 1991. Low nucleotide diversity in Man. *Genetics* **129**: 513–23.

Lipsitch, M. 1997*a*. Vaccination against colonizing bacteria with multiple serotypes. *Proc. Natl. Acad. Sci. USA* **94**: 6571–6.

Lipsitch, M. 1997*b*. Transmission rates and HIV virulence: comments to Massad. *Evolution* **51**: 319–20.

Lipsitch, M. and Levin, B.R. 1997*a*. The within-host population dynamics of anti-bacterial chemotherapy: conditions for the evolution of resistance. In *CIBA Foundation Symposium* **207**. J. Wiley, New York, pp. 112–27.

Lipsitch, M. and Levin, B.R. 1997*b*. The population dynamics of antimicrobial chemotherapy. *Antimicrob. Agents Chemother.* **41**: 363–73.

Lipsitch, M. and Levin, B.R. 1998. Population dynamics of tuberculosis treatment: mathematical models of the roles of non-compliance and bacterial heterogeneity in the evolution of drug resistance. *Int. J. Tuberculosis and Lung Dis.* **2(3)**: 187–99.

Lipsitch, M. and Moxon, E.R. 1997. Virulence and transmissibility of pathogens: What is the relationship? *Trends in Microbiol.* **5**: 31–7.

Lipsitch, M. and Nowak, M.A. 1994. The evolution of virulence in sexually transmitted HIV/AIDS. *J. Theor. Biol.* **174**: 427–40.

Lipsitch, M., Herre, E.A., and Nowak, M.A. 1995. Host population structure and the evolution of virulence: a law of diminishing returns. *Evolution* **49**: 743–8.

Lisker, R. and Motulsky, A.G. 1967. Computer simulation of evolutionary trends in an X linked trait. Application to glucose-6-phosphate dehydrogenase deficiency in man. *Acta. Genet. Stat. Med.* **17**: 465–74.

Lithell, H.O., McKeigue, P.M., Berglund, L., *et al.* 1996. Relation of size at birth to non-insulin dependent diabetes and insulin concentrations in men aged 50–60 years. *Br. Med. J.* **312**: 406–10.

Lithgow, G.J. 1996. Invertebrate gerontology: the age mutations of *Caenorhabditis elegans*. *BioEssays* **18**: 809–15.

Liu, R., Paxton, W.A., Choe, S., *et al.* 1996. Homozygous defect in HIV-1 coreceptor accounts for resistance of some multiply-exposed individuals to HIV-1 infection. *Cell* **86**: 367–77.

Livingstone, F.B. 1958. Anthropological implications of sickle cell gene distribution in West Africa. *Am. Anthropol.* **60**: 533–62.

Livingstone, F.B. 1984. The Duffy blood groups, vivax malaria and malarial selection in human populations. *Hum. Biol.* **56**: 413–25.

Livingstone, F.B. 1992. Gene flow in the Pleistocene. *Hum. Biol.* **64**: 649–57.

LLerena, A., Edman, G., Cobaleda, J., *et al.* 1993. Relationship between personality and debrisoquine hydroxylation capacity. *Acta Psychiatr. Scand.* **87**: 23–8.

Lockhart, A.B., Thrall, P.H., and Antonovics, J. 1996. Sexually transmitted diseases in animals: ecological and evolutionary implications. *Biol. Rev.* **71**: 415–71.

Loeb, L.A. 1991. Mutator phenotype may be required for multistage carcinogenisis. *Cancer Res.* **51**: 3075–9.

Longini, I.M. and Halloran, M.E. 1995. AIDS: modeling epidemic control. *Science* **267**: 1250–1.

Lorian, V. 1995. The need for surveillance for antimicrobial resistance. *Infect. Control Hosp. Epidemiol.* **16**: 638–41.

Louwagie, J., McCuthan, F.E., Peeters, M., *et al.* 1993. Phylogenetic analysis of gag genes from 70 international HIV-1 isolates provides evidence for multiple genotypes. *AIDS* **7**: 769–80.

Luc, G., Bard, J., Arveiler, D., *et al.* 1994. Impact of apolipoprotein E polymorphism on lipoproteins and risk of myocardial infarction: the ECTIM Study. *Arterioscler. Thromb.* **14**: 1412–19.

Luckinbill, L.S., Arking, R., Clare, M.J., Cirocco, W.C., and Buck, S.A. 1984. Selection for delayed senescence in *Drosophila melanogaster*. *Evolution* **38**: 996–1003.

Lum, J.K., Rickards, O., Ching, C., and Cann, R.L. 1994. Polynesian mitochondrial DNAs reveal three deep maternal lineage clusters. *Hum. Biol.* **66**: 567–90.

Lunt, D.H. and Hyman, B.C. 1997. Animal mitochondrial DNA recombination. *Nature* **387**: 247.

MacDonald, G. 1952. The analysis of equilibrium in malaria. *Trop. Dis. Bull.* **49**: 813–29.

Mace, R. 1996. Biased parental investment and reproductive success in Gabbra pastoralists. *Behav. Ecol. Sociobiol.* **38**: 75–81.

Mace, R. 1997. The co-evolution of human fertility and wealth inheritance patterns. *Phil. Trans. Roy. Soc. London B* **357**: 1–10.

Mace, R. and Sear, R. 1997. Birth interval and the sex of children: an evolutionary analysis. *J. Biosocial Sci.* **29**: 499–507.

Macilwain, C. 1996. Tribal groups attack ethics of genome diversity project. *Nature* **383**: 208.

MacLintock B. The significance of responses of the genome to challenge. Nobel Lecture, 8th December 1993.

MacSorley, K. 1964. An investigation into the fertility rates of mentally ill patients. *Ann. Hum. Genet.* **27**: 247–56.

Maddison, D.R., Ruvolo, M., and Swofford, D.L. 1992. Geographic origins of human mitochondrial DNA: phylogenetic evidence from control region sequences. *Syst. Biol.* **41**: 111–24.

Maddison, R.R. 1991. African origin of human mitochondrial DNA reexamined. *Syst. Zool.* **40**: 355–62.

Majesky, M.W. and Schwartz, S.M. 1997. An origin for smooth muscle cells for endothelium? *Circ. Res.* **80**: 601–3.

Maldonado, Y.A., Lawrence, E.C., *et al.* 1995. Early loss of measles antibody in infants of mothers with vaccine-induced immunity. *Pediatrics* **96**: 447–50.

Malécot, G. 1948. *Les mathématiques de l'hérédité.* Masson, Paris.

Malécot, G. 1966. *Probabilités et hérédité.* Presses Universitaires de France, Paris.

Malone, E.A., Inoue, T., and Thomas, J.H. 1996. Genetic analysis of the roles of daf-28 and age-1 in regulating *Caenorhabditis elegans* dauer formation. *Genetics* **143**: 1193–205.

Mangin, K.L., Lipsitch, M., and Ebert, D. 1995. Virulence and transmission modes of two microsporidia in *Daphnia magna*. *Parasitology* **111**: 133–42.

Manton, K.G., Corder, L., and Stallard, E. 1997. Chronic disability trends in elderly United States populations: 1982–1994. *Proc. Natl. Acad. Sci. USA* **94**: 2593–8.

Mao, E. F., Lane, L., Lee, J., and Miller, J.H. 1997. Proliferation of mutators in a cell population. *J. Bacteriol.* **179**: 417–22.

Marez, D., Legrand, M., Sabbagh, N., *et al.* 1997. Polymorphism of the cytochrome P450 CYP2D6 gene in a European population: characterization of 48 mutations and 53 alleles, their frequencies and evolution. *Pharmacogenetics* **7**: 193–202.

Marks, I.M. 1988. Blood-injury phobia: a review. *Am. J. Psych.* **145**: 1207–13.

Marks, I.M. and Nesse, R.M. 1994. Fear and fitness: an evolutionary analysis of anxiety disorders. *Ethol. Sociobiol.* **15**: 247–61.

Marks, J. 1995. *Human biodiversity.* Aldine de Gruyter, New York.

Marquet, S., Abel, L., Hillaire, D., *et al.* 1996. Genetic localization of a locus controlling the intensity of infection by *Schistosoma mansoni* on chromosome 5q31–q33. *Nature Genet.* **14**: 181–4.

Marsh, D.G., Neely, J.D., Breazeale, D.R., *et al.* 1994. Linkage analysis of IL4 and other chromosome 5q31.1 markers and total serum immunoglobulin E concentrations. *Science* **264**: 1152–6.

Martin, G.M. 1978. Genetic syndromes in man with potential relevance to the pathobiology of aging. *Birth Defects* **14**: 5–39.

Martin, G.M. 1992. Biological mechanisms of aging. In J.G. Evans. and T.F. Williams (eds) *Oxford textbook of geriatric medicine.* Oxford University Press, pp. 41–8.

Martin, G.M. 1993. Abiotrophic gene action in *Homo sapiens*, potential mechanisms and significance for the pathobiology of aging. *Genetica* **91**: 265–77.

Martin, G.M., Austad, S.N., and Johnson, T.E. 1996. Genetic analysis of ageing, role of oxidative damage and environmental stresses. *Nature Genet.* **13**: 25–34.

Martin, R.D. 1989 (1990 Princeton). *Primate origins and evolution: a phylogenetic reconstruction.* Chapman & Hall, London.

Martyn, C.N., Barker, D.J.P., and Osmond, C. 1996. Mothers' pelvic size, fetal growth and death from stroke and coronary heart disease in men in the UK. *Lancet* **348**: 1264–8.

Masimirembwa, C., Bertilsson, L., Johansson, I., *et al.* 1995. Phenotyping and genotyping of S-mephenytoin hydroxylase (cytochrome P450 2C19) in a Shona population of Zimbabwe. *Clin. Pharmacol. Ther.* **57**: 656–61.

Mason, J.W. 1975. Emotion as reflected in patterns of endocrine integration. I. Some theoretical aspects of psychoneuroendocrine studies of emotions. In L. Levi (ed.) *Emotions—their parameters and measurement.* Raven Press, New York.

Masoro, E.J. and Austad, S.N. 1996. The evolution of the antiaging action of dietary restriction: a hypothesis. *J. Gerontol.* **51A**: B387–B391.

Massad, E., Lundberg, S., and Yang, H.M. 1993. Modeling and simulating the evolution of resistance against antibiotics. *Int. J. Biomed. Comput.* **33**: 65–81.

Matic, I., Radman, M., Taddie, F,. *et al.* 1997. High variability of mutation rates in commensal and pathogenic *Escherichia coli. Science* **277**: 1833–4.

May, R.M. and Anderson, R.M. 1983. Epidemiology and genetics in the coevolution of parasites and hosts. *Proc. Roy. Soc. London B* **219**: 281–313.

May, R.M. and Nowak, M.A. 1995. Coinfection and the evolution of parasite virulence. *Proc. Roy. Soc. London B* **261**: 209–15.

Mayer, W.E., Jonker, M., Klein, D., *et al.* 1988. Nucleotide sequences of chimpanzee MHC class I alleles: evidence for trans-species mode of evolution. *EMBO J.* **7**: 2765–74.

Maynard Smith, J. 1958. The effects of temperature and of egg-laying on the longevity of *Drosophila subobscura. J. Exp.Biol.* **35**: 832–42.

Maynard Smith, J. 1962. Review lectures on senescence. I. The causes of ageing. *Proc. Roy. Soc. London B* **157**: 115–27.

Maynard Smith, J. 1964. Group selection and kin selection: a rejoinder. *Nature* **201**: 1145–7.

Maynard Smith, J. 1989. *Evolutionary genetics.* Oxford University Press.

Maynard Smith, J. 1991. The population genetics of bacteria. *Proc. Roy. Soc. London B* **245**: 37–41.

Maynard Smith, J. 1992. Analyzing the mosaic structure of genes. *J. Mol. Evol.* **34**: 126–9.

Maynard Smith, J., and Smith, N.H. 1998. Detecting recombination from gene trees. *Mol. Biol. Evol.* (in press).

Maynard Smith, J., Smith, N.H., O'Rourke, M., and Spratt, B.G. 1993. How clonal are bacteria? *Proc. Natl. Acad. Sci. USA* **90**: 4384–8.

Mayr, E. 1982. *The growth of biological thought: diversity, evolution, and inheritance.* Harvard University Press, Cambridge, Massachusetts.

McCullough, J.M. and York Barton, E. 1991. Relatedness and mortality risk during a crisis year: Plymouth Colony, 1620–1621. *Ethol. Sociobiol.* **12**: 195–209.

McGlynn, K.A., Rosvold, E.A., Lustbader, E.D., *et al.* 1995. Susceptibility to hepatocellular carcinoma is associated with genetic variation in the enzymatic detoxification of aflatoxin B1. *Proc. Natl. Acad. Sci. USA* **92**: 2384–7.

McGowan, J.E. 1983. Antimicrobial resistance in hospital organisms and its relation to antibiotic use. *Rev. Infect. Dis.* **5**: 1033–48

McGowan, J.E. 1986. Minimising antimicrobial resistance in hospital bacteria: can switching or cycling drugs help? *Infect. Control* **7**: 573–6.

McGuire, M., Marks, I., Nesse, R., and Troisi, A. 1992. Evolutionary biology: a basic science for psychiatry. *Acta Psychiat. Scand.* **86**: 89–96.

McGuire, M.T. and Troisi, A. 1998. *Darwinian psychiatry.* Harvard University Press, Cambridge, Massachusetts.

McGuire, W., Hill, A.V., Allsopp, C.E, *et al.* 1994. Variation in the TNF-alpha promoter region associated with susceptibility to cerebral malaria. *Nature* **371**: 508–10.

McKane, M. and Milkman, R. 1995. Transduction, restriction and recombination patterns in *Escherichia coli. Genetics* **139**: 35–43.

McKeigue, P.M., Shah, B., and Marmot, M.G. 1991. Relation of central obesity and insulin resistance with high diabetes prevalence and cardiovascular risk in South Asians. *Lancet* **337**: 382–6.

McKeown, T. 1979. *The role of modern medicine: dream, mirage or nemesis?* Basil Blackwell, Oxford.

McKusick, V.A. 1975. *Mendelian inheritance in man. Catalogs of autosomal dominant, autosomal recessive and X-linked phenotypes*, 4th edn. John Hopkins Press, Baltimore.

McLean, A.R. 1995. Vaccination, evolution and changes in the efficacy of vaccines: a theoretical framework. *Proc. Roy. Soc. London B* **261**: 389–93.

McLean, A.R. and Anderson, R.M. 1988. Measles in developing countries. Part II. The predicted impact of mass vaccination. *Epidemiol. Infect.* **100**: 419–42.

McLean, A.R. and Blower, S.M. 1993. Imperfect vaccines and herd immunity to HIV. *Proc. Roy. Soc. London B* **253**: 9–13.

McLean, A.R. and Blower, S.M. 1995. Modelling HIV vaccination. *Trends in Microbiol.* **3**: 458–63.

McLean, A.R. and Kirkwood, T.B.L. 1990. A model of human immunodeficiency virus infection in T helper cell clones. *J. Theor. Biol.* **147**: 177–203.

McLean, A.R. and Nowak, M.A. 1992. Models of interactions between HIV and other pathogens. *J. Theor. Biol.* **155**: 69–86.

McNeil, W. 1980. Migration patterns and infection in traditional societies. In N.F. Stanley and R.A. Joske (eds) *Changing disease patterns and human behavior.* Academic Press, London, pp. 28–36.

McNeil, W.H. 1976. *Plagues and peoples.* Anchor, Garden City, New York.

Medawar, P.B. 1952. *An unsolved problem of biology.* H.K. Lewis, London.

Medvedev, Z.A. 1990. An attempt at a rational classification of theories of ageing. *Biol. Rev.* **65**: 375–98.

Melton, T., Peterson, R., Redd, A.J., *et al.* 1995. Polynesian genetic affinities with Southeast Asian populations as identified by mtDNA analysis. *Am. J. Hum. Genet.* **57**: 403–14.

Meltzer, D.J. 1993. Pleistocene peopling of the Americas. *Evol. Anthropol.* **1**: 157–69.

Menozzi, P., Piazza, A., and Cavalli-Sforza, L.L. 1978. Synthetic maps of human gene frequencies in Europeans. *Science* **201**: 786–92.

Mentis, A., Blackwell, C.C., Weir, D.M., *et al.* 1991. ABO blood group, secretor status and detection of *Helicobacter pylori* among patients with gastric or duodenal ulcers. *Epidemiol. Infect.* **106**: 221–9.

Menzel, H-J., Kladetzky, R-G., and Assmann, G. 1983. Apolipoprotein E polymorphism and coronary artery disease. *Arteriosclerosis* **3**: 310–14.

Merbs, C.F. 1992. A New World of infectious disease. *Yearb. Phys. Anthropol.* **35**: 3–42.

Merimee, T.J., Rimoin, D.L., and Cavalli-Sforza, L.L. 1972. Metabolic studies in the African Pygmy. *J. Clin. Invest.* **51**: 395–401.

Merriwether, D.A., Hall, W.H., Vahlne, A., and Ferrell, R.E. 1996. mtDNA variation indicates Mongolia may have been the source for the founding population of the New World. *Am. J. Hum. Genet.* **59**: 204–12.

Merryweather-Clarke, A.T., Pointon, J.J., Shearman, J.D., and Robson K.J.H. 1997. Global prevalence of putative haemochromatosis mutations. *J. Med. Genet.* **34**: 275–8.

Meyer, C.G. and Kremsner, P.G. 1996. Malaria and onchocerciasis: on HLA and related matters. *Parasitol. Today* **12**: 179–86.

Meyer, U.A. 1991. Genotype or phenotype: the definition of a pharmacogenetic polymorphism. *Pharmacogenetics* **1**: 66–7.

Meyer, U.A. 1992. Drugs in special patient groups: clinical importance of genetics in drug effects. In K.F. Melmon and H.F. Morelli (eds) *Clinical pharmacology: basic principles of therapeutics.* McGraw-Hill, New York, pp. 875–94.

Meyer, U.A. and Zanger, U.M. 1997. Molecular mechanisms of genetic polymorphisms of drug metabolism. *Annu. Rev. Pharmacol. Toxicol.* **37**: 269–96.

Meynell, G.G. 1957. The applicability of the hypothesis of independent action to fatal infections in mice given *Salmonella typhimurium* by mouth. *J. Gen. Microbiol.* **16**: 396–404.

Michod, R.E. and Levin, B.R. 1987. *The evolution of sex. An examination of current ideas.* Sinauer Associates, Massachusetts.

Michod, R.E., Wojciechowski, M.F., and Hoelzer, M.A. 1988. DNA repair and the evolution of transformation in the bacterium *Bacillus subtilis*. *Genetics* **118**: 31–9.

Mikus, G., Bochner, F., Eichelbaum, M., *et al.* 1994. Endogenous codeine and morphine in poor and extensive metabolisers of the CYP2D6 (debrisoquine/sparteine) polymorphism. *J. Pharmacol. Exp. Ther.* **268**: 546–51.

Milic, A.B. and Adamsons, K. 1969. The relationship between anencephaly and prolonged pregnancy. *J. Obstet. Gynaecol. Br. Commonw.* **76**: 102–11.

Milkman, R. and Bridges, M.M. 1990. Molecular evolution of the *Escherichia coli* chromosome. III. Clonal frames. *Genetics* **126**: 505–17.

Miller, G.F. 1994. Evolution of the human brain through runaway sexual selection: the mind as a protean courtship device. *Dissert. Abstr. Int.: Sect. B: Sci.Engineer.* **54** (12B): 6466.

Miller, L.H., Mason, S.J., Clyde, D.F., and McGinniss, M.H. 1976. The resistance factor to *Plasmodium vivax* in blacks. The Duffy-blood-group genotype, FyFy.: *N. Engl. J. Med.* **295**: 302–4.

Miller, L.K., Lingg, A.J., and Bulla, L.A.J. 1983. Bacterial, viral and fungal insecticides. *Science* **219**: 715–21.

Milner, G.R., Humpf, D.A., and Harpending, H.C. 1989. Pattern matching of age-at-death distributions in paleodemographic analysis. *Am. J. Phys. Anthropol.* **80**: 49–58.

Milton, K. 1993. Diet and primate evolution. *Sci. Am.* **269** (Aug.): 86–93.

Mineka, S., Davidson, M., Cook, M., and Keir, R. 1984. Observational conditioning of snake fear in rhesus monkeys. *J. Abnorm. Psychol.* **93**: 355–72.

Ministry of Agriculture, Fisheries, and Foods. 1995. *Household food consumption and expenditure*

1990. With a study of trends over the period 1940–1990. HMSO, London, pp. 55–114.

Mirlesse, V., Frankenne, F., Alsat, E., *et al.* 1993. Placental growth hormone levels in normal pregnancy and in pregnancies with intrauterine growth retardation. *Pediat. Res.* **34**: 439–42.

Mirra, S.S., Heyman, A., McKeel, D., *et al.* 1991. The Consortium to Establish a Registry for Alzhiemer's Disease (CERAD). Part II. Standardization of the neuropathologic assessment of Alzheimer's disease. *Neurology* **41**: 479–86.

Mitchison, D.A. 1979. Basic mechanisms of chemotherapy. *Chest* **76**: 771–81.

Mitchison, D.A. 1984. Drug resistance in mycobacteria. *Br. Med. Bull.* **40**: 84–90.

Mittler, J.E., Levin, B.R., and Antia, R. 1996. T-cell homeostasis, competition and drift: AIDS as HIV-accelerated senescence of the immune repertoire. *J. AIDS Hum. Retovirol.* **12**: 233–48.

Miyake, T. 1960. Mutator factor in *Salmonella typhimurium*. *Genetics* **45**: 11–14.

Moberg, C.L. 1996. René Dubos: a harbinger of microbial resistance to antibiotics. *Microb. Drug Resist.* **2**: 287–97

Modi, R.I. and Adams, J. 1991. Coevolution in bacteria–plasmid populations. *Evolution* **45**: 656–67.

Møller, A.P. and Erritzøe, J. 1996. Parasite virulence and host immune defense: host immune response is related to nest reuse in birds. *Evolution* **50**: 2066–72.

Monath, T.P. 1994. Dengue: the risk to developed and developing countries. *Proc. Natl. Acad. Sci. USA* **85**: 7627–31.

Moore, T. and Haig, D. 1991. Genomic imprinting in mammalian development: a parental tug-of-war. *Trends Genet.* **7**: 45–9.

Morand, S., Manning, S.D., and Woolhouse, M.E.J. 1996. Parasite–host coevolution and geographic patterns of parasite infectivity and host susceptibility. *Proc. Roy. Soc. London B* **263**: 119–28.

Morgan, E. 1982. *The aquatic ape.* Souvenir Press, London.

Morral, N., Bertranpetit, J., Estivill, X., *et al.* 1994. The origin of the major cystic fibrosis mutation (DF508) in European populations. *Nature Genet.* **7**: 169–75.

Morris, J.Z., Tissenbaum, H.A., and Ruvkun, G. 1996. A phosphatidylinositol-3-OH kinase family member regulating longevity and diapause in *Caenorhabditis elegans*. *Nature* **382**: 536–9.

Morris, K., Morganlander, M., Coulehan, J., *et al.* 1990. Wood-burning stoves and lower respiratory tract infection in American Indian children. *Am. J. Dis. Child.* **144**: 105–8.

Morrison, N.A., Qi, J.C., Tokita, A., *et al.* 1994. Prediction of bone density from vitamin D receptor. *Nature* **367**: 284–7.

Morse, S.S. (ed.) 1994. *The evolutionary biology of viruses.* Raven Press, New York.

Morton, N.E. 1955. The inheritance of human birth weight. *Ann. Hum. Genet.* **20**: 123–34.

Morton, N.E. 1997. Genetic epidemiology. *Ann. Hum. Genet.* **61**: 1–13.

Morton, N.E., Yee, S., Harris, D.E., and Lew, R. 1972. Bioassay of kinship. *Theor. Pop. Biol.* **2**: 507–24.

Morton, S.R., Recher, H.F., Thompson, S.D., and Braithwaite, R.W. 1982. Comments on the relative advantages of marsupial and eutherian reproduction. *Am. Nat.* **120**: 128–34.

Mosher, W.D. and Pratt, W.F. 1990. Fecundity and infertility in the United States, 1965–1988. National Center for Health Statistics, Hyattsville, Maryland.

Motulsky, A.G. 1960. Metabolic polymorphisms and the role of infectious diseases in human evolution. *Hum. Biol.* **32**: 28–62.

Motulsky, A.G. 1995. Jewish diseases and origins. *Nature Genet.* **9**: 99–101.

Motulsky, A.G. 1996*a*. Human genetic variation in nutrition and chronic disease prevention. In M. Moya, G. Sawatzki, A. Motulsky, and J. Morán (eds) *Third International Symposium, Infant nutrition in the prevention of chronic pathology.* Ediciones Ergon S.A., Madrid.

Motulsky, A.G. 1996*b*. Invited editorial: Nutritional ecogenetics; homocysteine-related arteriosclerotic vascular disease, neural tube defects, and folic acid. *Am. J. Hum. Genet.* **58**: 17–20.

Motulsky, A.G. and Brunzell, J.D. 1992. The genetics of coronary atherosclerosis. In King *et al.* (1992), pp. 150–69.

Motulsky, A.G. and Deeb, S.S. 1995. Color vision and its genetic defects. In Scriver *et al.* (1995), Ch. 143.

Motulsky, A.G. and Stamatoyannopoulos, G. 1966. Clinical implications of glucose-6-phosphate dehydrogenase deficiency. *Ann. Intern. Med.* **65**: 1329–34.

Mountain, J.L., Hebert, J.M., Bhattacharyya, S., *et al.* 1995. Demographic history of India and mtDNA-sequence diversity. *Am. J. Hum. Genet.* **56**: 979–92.

Moxon, E.R. and Murphy, P.A. 1978. *Haemophilus influenzae* bacteraemia and meningitis resulting from survival of a single organism. *Proc. Natl. Acad. Sci. USA* **75**: 1534–6.

Moxon, E.R. and Thaler, D.S. 1997. The tinkerer's toolbox. *Nature* **387**: 659–62.

Moxon, E.R., Rainey, P.B., Nowak, M.A, and Lenski, R.E. 1994. Adaptive evolution of highly mutable loci in pathogenic bacteria. *Curr. Biol.* **4**: 24–33.

Mueller, L.D. 1987. Evolution of accelerated senescence in laboratory populations of *Drosophila. Proc. Natl. Acad. Sci. USA* **84**: 1974–7.

Muller, H.J. 1964. The relation of recombination to mutational advance. *Mutation Research* **1**: 2–9.

Munck, A., Guyre, P.M., and Holbrook, N.J. 1984. Physiological functions of glucocorticoids in stress and their relation to pharmacological actions. *Endocrine Rev.* **5**: 25–44.

Murphy, E.A. 1972. The normal, and the perils of the sylleptic argument. *Perspect. Biol. Med.* **15**: 566–82.

Murphy, P.M. 1993. Molecular mimicry and the generation of host defense protein diversity. *Cell* **72**: 823–6.

Murray, C. and Lopez, A. 1996. Evidence-based health policy-lessons from the global burden of disease study. *Science* **274**: 740–3.

Murray, C.J.L. and Lopez, A.D. 1997. Morality by cause for eight regions of the world: Global Burden of Disease Study. *Lancet* **349**: 1269–76.

Muskett, J.C., Reed, N.E., and Thornton, D.H. 1985. Increased virulence of an infectious bursal disease live virus vaccine after passage in chicks. *Vaccine* **3**: 309–12.

Musser, J.M., Kroll, J.S., Branoff, D.M., *et al.* 1990. Global genetic structure and molecular epidemiology of encapsulated *Haemophilus influenzae. Rev. Infect. Dis.* **12**: 75–111.

Nadel, S., Newport, M.J., Booy, R., and Levin, M. 1996. Variation in the tumor necrosis factor alpha gene promoter region may be associated with death from meningococcal disease. *J. Infect. Dis.* **174**: 878–80.

Nalbandov, A.V. 1976. *Reproductive physiology of mammals and birds,* 3rd edn. W. H. Freeman, San Francisco.

Nathanielsz, P.W. 1996. The timing of birth. *Am. Sci.* **84**: 562–9.

Nebert, D.W., McKinnon, R.A., and Puga, A. 1996. Human drug-metabolizing enzyme polymorphisms: effects on risk of toxicity and cancer. *DNA Cell Biol.* **15**: 273–80.

Neel, J.V. 1947. The clinical detection of the genetic carriers of inherited disease. *Medicine* **26**: 115–53.

Neel, J.V. 1962. Diabetes mellitus: a thrifty geno-type rendered detrimental by 'progress'? *Am. J. Hum. Genet.* **14**: 353–62.

Neel, J.V. 1994. *Physician to the gene pool.* John Wiley, New York.

Negri, M.C., Morosini, M.I., Loza, E., and Baquero, F. 1994. *In vitro* selective antibiotic concentrations of beta-lactams for penicillin-resistant *Streptococcus pneumoniae* populations. *Antimicrob. Agents Chemother.* **38**: 122–5.

Nei, M. and Graur, D. 1984. Extent of protein polymorphism and the neutral mutation theory. *Evol. Biol.* **17**: 73–118.

Nei, M. and Roychoudhury, A.K. 1993. Evolutionary relationships of human populations on a global scale. *Mol. Biol. Evol.* **10**: 927–43.

Nelson, D.L. 1993. Six human genetic disorders involving mutant trinucleotide repeats. In K.E. Davis and S.T. Warren (eds) *Genome analysis Vol. 7: Genome rearrangement and stability.* Cold Spring Harbor Laboratory Press, pp. 1–24.

Nelson, D.R., Koymans, L., Kamataki, T., *et al.* 1996. P450 superfamily: update on new sequences, gene mapping, accession numbers and nomenclature. *Pharmacogenetics* **6**: 1–42.

Nelson, K. and Selander, R.K. 1994. Intergeneric transfer and recombination of the 6-phosphogluconate dehydrogenase gene (gnd) in enteric bacteria. *Proc. Natl. Acad. Sci. USA* **91**: 10227–31.

Nesse, R.M. 1994. An evolutionary perspective on substance abuse. *Ethol. Sociobiol.* **15**: 339–48.

Nesse, R.M. 1999. What Darwinian medicine offers psychiatry. In W.R. Trevathan, J.J. McKenna, and E.O. Smith (eds) *Evolutionary medicine.* Oxford University Press (in press).

Nesse, R.M. and Berridge, K.C. 1997. Psychoactive drug use in evolutionary perspective. *Science* **278**: 63–6.

Nesse, R.M. and Klaas, R. 1994. Risk perception by patients with anxiety disorder. *J. Nerv. Ment. Dis.* **182**: 465–70.

Nesse, R.M. and Williams, G. 1994. Why we get sick: the new science of Darwinian medicine. Times Books, New York.

Neu, H.C. 1992. The crisis in antibiotic-resistance. *Science* **257**: 1064–73.

Neu, H.C. 1994. Emerging trends in antimicrobial resistance in surgical infections: a review. *Eur. J. Surg.* (Suppl.): 7–18.

Ni, Y. and Kemp, M.C. 1992. Strain-specific selection of genome segments in avian reovirus coinfections. *J. Gen. Virol.* **73**: 3107–13.

Nichol, S.T., Spiropoulou, C.F., Morzunov, S., *et al.* 1993. Genetic identification of a hantavirus associated with an outbreak of acute respiratory illness. *Science* **262**: 914–17.

Nissenen, A.P., Gronvoos, P., Huovinen, E., *et al.* 1995. Development of β-lactamase-mediated resistance to penicillin in middle-ear isolates of *Moraxella catarrhalis* in Finnish children 1978–1993. *Clin. Infect. Dis.* **21**: 1193–6.

Nitta, D.M., Jackson, M.A., Burry, V.F., and Olson, L.C. 1995. Invasive *Haemophilus influenzae* type f disease. *Pediat. Infect. Dis. J.* **14**: 157–60.

Norton, T.T. 1994. A new focus on myopia. *JAMA* **271**: 1363–4.

Novella, I.S., Duarte, E.A., Elena, S.F., *et al.* 1995. Exponential increase of RNA virus fitness during large population transmission. *Proc. Natl. Acad. Sci. USA* **92**: 5841–4.

Nowak, M.A. 1992. What is a quasispecies? *TREE* **7**: 118–21.

Nowak, M.A. and Bangham, C.R. 1996. Population dynamics of immune responses to persistent viruses. *Science* **272**: 74–9.

Nowak, M.A. and May, R.M. 1993. AIDS pathogenesis: mathematical models of HIV and SIV infections. *AIDS* **7**: S3–S18.

Nowak, M.A. and May, R.M. 1994. Superinfection and the evolution of parasite virulence. *Proc. Roy. Soc. London B* **255**: 81–9.

Nowak, M.A., Anderson, R.M., McLean, A.R., *et al.* 1991. Antigenic diversity thresholds and the development of AIDS. *Science* **254**: 963–9.

O'Dea, K. 1984. Marked improvement in the carbohydrate and lipid metabolism in diabetic Australian Aborigines after temporary reversion to traditional lifetsyle. *Diabetes* **33**: 596–603.

O'Dea, K. 1991. Traditional diet and food preferences of Australian Aboriginal hunter-gatherers. *Proc. Roy. Soc. London B* **334**: 233–41.

O'Gara, B.W. 1969. Unique aspects of reproduction in the female pronghorn (*Antilocapra americana* Ord). *Am. J. Anat.* **125**: 217–32.

O'Rourke, M. and Stevens, E. 1993. Genetic structure of *Neisseria gonorrhoeae* populations: a non-clonal pathogen. *J. Gen. Microbiol.* **139**: 2603–11.

Obaro, S.K., Adegbola, R.A., Banya, W.A., and Greenwood, B.M. 1996. Carriage of pneumococci after pneumococcal vaccination. *Lancet* **348**: 271–2.

Ober, C., Weitkamp, L.R., Cox, N., *et al.* 1997. HLA and mate choice in humans. *Am. J. Hum. Genet.* **61**: 497–505.

Obrebski, S. 1975. Parasite reproductive strategy

and evolution of castration of hosts by parasites. *Science* **188**: 1314–16.

Oftedal, O.T., Bowen, W.D., and Boness, D.J. 1993. Energy transfer by lactating hooded seals and nutrient deposition in their pups during the four days from birth to weaning. *Phys. Zool.* **66**: 412–36.

Öhman, A., Dimberg, U., and Ost, L. 1985. Animal and social phobias: biological constraints on learned fear responses. In S. Reiss and R. Bootzin (eds) *Theoretical issues in behavioral therapy.* Academic Press, Orlando, pp. 123–75.

Olsson, M., Shine, R., Madsen, T., *et al.* 1996. Sperm selection by females. *Nature* **383**: 585.

Oppliger, A., Christe, P., and Richner, H. 1996. Clutch size and malaria resistance. *Nature* **381**: 565.

Orskov, F. and Orskov, I. 1983. Summary of a workshop on the clone concept in the epidemiology, taxonomy and evolution of the Enterobacteriaceae and other bacteria. *J. Infect. Dis.* **148**: 346–57.

Oshima, J., Yu, C-E., Piussan, C., *et al.* 1996. Homozygous and compound heterozygous mutations at the Werner syndrome locus. *Hum. Mol. Genet.* **5**: 1909–13.

Ott, W.J. 1993. Intrauterine growth retardation and preterm delivery. *Am. J. Obstet. Gynecol.* **168**: 1710–17.

Overstreet, J.W. 1983. Transport of gametes in the reproductive tract of the female mammal. In J.F. Hartmann (ed.) *Mechanism and control of animal fertilization.* Academic Press, New York.

Palermo, G., Jons, H., Devroey, P., and van Steirteghem, A.C. 1992. Pregnancies after intracytoplasmic injection of single spermatozoon into an oocyte. *Lancet* **340**: 17–18.

Palinski, W., Ord, V.A., Plump, A.S., *et al.* 1994. ApoE-deficient mice are a model of lipoprotein oxidation in atherogenesis. Demonstration of oxidation-specific epitopesin lesions and high titers of autoantibodies to malondialdehyde-lysine in serum. *Arterioscler. Thromb.* **14**: 605–16.

Palmer, D.A. and Bauchner, H. 1997. Parents' and physicans' views on antibiotics. *Pediatrics* **99**: P6 (electronic version www.pediatrics.org).

Pamilo, P. and Nei., M 1988. Relationship between gene trees and species trees. *Mol. Biol. Evol.* **5**: 568–83.

Panterbrick, C. 1993. Seasonal and sex variation in physical activity levels among agro-pastoralists in Nepal. *Am. J. Phys. Anthropol.* **100**: 7–21.

Parker, G.A. and Maynard Smith, J. 1991. Optimality theory in evolutionary biology. *Nature* **348**: 27–33.

Parker, M.A. 1985. Local population differentiation for compatibility in an annual legume and its host-specific fungal pathogen. *Evolution* **39**: 713–23.

Partridge, L. and Barton, N.H. 1993. Optimality, mutation and the evolution of ageing. *Nature* **362**: 305–11.

Partridge, L. and Barton, N.H. 1996. On measuring the rate of ageing. *Proc. Roy. Soc. London B* **263**: 1365–71.

Partridge, L. and Fowler, K. 1992. Direct and correlated responses to selection on age at reproduction in *Drosophila melanogaster*. *Evolution* **46**: 76–91.

Pate, J., Pumariega, A., Hester, C., and Garner, D. 1992. Cross-cultural patterns in eating disorders: a review. *J. Am. Acad. Child Adolesc. Psychiatry* **31**: 802–9.

PDAY Research Group 1993. Natural history of aortic and coronary atherosclerotic lesions in youth. Findings from the PDAY Study. *Arterioscler. Thromb.* **13**: 1291–8.

Pedersen, J.C. and Berg, K. 1989. Interaction between low density lipoprotein receptor (LDLR) and apolipoprotein E (apoE) alleles contributes to normal variation in lipid level. *Clin. Genet.* **35**: 331–7.

Pennington, R. 1992. Did food increase fertility—evaluation of Kung and Herero history. *Hum. Biol.* **64**: 497–521.

Penrose, L.S. 1954. Some recent trends in human genetics. *Carylogia* **6** (Suppl.): 521–30.

Pereira, M.E. and Pond, C.M. 1995. Organization of white adipose tissue in Lemuridae. *Am. J. Primatol.* **35**: 1–13.

Perelson, A. S., Essunger, P., Cao, Y. *et al.*, 1997. Decay characteristics of HIV-1-infected compartments during combination therapy. *Nature* **387**: 188–91.

Perelson, A.S., Neumann, A.U., Markowitz, M., *et al.* 1996. HIV-1 dynamics *in vivo*: virion clearance rate, infected cell life-span, and viral generation time. *Science* **271**: 1582–6.

Pérez-Lezaun, A., Calafell, F., Mateu, E., *et al.* 1997. Microsatellite variation and the differentiation of modern humans. *Hum. Genet.* **99**: 1–7.

Pericak-Vance, M.A. and Haines, J.L. 1995. Genetic susceptibility to Alzheimer disease. *Trends Genet.* **11**: 504–8.

Peters, W. 1987. *Chemotherapy and drug resistance in malaria.* Academic Press, New York

Peto, J., Easton, D.F., Matthews, F.E., *et al.* 1996. Cancer mortality in relatives of women with breast cancer: the OPCS study. *Int. J. Cancer* **65**: 275–83.

Phillips, D.I.W. 1996. Insulin resistance as a programmed response to fetal undernutrition. *Diabetologia* **39**: 1119–22.

Pichichero, M.E., Francis, A.B., Blatter, M.M., *et al.* 1992. Acellular pertussis vaccination of 2-month-old infants in the United States. *Pediatrics* **89**: 882–7.

Piersma, T. and Lindstrom, A. 1997. Rapid reversible changes in organ size as a component of adaptive behaviour. *TREE* **12**: 134–8.

Pio, A., Leowski, J., and Ten Dam, H.G. 1985. The magnitude of the problem of acute respiratory infections. In R.M. Douglas and E. Kerby-Eaton (eds) *Acute respiratory infections in childhood.* University of Adelaide Press, pp. 3–16.

Plummer, F.A., Simonson, J.H., Chubb, H., *et al.* 1989. Epidemiological evidence for the development of serovar-specific immunity after gonococcal infection. *J. Clin. Invest.* **83**: 1472–6.

Poloni, E.S., Excoffier, L., Mountain, J.L., *et al.* 1995. Nuclear DNA polymorphism in a Mandenka population from Senegal: comparison with eight other human populations. *Ann. Hum. Genet.* **59**: 43–61.

Poloni, E.S., Passarino,G., Santachiara Benerecetti, A.S., *et al.* 1997. Human genetic affinities for Y chromosome p49a,f/Taq I hapotypes show strong correspondence with linguistics. *Am. J. Hum. Genet.* (in press).

Pomerleau, C. 1997. Cofactors for smoking and evolutionary psychobiology. *Addiction* **92**: (in press).

Pond, C.M. 1968. Morphological aspects and the ecological and mechanical consequences of fat deposition in wild vertebrates. *Annu. Rev. Ecol. Syst.* **9**: 519–70.

Pond, C.M. 1997. The biological origins of adipose tissue in humans. In M.E. Morbeck, A. Galloway, and A.L. Zihlman (eds) *The evolving female.* Princeton University Press, pp. 147–62.

Pond, C.M. and Mattacks, C.A. 1987. The anatomy of adipose tissue in captive *Macaca* monkeys and its implications for human biology. *Folia Primatol.* **48**: 164–85.

Pope, C.A., Thun M.J., Namboodiri M.M., *et al.* 1995. Particulate air pollution as a predictor of mortality in a prospective study of U.S. Adults. *Am. J. Resp. Crit. Care Med.* **151**: 669–74.

Potts, W.K. and Wakeland, E.K. 1993. Evolution of MHC genetic diversity: a tale of incest, pestilence and sexual preference. *Trends Genet.* **9**: 408–12.

Potts, W.K., Manning, C.J., Wakeland, E.K., 1991. Mating patterns in seminatural populations of mice influenced by MHC genotype. *Nature* **352**: 619–21.

Power, J.P., Lawlor, E., Davidson, F., *et al.* 1995. Molecular epidemiology of an outbreak of infection with hepatitis C virus in recipients of anti-D immunoglobulin. *Lancet* **345**: 1211–13.

Price, P.N., Duncan, S.L.B., and Levin, R.J. 1981. Oxygen consumption of human endometrium during the menstrual cycle measured *in vitro* using an oxygen electrode. *J. Reprod. Fert.* **63**: 185–92.

Profet, M. 1993. Menstruation as a defense against pathogens transported by sperm. *Q. Rev. Biol.* **68**: 335–86.

Promislow, D.E.L., Tatar, M., Khazaeli, A.A., and Curtsinger, J.W. 1996. Age-specific patterns of genetic variance in *Drosophila melanogaster*. I. Mortality. *Genetics* **143**: 839–48.

Radman, M., Matic, I., Halliday, J., and Taddei, F. 1995. Editing DNA replication and recombination by mismatch repair: from bacterial genetics to mechanisms of predisposition to cancer in humans. *Phil. Trans. Roy. Soc. London B* **347**: 97–103.

Rahman, M. and DaVanzo, J. 1993. Gender preferences and birth spacing in Matlab, Bangladesh. *Demography* **30**: 315–24.

Rainey, P.B., Moxon, E.R., and Thompson, I.P. 1993. Intraclonal polymorphism in bacteria. *Adv. Microb. Ecol.* **13**: 263–300.

Rasa, A., Vogel, C., and Voland, E. (eds). 1989. *The sociobiology of sexual and reproductive strategies.* Chapman & Hall, London.

Rawls, J. 1971. *A theory of justice.* Harvard University Press, Cambridge, Massachusetts.

Raza, M.W., Blackwell, C.C., Molyneaux, P., *et al.* 1991. Association between secretor status and respiratory viral illness. *Br. Med. J.* **303**: 815–18.

Read, A.F. 1994. The evolution of virulence. *Trends in Microbiol.* **2**: 73–6.

Read, A.F. and Viney, M.E. 1996. Helminth immunogenetics: why bother? *Parasitol. Today* **12**: 337–43.

Read, A.F., Anwar, M., Shutler, D., and Nee, S. 1995. Sex allocation and population structure in malaria and related parasitic protozoa. *Proc. Roy. Soc London B* **260**: 359–63.

Redd, A.J., Takezaki, N., Sherry, S.T., *et al.* 1995.

Evolutionary history of the COII/tRNA intergenic 9 base pair deletion in human mitochondrial DNAs from the Pacific. *Mol. Biol. Evol.* **12**: 604–15.

Reder, J. 1918. Die Bedeutung der Masern in Sammelniederlassungen nach den in k.k. Flüchtlingslager Gmünd gemachten Wahrnehmungen. *Z. Kinderheilk* **18**: 355–70.

Redfield, R. J. 1993. Genes for breakfast—the have-your-cake-and-eat-it-too of bacterial transformation. *J. Hered.* **84**: 400–14.

Redman, C. 1989. Hypertension in pregnancy. In M. de Swiet (ed.) *Medical disorders in obstetric practice*, 2nd edn. Blackwell Scientific, Oxford, pp. 249–305.

Reilly, S.L., Ferrell, R.E., Kottke, B.A., *et al.* 1991. The gender-specific apolipoprotein E genotype influence on the distribution of lipids and apolipoproteins in the populations of Rochester, MN. I. Pleiotropic effects on means and variances. *Am. J. Hum. Genet.* **49**: 1155–66.

Reilly, S.L., Ferrell, R.E., Kottke, B.A., and Sing, C.F. 1992. The gender specific apolipoprotein E genotype influence on the distribution of plasma lipids and apolipoproteins in the population of Rochester, Minnesota. II. Regression relationships with concomitants. *Am. J. Hum. Genet.* **51**: 1311–24.

Reilly, S.L., Ferrell, R.E., and Sing, C.F. 1994. The gender-specific apolipoprotein E genotype influence on the distribution of plasma lipids and apolipoproteins in the population of Rochester, MN. III. Correlations and covariances. *Am. J. Hum. Genet.* **55**: 1001–18.

Reilly, S.L., Haviland, M.B., Peyser, P., *et al.* 1995. Apolipoprotein E genotype as a predictor of presence of coronary artery calcification (CAC). The American Society of Human Genetics. 45th Annual Meeting. Minneapolis, MN. *Am. J. Hum. Genet.* **57**: A11.

Renfrew, C. 1987. *Archaeology and language. The puzzle of Indo-European origins.* Jonathan Cape, London.

Renfrew, C. 1991. Before Babel: speculations on the origins of linguistic diversity. *Cambridge Archaeol. J.* **1**: 3–23.

Rich-Edwards, J., Stampfer, M., Manson, J., *et al.* 1995. Birthweight, breastfeeding, and the risk of coronary heart disease in the Nurses' Health Study. *Am. J. Epidemiol.* **141**: S78.

Richards, R.L., Kinner, D.K., Lunde, I., and Benet, M. 1988. Creativity in manic-depressives, cyclothymes and their normal first-degree relatives: a preliminary report. *J. Abnorm. Psychol.* **97**: 281–8.

Rico-Hesse, R. 1990. Molecular evolution and distribution of dengue viruses type-1 and type 2 in nature. *Virology* **174**: 479–93.

Riddle, D.L. 1988. The dauer larva. In W.B. Wood (ed.) *The nematode* Caenorhabditis elegans. Cold Spring Harbor Laboratory Press, New York, pp. 393–412.

Ridley, M. 1993. *Evolution.* Blackwell Scientific, Boston.

Riley, M. 1993. Functions of the gene products of *Escherichia coli. Microbiol. Rev.* **57**: 862–952.

Risch, A., Wallace, D.M.A., Bathers, S., and Sim, E. 1995. Slow *N*-acetylation genotype is a susceptibility factor in occupational and smoking related bladder cancer. *Hum. Mol. Genet.* **4**: 231–36.

Risch, N. and Merikangas, K. 1996. The future of genetic studies of complex human diseases. *Science* **273**: 1516–17.

Risch, N., de Leon, D., Ozelius, L., *et al.* 1995. Genetic analysis of idiopathic torsion dystonia in Ashkenazi Jews and their recent descent from a small founder population. *Nature Genet.* **9**: 152–9.

Rist, N. 1964. Nature and development of resistance of tubercle bacilli to chemotherapeutic agents. In V.C. Barry (ed.) *Chemotherapy of tuberculosis.* Butterworths, London, pp. 192–227.

Roberts, A.G., Whatley, S.D., Morgan, R.R., *et al.* 1997. Increased frequency of the haemochromatosis Cys282Tyr mutation in sporadic porphyria cutanea tarda. *Lancet* **349**: 321–3.

Roberts, C. and Manchester, K. 1995. *The archaeology of disease*, 2nd edn. Cornell University Press, Ithaca.

Roberts, D.J., Craig, A.G., Berend, A.R., *et al.* 1992. Rapid switching to multiple antigenic and adhesive phenotypes in malaria. *Nature* **357**: 689–92.

Roberts, M. S. and Cohan, F.M. 1993. The effect of DNA sequence divergence on sexual isolation in *Bacillus. Genetics* **134**: 401–8.

Robertson, B.D. and Meyer, T.F. 1992. Genetic variation in pathogenic bacteria. *Trends Genet.* **8**: 422–7.

Robertson, D.L., Sharp, P.M., McCuthan. F.E., and Hahn, B.H. 1995. Recombination in HIV-1. *Nature* **374**: 124–6.

Robertson, T.L., Kato, H., Rhodes, G.G., *et al.* 1977. Epidemiologic studies of coronary heart disease and stroke in Japanese men living in Japan, Hawaii and California: incidence of myocardial infarction and death from coronary heart disease. *Am. J. Cardiol.* **39**: 239–43.

Roche, R.J., High, N.J., and Moxon, E.R. 1994. Phase-variation of *Haemophilus influenzae* lipopolysaccharide: characterisation of lipopolysaccharide from individual colonies. *FEMS Microbiol. Letters* **120**: 279–84.

Rode, A. and Shephard, R.J. 1994. Physiological consequences of acculturation: a 20-year study of fitness in an Inuit community. *Eur. J. Appl. Physiol.* **69**: 516–24.

Rogers, A. 1995a. For love or money: the evolution of reproductive and material motivations. In R.I.M. Dunbar (ed.) *Human reproductive decisions.* MacMillan, London, pp. 76–95.

Rogers, A. 1995b. Genetic evidence for a Pleistocene population explosion. *Evolution* **49**: 608–15.

Rogers, A.R. 1993. Why menopause? *Evol. Ecol.* **7**: 406–20.

Rogers, A.R. and Harpending, H. 1992. Population growth makes waves in the distribution of pairwise genetic differences. *Mol. Biol. Evol.* **9**: 552–69.

Rook, G.A. 1988. The role of vitamin D in tuberculosis. *Am. Rev. Respir. Dis.* **138**: 768–70.

Roosevelt, A.C. 1984. Population, health, and the evolution of subsistence: conclusions from the conference. In Cohen and Armelagos (1984a), pp. 559–81.

Roper, C., Pignatelli, P., and Partridge, L. 1993. Evolutionary effects of selection on age at reproduction in larval and adult *Drosophila melanogaster. Evolution* **47**: 445–55.

Rose, M.R. 1984. Laboratory evolution of postponed senescence in *Drosophila melanogaster. Evolution* **38**: 1004–10.

Rose, M.R. 1991. *The evolutionary biology of ageing.* Oxford University Press.

Rose, M.R. and Lauder, G.V. (eds). 1996. *Adaptation.* Academic Press, New York.

Rosenberg, S.M., Longerich, S., Gee, P., and Reuben, S.H. 1994. Adaptive mutation by deletions in small mononucleotide repeats. *Science* **265**: 405–7.

Rota, J.S., Hummel, K.B., Rota, P.A., and Bellini, W.J. 1992. Genetic variability of the glycoprotein genes of current wild-type measles isolates. *Virology* **188**: 135–42.

Roth, E.A. 1985. A note on the demographic concomitants of sedentism. *Am. Anthropol.* **87**: 380–2.

Roth, E.F. Jr., Raventos-Suarez, C., Rinaldi, A., and Nagel, R.L. 1983. Glucose-6-phosphate dehydrogenase deficiency inhibits *in vitro* growth of *Plasmodium falciparum. Proc. Natl. Acad. Sci. USA* **80**: 298–9.

Rotter, J.I. and Diamond, J.M. 1987. What maintains the frequencies of human genetic diseases? *Nature* **329**: 289–90

Roy, S., McGuire, W., Mascie-Taylor, C.G.N., et al. 1997. Association of a tumor necrosis factor polymorphism with lepromatous but not tuberculoid leprosy. *J. Infect. Dis.* **176**: 530–2.

Rubin, L.G. 1987. Bacterial colonisation and infection resulting from multiplication of a single organism. *Rev. Infect. Dis.* **9**: 488–93.

Rubinsztein, D.C., Amos, W., Leggo, J., et al. 1994. Mutational bias provides a model for the evolution of Huntington's disease and predicts a general increase in disease prevalence. *Nature Genet.* **7**: 525–30.

Ruff, C.B. 1992. Biomechanical analyses of archaeological human material. In S.R. Saunders and A. Katzenberg (eds) *The skeletal biology of past peoples.* Alan R. Liss, New York, pp. 41–62.

Ruff, C.B., Trinkaus, E., Walker, A., and Larsen, C.S. 1993. Postcranial robusticity in *Homo.* I: temporal trends and mechanical interpretations. *Am. J. Phys. Anthropol.* **91**: 21–53.

Ruhlen, M. 1991. *A guide to the world's languages. Vol. 1: Classification.* Edward Arnold, London.

Rülicke, T., Chapuisat, M., Homberger, F.R. et al. 1998. MHC-genotype of progeny influenced by parental infection. *Proc. Roy. Soc. London B* **265**: 711–16.

Runger, T.M., Bauer, C., Dekant, B., et al. 1994. Hypermutable ligation of plasmid DNA ends in cells from patients with Werner syndrome. *J. Invest. Dermatol.* **102**: 45–8.

Rush, A.J., Kain, J.W., and Rease, J. 1991. Neurological bases for psychiatric disorders. In R.N. Rosenberg (ed.) *Comprehensive neurology.* Raven Press, New York, pp. 555–603.

Russo, A.R., Ausman, L.M., Gallina, D.L., and Hegsted, D.M. 1980. Developmental body composition of the squirrel monkey (*Saimiri sciureus*). *Growth* **44**: 271–86.

Ruwende, C., Khoo, S.C., Snow, R.W., et al. 1995. Natural selection of hemi- and heterozygotes for G6PD deficiency in Africa by resistance to severe malaria. *Nature* **376**: 246–9.

Ryan, F. 1993. *The forgotten plague.* Little Brown, Boston.

Ryman, N., Chakraborty, R., and Nei, M. 1983. *Hum. Hered.* **33**: 93–102.

Sabin, A.B. 1993. HIV vaccination dilemma. *Nature* **362**: 212.

Sacher, G.A. 1978. Evolution of longevity and survival characteristics in mammals. In E.L.

Schneider (ed) *The genetics of aging.* Plenum Press, New York, pp. 151–67.

Saitou, N. 1996. Contrasting gene trees and population trees of the evolution of modern humans. In Boyce and Mascie-Taylor (1996), pp. 265–82.

Sajantila, A., Lahermo, P., Anttinen, T., *et al.* 1995. Genes and languages in Europe: analysis of mitochondrial lineages. *Genome Res.* **5**: 42–52.

Salk, D., Au, K., Hoehn, H., and Martin, G.M. 1981. Cytogenetics of Werner's syndrome cultured skin fibrobasts, variegated translocation mosaicism. *Cytogenet. Cell Genet.* **30**: 92–107.

Samson, M., Libert, F., Doranz, B.J., *et al.* 1996. Resistance to HIV-1 infection in Caucasian individuals bearing mutant alleles of the CCR-5 chemokine receptor gene. *Nature* **382**: 722–5.

Sanchez-Mazas, A. and Langaney, A. 1988. Common genetic pools between human populations. *Hum. Genet.* **78**: 161–6.

Sanchez-Mazas, A. and Pellegrini, B. 1990. Polymorphismes Rhésus, Gm et HLA et histoire de l'Homme moderne. *Bull. Mém. Société d'Anthrop. Paris* **2** (1): 57–76.

Sanchez-Mazas, A., Dard, P., Excoffier, L., and Langaney, A. 1995. Selective neutrality tested on HLA. *Eur. J. Immunogen.* **22**: 131 (abstract).

Sanger, F. *et al.* 1977. Nucleotide sequencing of bacteriophage phi X174 DNA. *Nature* **246**: 687–95.

Sanger, F., Coulson, A.R., Hong, G.F., *et al.* 1982. Nucleotide sequence of bacteriophage lambda DNA. *J. Mol. Biol.* **162**: 729–73.

Santos, M., Schilham, M.W., Rademakers, L.H.P.M., *et al.* 1996. Defective iron homeostasis in beta 2-microglobulin knockout mice recapitulates hereditary hemochromatosis in man. *J. Exp. Med.* **184**: 1975–85.

Sapolsky, R.M. 1992. Stress, the aging brain, and the mechanisms of neuron death. MIT Press, Cambridge, Massachusetts.

Satsangi, J., Parkes, M., Louis, L., *et al.* 1996. Two stage genome-wide search in inflammatory bowel disease provides evidence for susceptibility loci on chromosomes 3, 7 and 12. *Nature Genet.* **14**: 199–202.

Satta, Y., O'Huigin, C., Takahata, N., and Klein, J. 1994. Intensity of natural selection at the major histocompatibility complex loci in primates. *Proc. Natl. Acad. Sci. USA* **91**: 7184–8.

Saunders, J.R. 1995. Population genetics of phase variable antigens. In Baumberg *et al.* (1995), pp. 247–68.

Sawyer, S.A. 1989. Statistical tests for detecting gene conversion. *Mol. Biol. Evol.* **6**: 526–38.

Schächter, F., Cohen, D., and Kirkwood T.B.L. 1993. Prospects for the genetics of human longevity. *Hum. Genet.* **91**: 519–26.

Schiefenhövel, W. 1989. Reproduction and sex-ratio manipulation through preferential female infanticide among the Eipo, in the highlands of West New Guinea. In Rasa *et al.* (1989), pp. 170–93.

Schlesinger, L.S. 1996. Entry of *Mycobacterium tuberculosis* into mononuclear phagocytes. *Curr. Top. Microbiol. Immunol.* **215**: 71–96.

Schmidt-Nielsen, K. 1990. *Animal physiology: adaptation and environment.* Cambridge University Press.

Schrag, S. and Perrot, V. 1996. Reducing antibiotic resistance. *Nature* **381**: 120–1

Schrag, S., Perrot, V., and Levin, B.R. 1997. Adaptation to the fitness cost of antibiotic resistance in *Escherichia coli. Proc. Roy. Soc. London B* (in press).

Schuitemaker, H., Koot, M., Kootstra, N.A., *et al.* 1992. Biological phenotype of human immunodeficiency virus type 1 clones at different stages of infection: progression of disease is associated with a shift from monocytotropic to T-cell-tropic virus populations. *J. Virol.* **66**: 1354–60.

Schultz, A.H. 1926. Fetal growth of man and other primates. *Q. Rev. Biol.* **1**: 465–521.

Schultz, A.H. 1936. Characters common to higher primates and characters specific for man. *Q. Rev. Biol.* **11**: 259–83.

Schultz, A.H. 1940. Growth and development of the chimpanzee. Carnegie Inst. Washington Publ. 518. *Contrib. Embryol.* **28**: 1–63.

Schultz, A.H. 1969. *The life of primates.* Universe Books, New York.

Schuster, P. 1993. RNA based evolutionary optimization. *Origins Life Evol. Biosphere* **23**: 373–91.

Schuster, P., Fontana, W., Stadler, P.F., and Hofacker, I.L. 1994. From sequences to shapes and back: a case study in RNA secondary structures. *Proc. Roy. Soc. London B* **255**: 279–84.

Schwarcz, H.P., Grün, R., Vandermeersch, B., *et al.* 1988. ESR dates for the hominid burial site of Qafzeh in Israel. *J. Hum. Evol.* **17**: 733–7.

Schwartz, S.M., deBlois, D., and OBrien, E.R.M. 1995*a*. The intima: soil for atherosclerosis and restenosis. *Circ. Res.* **77**: 445–65.

Schwartz, S.M., Majesky, M.W., and Murry, C.E. 1995*b*. The intima: development and monoclonal responses to injury. *Atherosclerosis* **118**: S125–S140.

Scofield, V.L., Schlumpberger, J.M., West, L.A., and Weissman, I.L. 1982. Protochordate allorecognition is controlled by a MHC-like gene system. *Nature* **295**: 499–502.

Scriver, C.R., Beaudet, A.L, Sly, W.S., and Valle, D. 1995. *The metabolic and molecular basis of inherited diseases*, 7th edn. McGraw-Hill, New York.

Seckler, D. 1980. 'Malnutrition': an intellectual odyssey. *West. J. Agric. Econ.* **5**: 219–27.

Secor, S.M. and Diamond, J. 1995. Adaptive responses to feeding in Burmese pythons: pay before pumping. *J. Exp. Biol.* **198**: 1313–25.

Selander, R.K. and Levin, B.R. 1980. Genetic diversity and structure in *Escherichia coli* populations. *Science* **210**: 545–7.

Selander, R.K., Musser, J.M., Caugant, D.A., *et al.* 1987. Population genetics of pathogenic bacteria. *Microb. Pathog.* **3**: 1–7.

Selander, R.K., Li, J., and Nelson, K. 1996. Evolutionary genetics of *Salmonella enterica*. In F.C. Neidhardt, R. Curtiss, J.L. Ingraham, *et al.* (eds) Escherichia coli *and* Salmonella. ASM Press, Washington DC.

Seppala, H., Llaukka, T., Vuopio-Varkila, J., *et al.* 1997. The effects of changes in the consumption of macrolide antibiotics of erythromycin resistance in group A streptocci in Finland. *N. Engl. J. Med.* **337**: 441–6.

Sergent, E. and Sergent, E. 1921. Étude experimentale du paludisme. Paludisme des oiseaux (*Plasmodium relictum*). *Bull. Soc. Pathol. Exotique* **14**: 72–8.

Serjeantson, S.W. 1989. HLA genes and antigens. In A.V.S. Hill and S.W. Serjeantson (eds) *The colonization of the Pacific: a genetic trail*. Clarendon Press, Oxford, pp. 120–73.

Serjeantson, S.W. and Gao, X. 1995. *Homo sapiens* is an evolving species: origins of the Austronesians. In P. Bellwood, J.J. Fox, and D. Tryon (eds) *The Austronesians: historical and comparative perspectives*. Canberra ACT, pp. 165–80.

Shanley, D.P. and Kirkwood, T.B.L. 1998. Is the extended lifespan of calorie-restricted mice an evolutionary adaptation? *Lifespan* (in press).

Sharp, P.M., Robertson, D.L., and Hahn, B.H. 1995. Cross-species transmission and recombination of AIDS viruses. *Phil. Trans. Roy. Soc. London B* **349**: 41–7.

Shavit, Y., Fischer, C. S., and Koresh, Y. 1994. Kin and nonkin under collective threat: Israeli networks during the Gulf War. *Social Forces* **72**: 1197–215.

Sheinfeld, J., Schaeffer, A.J., Cordon-Cardo, C., *et al.* 1989. Association of the Lewis blood-group phenotype with recurrent urinary tract infections in women. *N. Engl. J. Med.* **320**: 773–7.

Shimamoto, T., Komachi, Y., Inada, H., *et al.* 1989. Trends for coronary heart disease and stroke and their risk factors in Japan. *Circulation* **79**: 503–15.

Short, R. 1976. The evolution of human reproduction. *Proc. Roy. Soc. London B* **195**: 3–24.

Shpaer, E.G. and Mullins, J.I. 1993. Rates of amino acid change in the envelope protein correlate with pathogenicity of primate lentiviruses. *J. Mol. Evol.* **37**: 57–65.

Shprentz, D.S. 1996. *Breathtaking: premature mortality due to particulate air pollution in 239 American cities*. Natural Resources Defense Council, New York.

Siegel, E.C. and Bryson, V. 1967. Mutator gene of *Escherichia coli* B. *J. Bacteriol.* **94**: 38–47.

Siegwart, J.T. and Norton, T.T. 1993. Refractive and ocular changes in tree shrews raised with plus or minus lenses. *Invest. Ophthalmol. Vis. Sci.* **34**: 1208.

Simmonds, P., Holmes, E.C., Cha, T.A., *et al.* 1993. Classification of hepatitis C virus into 6 major genotypes and a series of subtypes by phylogenetic analysis of the NS-5 region. *J. Gen. Virol.* **74**: 2391–9.

Simmonds, P., Mellor, J.M., Sakuldamrongpanich, T., *et al.* 1996. Evolutionary analysis of variants of hepatitis C virus found in South East Asia: comparison with classification based on sequence similarity. *J. Gen. Virol.* **77**: 3013–24.

Simon, M., Zieg, J., Silverman, M., *et al.* 1980. Phase variation in the evolution of a controlling element. *Science* **209**: 1370–4.

Simoons, F.J. 1978. The geographic hypothesis and lactose malabsorption. A weighing of the evidence. *Dig. Dis.* **23**: 963–80.

Simopoulos, A.P. and Childs, B. 1990. *Genetic variation and nutrition*. Karger, Basel.

Sindrup, S.H. and Brøsen, K. 1995. The pharmacogenetics of codeine hypoalgesia. *Pharmacogenetics* **5**: 335–46.

Sindrup, S.H., Poulsen, L., Brosen, K., *et al.* 1993. Are poor metabolisers of sparteine/debrisoquine less pain tolerant than extensive metabolisers. *Pain* **53**: 335–9.

Sing, C.F. and Davignon, J. 1985. Role of apolipoprotein E genetic polymorphism in determining normal plasma lipid and lipoprotein variation. *Am. J. Hum. Genet.* **37**: 268–85.

Sing, C.F. and Hanis, C.L. (eds). 1993. *Genetics of cellular, individual, family and population variability*. Oxford University Press.

Sing, C. and Harris, C. (eds). 1993. *Genetics of cellular, individual, family, and population viability.* Oxford University Press, New York.

Sing, C.F. and Moll, P.P. 1989. Genetics of variability of CHD risk. *Int. J. Epidemiol.* **18** (Suppl. 1): S183–S195.

Sing, C.F. and Reilly, S.L. 1993. Genetics of common diseases that aggregate, but do not segregate, in families. In Sing and Hanis (1993), pp. 140–61.

Sing, C.F., Boerwinkle, E., Moll, P.P., and Templeton, A.R. 1988. Characterization of genes affecting quantitative traits in humans. In B.S. Weir, E.J. Eisen, M.M. Goodman, and G. Namkoong (eds) *Proceedings of the Second International Conference on Quantitative Genetics.* Sinauer, Sunderland, Massachusetts, pp. 250–69.

Sing, C.F., Haviland, M.B., Templeton, A.R., *et al.* 1992. Biological complexity and strategies for finding DNA variations responsible for interindividual variation in risk of a common chronic disease, coronary artery disease. *Ann. Med.* **24**: 539–47.

Sing, C.F., Haviland, M.B., and Reilly, S.L. 1996. Genetic architecture of common multifactorial diseases. In *Variation in the human genome (Ciba Foundation Symposium* **197**). John Wiley, Chichester, pp. 211–32.

Singh, D. and Young, R.K. 1995. Body weight, waist-to-hip ratio, breasts, and hips: role in judgments of female attractiveness and desirability for relationships. *Ethol. Sociobiol.* **16**: 483–507.

Singh, N., Agrawal, S., and Rastogi, A.K. 1997. Infectious diseses and immunity: special reference to major histocompatibility complex. *Emerg. Infect. Dis.* **3**: 41–9.

Singh, S.P., Mehra, N.K., Dingley, H.B., *et al.* 1983. Human leukocyte antigen (HLA)-linked control of susceptibility to pulmonary tuberculosis and association with HLA-DR types. *J. Infect. Dis.* **148**: 676–81.

Siniscalco, M., Bernini, L., Latte, B., and Motulsky, A.G. 1961. Favism and thalassaemia and their relationship to malaria. *Nature* **190**: 1179–80.

Slatkin, M. and Hudson, R.R. 1991. Pairwise comparisons of mitochondrial DNA sequences in stable and exponentially growing populations. *Genetics* **129**: 555–62.

Slavin, M. and Kriegman, D. 1992. *The adaptive design of the human psyche: psychoanalysis, evolutionary biology, and the therapeutic process.* Guilford Press, New York.

Small, P.M. and Moss, A. 1993. Molecular epidemiology and the new tuberculosis. *Infect. Agents Dis.* **2**: 132–8.

Smith, A.H., Butler, T.M., and Pace, N. 1975. Weight growth of colony-reared chimpanzees. *Folia Primatol.* **24**: 29–59.

Smith, B.H. and Tompkins, R.L. 1995. Toward a life history of the Hominidae. *Annu. Rev. Anthropol.* **24**: 257–79.

Smith, D.A. 1991. Species differences in metabolism and pharmacokinetics: are we close to an understanding? *Drug Met. Rev.* **23**: 355–73.

Smith, D.W.E. 1994. *Human longevity.* Oxford University Press.

Smith, E.A. 1992*a*. Human behavioral ecology: I. *Evol. Anthropol.* **1**: 20–25.

Smith, E.A. 1992*b*. Human behavioral ecology: II. *Evol. Anthropol.* **1**: 50–55.

Smith, G., Stanley, L.A., Sim, E., *et al.* 1995*a*. *Metabolic polymorphisms and cancer susceptibility. Cancer Surveys XXV. Genetics and cancer: a second look.* ICRF, Lyon.

Smith, H.O., Tomb, J.F., Dougherty, B.A., *et al.* 1995*b*. Frequency and distribution of DNA uptake sequence sites in the *Haemophilus infleunzae* Rd genome. *Science* **269**: 1–3.

Smith, N.H., Maynard Smith, J., and Spratt, B.G. 1995*c*. Sequence evolution of the porB gene of *Neisseria gonorrhoeae* and *Neisseria meningitidis;* evidence of positive Darwinian evolution. *Mol. Biol. Evol.* **12**: 363–370.

Smith, P., Bar-Yosel, O., and Sillen, A. 1984. Archaeological and skeletal evidence for dietary change during the late Pleistocene/early Holocene in the Levant. In Cohen and Armelagos (1984), pp. 101–36.

Smith, T., Lehmann, D., Montgomery, J., *et al.* 1993. Acquisition and invasiveness of different serotypes of *Streptococcus pneumoniae* in young children. *Epidemiol. Infect.* **111**: 27–39.

Sniegowski, P.D., Gerrish, P.I., and Lenski, R.E. 1997. Evolution of high mutation rates in experimental population of *E. coli. Nature* **387**: 703–5.

Sokal, R.R. 1988. Genetic, geographic, and linguistic distances in Europe. *Proc. Natl. Acad. Sci. USA* **85**: 1722–6.

Sokal, R.R. and Thomson, B.A. 1997. Spatial genetic structure of human populations in Japan. *Hum. Biol.* (in press).

Sokal, R.R., Oden, N.L., and Wilson, C. 1991. Genetic evidence for the spread of agriculture in Europe by demic diffusion. *Nature* **351**: 143–5.

Soloman, S.J., Kurzer, M.S., and Calloway, D.H.

1982. Menstrual cycle and basal metabolic rate in women. *Am. J. Clin. Nutr.* **36**: 611–16.

Sorensen, T.I., Nielsen, G.G., Andersen, P.K., and Teasdale T.W. 1988. Genetic and environmental influences on premature death in adult adoptees. *N. Engl. J. Med.* **318**: 727–32

Soto-Ramirez, L.E., Renjifo, B., McLane, M.F., *et al.* 1996. HIV-1 Langerhans' cell tropism associated with heterosexual transmission of HIV. *Science* **271**: 1291–3.

Southgate, D.A.T. and Hey, E.N. 1976. Chemical and biochemical development of the human fetus. In D.F. Roberts and A.M. Thomson (eds) *The biology of human fetal growth. Symp. Soc. Study Hum. Biol*, Vol. 15. Taylor & Francis, London, pp. 195–209.

Spence, J. 1954. *One thousand families in Newcastle upon Tyne.* Oxford University Press.

Speth, J.D. 1987. Early hominid subsistence strategies in seasonal habitats. *J. Archaeol. Sci.* **14**: 13–29.

Spicer, D.V. and Pike, M.C. 1993. Breast cancer prevention through modulation of endogenous hormones. *Breast Cancer Res. Treat.* **28**: 179–93.

Spielmann, R.S., Fajans, S.S., Neel, J.V., *et al.* 1982. Glucose tolerance in two unacculturated Indian tribes of Brazil. *Diabetologia* **23**: 90–3.

Spina, E., Gitto, C., Avenoso, A., *et al.* 1997. Relationship between plasma desipramine levels, CYP2D6 phenotype and clinical response to desipramine, a prospective study. *Eur. J. Clin. Pharmacol.* **51**: 395–8.

Spratt, B.G., Zhang, Q.-Y., Jones, D.M., *et al.* 1989. Recruitment of a penicillin-binding gene from *Neisseria flavescens* during the emergence of penicillin resistance in *Neisseria meningitidis*. *Proc. Natl. Acad. Sci. USA* **86**: 8988–92.

Squitieri, F., Andrew, S.E., Goldberg, Y.P., *et al.* 1994. DNA haplotype analysis of Huntington disease reveals clues to the origins and mechanisms of CAG expansion and reasons for geographic variations of prevalence. *Hum. Mol. Genet.* **3**: 2103–14.

Staddon, J.E.R. 1983. *Adaptive behavior and learning.* Cambridge University Press.

Staeheli, P., Haller, O., Boll, W., *et al.* 1986. Mx protein: constitutive expression in 3T3 cells transformed with cloned Mx cDNA confers selective resistance to influenza virus. *Cell* **44**: 147–58.

Stearns, S.C. 1987*a*. The evolution of sex and its consequences. Birkhäuser Verlag, Basel.

Stearns, S.C. 1987*b*. The selection-arena hypothesis. In Stearns (1978*a*), pp. 337–80.

Stearns, S.C. 1992. *The evolution of life histories.* Oxford University Press.

Stefaneanu, L.L., Kovacs, K., Lloyd, R.V., *et al.* 1992. Pituitary lactotrophs and somatotrophs in pregnancy: a correlative *in situ* hybridization and immunocytochemical study. *Virchows Archiv B Cell Path.* **62**: 291–6.

Stein, C.E., Fall, C.H.D., Kumaran, K., *et al.* 1996. Fetal growth and coronary heart disease in South India. *Lancet* **348**: 1269–73.

Stein, Z., Susser, M., Saenger, G., and Marolla, F. 1975. *Famine and human development.* Oxford University Press, New York.

Steinberg, D. 1997. Lewis A. Conner Memorial Lecture: Oxidative modification of LDL and atherogenesis. *Circulation* **95**: 1062–71.

Steiner, M.C. and Kuhn, S.L. 1992. Subsistence technology and adaptive variation in Middle Paleolithic Italy. *Am. Anthropol.* **94**: 306–39.

Stengård, J.H., Zerba, K.E., Pekkanen, J., *et al.* 1995. Apolipoprotein E polymorphism predicts death from coronary heart disease in a longitudinal study of elderly Finnish men. *Circulation* **91**: 265–9.

Stengård, J.H., Pekkanen, J., Ehnholm, C., *et al.* 1996. Genotypes with the apolipoprotein E4 allele are predictors of coronary heart disease mortality in a longitudinal study of elderly Finnish men. *Hum. Genet.* **97**: 677–84.

Stengård, J.H., Pekkanen, J., Ehnholm, C., *et al.* Utility of the predictors of coronary heart disease mortality in a longitudinal study is dependent on Apo E genotype in elderly Finnish men aged 65 to 84 years. *Circulation* (in press).

Stephens, J. 1985. Statistical methods of DNA sequence analysis: detection of intragenic recombination or gene conversion. *Mol. Biol. Evol.* **2**: 539–56.

Stern, A., Brown, M., Nickel, P., and Meyer, T.F. 1986. Opacity genes in *Neisseria gonorrhoeae*: control of phase and antigenic variation. *Cell* **47**: 61–7.

Stevens, A. and Price, J. 1996. *Evolutionary psychiatry: A new beginning.* Routledge, London.

Stewart, F.M. and Levin, B.R. 1977. The population biology of bacterial plasmids: a priori conditions for the existence of conjugationally transmitted factors. *Genetics* **87**: 209–28.

Stewart, F.M., Antia, R., Levin, B.R., *et al.* 1997. The population dynamics of antibiotic resistance. II: Analytic theory for sustained populations of bacteria in a population of hosts. *Theor. Pop. Biol* (in press).

Stock, E.P., Stock, A.M., and Mottonen, J.M. 1990. Signal transduction in bacteria. *Nature* **344**: 395–400.

Stockley, P. 1997. Sexual conflict resulting from adaptations to sperm competition. *TREE* **12**: 154–9.

Stoltzfus, A., Leslie, J.F., and Milkman, R. 1988. Molecular evolution of the *Escherichia coli* chromosome. I. Analysis of structure and natural variation in a previously uncharacterised region between trp and tonB. *Genetics* **120**: 345–58.

Strassmann, B.I. 1992. The function of menstrual taboos among the Dogon: defense against cukoldry? *Hum. Nat.* **3**: 89–131.

Strassmann, B.I. 1996a. Menstrual hut visits by Dogon women: a hormonal test distinguishes deceit from honest signaling. *Behav. Ecol.* **7**: 304–15.

Strassmann, B.I. 1996b. The evolution of endometrial cycles and menstruation. *Q. Rev. Biol.* **71**: 181–220.

Strassmann, B.I. 1997. The biology of menstruation in *H. sapiens*: total lifetime menses, fecundity, and nonsynchrony in a natural fertility population. *Curr. Anthropol.* **38**: 123–9.

Strauss, J.F., Gåfvels, M., and King, B.F. 1995. Placental hormones. In L.J. DeGroot (ed.) *Endocrinology*, 3rd edn. W.B. Saunders, Philadelphia, pp. 2171–206.

Streisinger, G., Okada, Y., Emrich, J., *et al.* 1966. Frameshift mutations and the genetic code. *Cold Spring Harbor Symp. Quant. Biol.* **31**: 77.

Stringer, C.B. and Andrews, P. 1988. Genetic and fossil evidence for the origin of modern humans. *Science* **239**: 1263–8.

Stuyt, P.M.J., Brenninkmeijer, B.J., Demacker, P.N.M., *et al.* 1991. Apolipoprotein E phenotypes, serum lipoproteins and apolipoprotein in angiographically assessed coronary heart disease. *Scand. J. Clin. Lab. Invest.* **51**: 425–35.

Suchindran, C.M. and Lachenbruch, P.A. 1975. Estimates of fecundability from a truncated distribution of conception times. *Demography* **12**: 291–301.

Sulloway, F.J. 1996. *Born to rebel.* Pantheon, New York.

Summerfield, J.A., Ryder, S., Sumiya, M., *et al.* 1995. Mannose binding protein gene mutations associated with unusual and severe infections in adults. *Lancet* **345**: 886–9.

Suzuki, H., Kurihara, Y., Takeya, M., *et al.* 1997. A role for macrophage scavenger receptors in atherosclerosis and susceptibility to infection. *Nature* **386**: 292–6.

Swartz, M.N. 1994. Hospital-acquired infections: diseases with increasingly limited therapies. *Proc. Natl. Acad. Sci. USA* **91**: 2420–7.

Swisher, C.C., Curtis, G.H., Jacob, T., *et al.* 1994. Age of the earliest known hominids in Java, Indonesia. *Science* **263**: 1118–21.

Symons, D. 1979. *The evolution of human sexuality.* Oxford University Press.

Taddei, F., Radman, M., Maynard Smith, J., *et al.* 1997. Role of mutator alleles in adaptive evolution. *Nature* **387**: 700–2.

Tajima, F. 1983. Evolutionary relationship of DNA sequences in finite populations. *Genetics* **105**: 437–60.

Takahata, N. 1991. Trans-species polymorphism of HLA molecules, founder principle, and human evolution. In J. Klein and D. Klein (eds) *Molecular evolution of the major histocompatibility complex.* Springer-Verlag, Heidelberg, pp. 29–49.

Takahata, N. 1993. Evolutionary genetics of human paleo-populations. In N. Takahata and A.G. Clark (eds) *Mechanisms of molecular evolution.* Sinauer, Sunderland, Massachusetts, pp. 1–21.

Takahata, N. and Nei, M. 1990. Allelic genealogy under overdominant and frequency-dependent selection and polymorphism of major histocompatibility complex loci. *Genetics* **124**: 967–78.

Takahata, N., Satta, Y., and Klein, J. 1995. Divergence time and population size in the lineage leading to modern humans. *Theor. Pop. Biol.* **48**: 198–221.

Tamin, A., Rota, P.A., Wang, Z.D., *et al.* 1994. Antigenic analysis of current wild type and vaccine strains of measles virus. *J. Infect. Dis.* **170**: 795–801

Tamura, K. and Nei, M. 1993. Estimation of the number of nucleotide substitutions in the control region of mitochondrial DNA in humans and chimpanzees. *Mol. Biol. Evol.* **10**: 512–26.

Tanner, J.M. 1973. The regulation of human growth. *Child Devel.* **34**: 817–47.

Targett, G.A. 1997. The search for an effective malaria vaccine. *J. Med. Microbiol.* **46**: 357–9.

Tatar, M., Promislow, D.E.L., Kaezaeli, A.A., and Curtinger, J.W. 1996. Age-specific patterns of genetic variance in *Drosophila melanogaster*. II. Fecundity and its genetic covariance with age-specific mortality. *Genetics* **143**: 849–58.

Tatusov, R.L., Mushegian, A.R., Bork, P., *et al.* 1996. Metabolism and evolution of *Haemophilus influenzae* deduced from whole-genome comparison with *Escherichia coli*. *Curr. Biol.* **6**: 279–91.

Tchuem Tchuenté, L.A., Southgate, V.R., Imbert-

Establet, D., and Jourdane, J. 1995. Change of mate and mating competition between males of *Schistosoma intercalatum* and *S. mansoni*. *Parasitology* **110**: 45–52.

Tchuem Tchuenté, L.A., Southgate, V.R., Combes, C., and Jourdane, J. 1996. Mating behaviour in schistosomes: are paired worms always faithful? *Parasitol. Today* **12**: 231–6.

Templeton, A.R. 1992. Human origins and analysis of mitochondrial DNA sequences. *Science* **255**: 737.

Templeton, A.R. 1993. The 'Eve' hypotheses: a genetic critique and reanalysis. *Am. Anthropol.* **95**: 51–72.

Templeton, A.R. and Sing, C.F. 1993. A cladistic analysis of phenotypic associations with haplotypes inferred from restriction endonuclease mapping. IV. Nested analyses with cladogram uncertainty and recombination. *Genetics* **134**: 659–69.

Templeton, A.R., Boerwinkle, E., and Sing, C.F. 1987. A cladistic analysis of phenotypic associations with haplotypes inferred from restriction endonuclease mapping. I. Basic theory and an analysis of alcohol dehydrogenase activity in *Drosophila*. *Genetics* **117**: 343–51.

Templeton, A.R., Crandall, K.A., and Sing, C.F. 1992. A cladistic analysis of phenotypic associations with haplotypes inferred from restriction endonuclease mapping and DNA sequence data. III. Cladogram estimation. *Genetics* **132**: 619–33.

Tenner, E. 1996. *Why things bite back: technology and the revenge of unintended consequences*. Knopf, New York.

Tenover, F.C. and Hughes, J.M. 1996. The challenges of emerging infectious diseases: development and spread of multiply-resistant bacterial pathogens. *JAMA* **275**: 300–4

Thaler, D.S. 1994. Sex is for sisters: intra genomic recombination of homology-dependent mutation as sources of evolutionary variation. *TREE* **9**: 108–10.

Thompson, E.A. and Neel, J.V. 1997. Allelic disequilibrium as a function of social and demographic history. *Am. J. Hum. Genet.* **60**: 197–204.

Thursz, M.R., Kwiatkowski, D., Allsopp, C.E., *et al.* 1995. Association between an MHC class II allele and clearance of hepatitis B virus in the Gambia. *N. Engl. J. Med.* **332**: 1065–9.

Tiercy, J.-M., Sanchez-Mazas, A., Excofffier, L., *et al.* 1992. HLA-DR polymorphism in a Senegalese Mandenka population: DNA oligotyping and population genetics of DRB1 specificities. *Am. J. Hum. Genet.* **51**: 592–602.

Tietze, C. 1957. Reproductive span and rate of reproduction among Hutterite women. *Fertil. Steril.* **8**: 89–97.

Tietze, C. 1959. Differential fecundity and effectiveness of contraception. *Eugenics Rev.* **50**: 231–7.

Tietze, C. 1968. Fertility after discontinuation of intrauterine and oral contraception. *Int. J. Fertil.* **13**: 385–9.

Tinbergen, N. 1963. On the aims and methods of ethology. *Zeit. Tierpsy.* **20**: 410–63.

Todd, J.A., Bell, J.I., and McDevitt, H.O. 1987. HLA-DQ beta gene contributes to susceptibility and resistance to insulin-dependent diabetes mellitus. *Nature* **329**: 599–604.

Todd, J.R., West, B.C., and McDonald, J.C. 1990. Human leukocyte antigen and leprosy: study in northern Louisiana and review. *Rev. Infect. Dis.* **12**: 63–74.

Tooby, J. and Cosmides, L. 1989. Evolutionary psychology and the generation of culture. Part I: Theoretical considerations. *Ethol. Sociobiol.* **10**: 29–49.

Tournamille, C., Colin, Y., Cartron, J.P., and Le-Van-Kim, C. 1995. Disruption of a GATA motif in the Duffy gene promoter abolishes erythroid gene expression in Duffy-negative individuals. *Nature Genet.* **10**: 224–8.

Trabuchet, G., Elion, J., Baudot, B., *et al.* 1991. Origin and spread of β-globin gene mutations in India, Africa, and Mediterranean: analysis of the 5′ flanking and intergenic sequences of βS and βC genes. *Hum. Biol.* **63**: 241–52.

Treffers, H.P., Spinelli, V., and Belser, N.O. 1954. A factor (or mutator gene) influencing mutation rates in *Escherichia coli*. *Proc. Natl. Acad. Sci. USA* **40**: 1064–71.

Trivers, R.L. 1972. Parental investment and sexual selection. In B. Campbell (ed.) *Sexual selection and the descent of man*. Aldine, Chicago, pp. 136–79.

Trivers, R.L. 1974. Parent–offspring conflict. *Am. Zool.* **14**: 249–64.

Trivers, R.L. and Willard, D.E. 1973. Natural selection of parental ability to vary the sex ratio of offspring. *Science* **179**: 90–2.

Troilo, D. and Judge, S.J. 1993. Ocular development and visual deprivation myopia in the common marmoset (*Callithrix jacchus*). *Vision Res.* **33**: 1311–24.

Tsuang, M.T. and Faraone, S.V. 1996. Genetics of Alzheimer's disease. *J. Formosan Med. Assoc.* **95**: 733–40.

Tyler Smith, W. 1856. Causes of abortion. *Lancet* **i**: 275–8.

Utermann, G., Langenback, V., Beisiegel, U., and Weber, W. 1980. Genetics of the apolipoprotein E system in man. *Am. J. Hum. Genet.* **32**: 339.

Utermann, G., Hardewig, A., and Zimmer, F. 1984. Apo E phenotypes in patients with myocardial infarction. *Hum. Genet.* **65**: 237–41.

Valladas, H., Reyss, J.L., Valladas, G., *et al.* 1988. Thermoluminescence dating of the Mousterian 'Proto-Cro-Magnon' remains from Israel and the origin of modern man. *Nature* **331**: 614–16.

Valverde, P., Healy, E., Jackson, I., *et al.* 1995. Variants of the melanocyte-stimulating hormone receptor gene are associated with red hair and fair skin in humans. *Nature Genet.* **11**: 328–30.

Van Assche, F.A. and Aerts, L. 1985. Long-term effect of diabetes and pregnancy in the rat. *Diabetes* **34**: 116–18.

van Baalen, M. and Sabelis, M.W. 1995*a*. The dynamics of multiple infection and the evolution of virulence. *Am. Nat.* **146**: 881–910.

van Baalen, M. and Sabelis, M.W. 1995*b*. The scope for virulence management—a comment on Ewald's view on the evolution of virulence. *TMB* **3**: 414–16.

van der Horst, C.J. and Gillman, J. 1941. The number of eggs and surviving embryos in *Elephantulus. Anat. Rec.* **80**: 443–52.

van Holde, K.E. 1989. *Chromatin.* Springer-Verlag, New York.

van Voorhies, W.A. 1992. Production of sperm reduces nematode lifespan. *Nature* **360**: 456–8.

Vander, A., Sherman, J.H., and Luciano, D.S. 1985. *Human physiology: the mechanisms of body function*, 5th edn. McGraw Hill, New York.

Vansina, J. 1984. Western Bantu expansion. *J. Afr. Hist.* **25**: 129–45.

Varkey, J.P., Muhirad, P.J., Minniti, A.N., *et al.* 1996. The *Caenorhabditis elegans* spe-26 gene is necessary to form spermatids and encodes a protein similar to the actin-associated proteins kelch and scruin. *Genes Dev.* **9**: 1074–86.

Vatsis, K.P., Wendell, W.W., Douglas, A.B., *et al.* 1995. Nomenclature for *N*-acetyltransferases. *Pharmacogenetics* **5**: 1–17.

Vaupel, J.W., Manton, K.G., and Stallard, E. 1979. The impact of heterogeneity in individual frailty on the dynamics of mortality. *Demography* **16**: 439–54.

Venter, J.C., Smith, H.O., and Hood, L. 1996. A new strategy for genome sequencing. *Nature* **381**: 364–6.

Vesell, E.S. 1990. Pharmacogenetic perspectives gained from twin and family studies. *Pharmacol. Ther.* **41**: 535–52.

Vidal, S., Tremblay, M.L., Govoni, G., *et al.* 1995. The Ity/Lsh/Bcg locus: natural resistance to infection with intracellular parasites is abrogated by disruption of the Nramp1 gene. *J. Exp. Med.* **182**: 655–66.

Vigilant, L., Pennington, R., Harpending, H., *et al.* 1989. Mitochondrial DNA sequences in single hairs from a southern African population. *Proc. Natl. Acad. Sci. USA* **86**: 9350–4.

Vigilant, L., Stoneking, M., Harpending, H., *et al.* 1991. African populations and the evolution of mitochondrial DNA. *Science* **253**: 1503–7.

Vineis, P., Bartsch, H., Caporaso, N., *et al.* 1994. Genetically based *N*-acetyltransferase metabolic polymorphism and low-level environmental exposure to carcinogens. *Nature* **369**: 154–6.

Vining, D.R. 1986. Social versus reproductive success—the central theoretical problem of human sociobiology. *Behav. Brain Sci.* **9**: 167–260.

Vogel, F. and Motulsky, A. 1996. *Human genetics: problems and approaches*, 3rd edn. Springer-Verlag, Heidelberg.

Voland, E. 1988. Differential infant and child mortality in evolutionary perspective: data from late 17th to 19th century Ostfriesland (Germany). In Betzig *et al.* (1989), pp. 253–61.

Voland, E. and Dunbar, R.I.M. 1995. Resource competition and reproduction: the relationship between economic and parental strategies in the Krummhörn population (1720–1874). *Hum. Nat.* **6**: 33–49.

Voland, E., Dunbar, R.I.M., Engel, C., and Stephan, P. 1997. Population increase and sex-biased parental investment in humans: evidence from 18th- and 19th-Century Germany. *Curr. Anthropol.* **38**: 129–35.

Vulliamy, T., Beutler, E., and Luzzatto, L. 1993. Variants of glucose-6-phosphate dehydrogenase are due to missense mutations spread throughout the coding region of the gene. *Hum. Mutat.* **2**: 159–67.

Waddle, D.M. 1994. Matrix test correlation tests support a single origin for modern humans. *Nature* **368**: 452–4.

Wainscoat, J.S., Hill, A.V.S., Boyce, A.L., *et al.* 1986. Evolutionary relationships of human populations from an analysis of nuclear DNA polymorphisms. *Nature* **319**: 491–3.

Wald, N.J. and Kennard, A. 1996. Prenatal screening for neural tube defects and Down syndrome. In D.L. Rimoin, J.M. Connor, and

R.E. Pyeritz (eds) *Emery and Rimoin's principles and practice of medical genetics*. Churchill Livingstone, San Francisco, pp. 545–62.

Wallman, J. 1994. Nature and nurture of myopia. *Nature* **371**: 201–2.

Wallman, J. and McFadden, S. 1995. Monkey eyes grow into focus. *Nature Med.* **1**: 737–9.

Ward, C.W. 1993. Progress towards a higher taxonomy of viruses. *Res. Virol.* **144**: 419–53.

Watson, E., Bauer, K., Aman, R., *et al.* 1996. mtDNA sequence diversity in Africa. *Am. J. Hum. Genet.* **59**: 437–44.

Watterson, G.A. 1978. The homozygosity test of neutrality. *Genetics* **88**: 405–17.

Watterson, G.A. 1986. The homozygosity test after a change in population size. *Genetics* **112**: 899–907.

Wedekind, C. 1994. Mate choice and maternal selection for specific parasite resistances before, during and after fertilization. *Phil. Trans. Roy. Soc. London B* **346**: 303–11.

Wedekind, C. and Folstad, I. 1994. Adaptive or non-adaptive immunosuppression by sex hormones? *Am. Nat.* **143**: 936–8.

Wedekind, C. and Füri, S. 1997. Body odour preference in men and women: aims for specific MHC-combinations or simply heterozygosity? *Proc. Roy. Soc. London B.* **264**: 1471–9.

Wedekind, C., Seebeck, T., Bettens, F., and Paepke, A.J. 1995. MHC-dependent mate preferences in humans. *Proc. Roy. Soc. London B* **260**: 245–9.

Wedekind, C., Chapuisat, M., Macas, E., and Rülicke, T. 1996. Non-random fertilization in mice correlates with the MHC and something else. *Heredity* **77**: 400–9.

Wei, X., Ghosh, S.K., Taylor, M.E., *et al.* 1995. Viral dynamics in human immunodeficiency virus type I infection. *Nature* **373**: 117–22.

Weidt, G., Deppert, W., Utermohlen, O., *et al.* 1995. Emergence of virus escape mutants after immunization with epitope vaccine. *J. Virol.* **69**: 7147–51.

Wein, L.M., Zenios, S.A., and Nowak, M.A. 1997. Dynamic multi-drug therapies for HIV: a control theoretic approach. *J. Theor. Biol.* **185**: 15–29.

Weinberg, E.D. 1984. Iron withholding: a defense against infection in neoplasia. *Physiol. Rev.* **64**: 65–102.

Weindruch, R. 1996. Caloric restriction and aging. *Sci. Am.* (Jan.): 32–8.

Weindruch, R. and Walford R.L. 1988. *The retardation of aging and disease by dietary restriction*. Charles C. Thomas, Springfield.

Weiser, J.N., Love, J.M., and Moxon, E.R. 1989. The molecular mechanism of phase-variation of *H. influenzae* lipopolysaccharide. *Cell* **59**: 657–65.

Weiss, G. and von Haeseler, A. 1996. Estimating the age of the common ancestor of men from the ZFY intron. *Science* **272**: 1359.

Weiss, K.M. 1995. *Genetic variation and human disease. Principles and evolutionary approaches*. Cambridge University Press.

Weiss, K.M. 1996. Is there a paradigm shift in genetics? Lessons from the study of human disease. *Mol. Phylogenet. Evol.* **5**: 1–7.

Wenegrat, B. 1990. *Sociobiological psychiatry: a new conceptual framework*. Lexington, Lexington, Mississippi.

Westendorp, R.G.J., Langermans, J.A.M., Huizinga, T.W.J., *et al.* 1997. Genetic influence on cytokine production and fatal meningococcal disease. *Lancet* **349**: 170–3.

White, N.J. 1997. Assessment of the pharmaco-dynamic properties of antimalarial drugs *in vitro* (submitted).

White, T.D., Suwa G., and Asfaw, B. 1994. *Australopithecus ramidus*, a new species of early hominid from Aramis, Ethiopia. *Nature* **371**: 306–12.

Whitfield, L.S., Sulston, J.E., and Goodfellow, P.N. 1995. Sequence variation of the human Y chromosome. *Nature* **378**: 379–80.

Whittam, T.S. 1995. Genetic population structure and pathogenicity. In Baumberg *et al.* (1995), pp. 217–45.

WHO DIAMOND Project Group. 1990. WHO multinational project for childhood diabetes. *Diabetes Care* **13**: 1062.

Widdowson, E.M. 1950. Chemical composition of newly born mammals. *Nature* **166**: 626–8.

Widdowson, E.M. 1980. Chemical composition and nutritional needs of the fetus at different stages of gestation. In H. Aebi and R. Whitehead (eds) *Maternal nutrition during pregnancy and lactation*. Hans Huber, Bern, pp. 39–48.

Widdowson, E.M. 1991. Contemporary human diets and their relation to health and growth: overview and conclusions. *Proc. Roy. Soc. London B* **334**: 289–95.

Wilcox, A.J., Weinberg, C.R., O'Connor, J.F., *et al.* 1988. Incidence of early loss of pregnancy. *N. Engl. J. Med.* **319**: 189–94.

Wilkinson, G.R., Guengerich, F.P., and Branch, R.A. 1992. Genetic polymorphism of *S*-mephenytoin hydroxylation. In Kalow (1992) pp. 657–85.

Willcox, M., Bjorkman, A., Brohult, J., *et al.* 1983. A case–control study in northern Liberia of *Plasmodium falciparum* malaria in haemoglobin S and beta-thalassaemia traits. *Ann. Trop. Med. Parasitol.* **77**: 239–46.

Williams, B. 1995. Westernized Asians and cardiovascular disease: nature or nurture? *Lancet* **345**: 401–2.

Williams, G.C. 1957. Pleiotropy, natural selection and the evolution of senescence. *Evolution* **11**: 398–411.

Williams, G.C. 1966. *Adaptation and natural selection.* Princeton University Press.

Williams, G.C. 1992. *Natural selection. Domains, levels, and challenges.* Oxford University Press.

Williams, G.C. 1997. *Plan and purpose in nature.* Basic Books, New York.

Williams, G.C. and Nesse, R.M. 1991. The dawn of Darwinian medicine. *Q. Rev. Biol.* **66**: 1–22.

Williams, T., Maitland, K., Bennett, S., *et al.* 1996. High incidence of malaria in a thalassaemic children. *Nature* **383**: 522–5.

Wills, C. 1989. *The wisdom of genes: new pathways in evolution.* Oxford University Press.

Wilson, D., Clark, A., Coleman, K., and Dearstyne, T. 1994. Shyness and boldness in humans and other animals. *TREE* **9**: 442–6.

Wilson, D.S. and Sober, E. 1994. Reintroducing group selection to the human behavioral sciences. *Behav. Brain Sci.* **17**: 585–607.

Wilson, J.G., Murphy, E.E., Wong, W.W., *et al.* 1986. Identification of a restriction fragment length polymorphism by a CR1 cDNA that correlates with the number of CR1 on erythrocytes. *J. Exp. Med.* **164**: 50–9

Wilson, J.G., Wong, W.W., Murphy, E.E., *et al.* 1987. Deficiency of the C3b/C4b receptor (CR1) of erythrocytes in systemic lupus erythematosus: analysis of the stability of the defect and of a restriction fragment length polymorphism of the CR1 gene. *J. Immunol.* **138**: 2708–10.

Wilson, M. and Daly, M. 1985. Competitiveness, risk taking, and violence: The young male syndrome. *Ethology and Sociobiology* **6**: 59–73.

Witte, W. 1997. Impact of antibiotic use in animal feeding on resistance of bacterial pathogens in humans. In D. J. Chadwick and J. Goode (eds) *Antibiotic resistance: origins, evolution, selection and spread.* John Wiley, Chichester, pp. 61–71.

Wolpert, L. 1992. *The unnatural nature of science.* Oxford University Press.

Wolpoff, M.H. 1989. Multiregional evolution. The fossil alternative to Eden. In P. Mellars and C.B. Stringer (eds) *The human revolution.* Princeton University Press, pp. 62–108.

Wolpoff, M.H. 1995. *Human evolution.* McGraw-Hill, New York.

Wong, A., Boutis, P., and Hekimi, S. 1995. Mutations in the clk-1 gene of *Caenorhabditis elegans* affect developmental and behavioral timng. *Genetics* **139**: 1247–59.

Wood Jones, F. 1929. *Man's place among the mammals.* Longmans, Green, & Co., New York.

Wood, J.W., Milner, G.R., Harpending, H.C., and Weiss, K.M. 1992. The osteological paradox: problems of inferring prehistoric health from skeletal samples. *Curr. Anthropol.* **33**: 343–70.

Wright, S. 1932. The roles of mutation, inbreeding, crossbreeding, and selection in evolution. *Proc. 6th Int. Congr. Genet.* **1**: 356–66.

Wright, S. 1934. Physiological and evolutionary theories of codominance. *Am. Nat.* **68**: 25–53.

Wright, S. 1982. Character change, speciation, and the higher taxa. *Evolution* **36**: 427–43.

Xhignesse, M., Lussier-Cacan, S., Sing, C.F., *et al.* 1991. Influences of common variants of apolipo-protein E on measures of lipid metabolism in a sample selected for health. *Arterioscler. Thromb.* **11**: 1100–10.

Yamazaki, K., Boyse, E.A., Miké, V., *et al.* 1976. Control of mating preference in mice by genes in the major histocompatibility complex. *J. Exp. Med.* **144**: 1324–35.

Yamazaki, K., Yamaguchi, M., Baranoski, L., *et al.* 1979. Recognition among mice. Evidence from the use of a Y-maze differentially scented by congenic mice of different major histo-compatibility types. *J. Exp. Med.* **150**: 755–60.

Yamazaki, K., Beauchamp, G.K., Ecorov, I.K., *et al.* 1983. Sensory distinction between H-2b and H-2bml mutant mice. *Proc. Natl. Acad. Sci. USA* **80**: 5685–8.

Yanagisawa, H., Fujii, K., Nagafuchi, S., *et al.* 1996. A unique origin and multistep process for the generation of expanded DRPLA triplet repeats. *Hum. Mol. Genet.* **5**: 373–9.

Yates, S.N.R. 1995. Human genetic diversity and selection by malaria in Africa. DPhil thesis, University of Oxford.

Ye, L., Miki, T., Nakura, J., *et al.* 1997. Association of a polymorphic variant of the Werner helicase gene with myocardial infarction in Japanese population. *Am. J. Med. Genet.* **68**: 494–9.

Yu, C-E., Oshima, J., Fu, Y.H., *et al.* 1996. Positional cloning of the Werner's syndrome gene. *Science* **272**: 193–4.

Yu, C-E., Oshima, J., Wijsman, E.M., *et al.* 1997. Mutations in the concensus helicase domains of the Werner syndrome gene. *Am. J. Hum. Genet.* **60**: 330–41.

Zahavi, A. 1975. Mate selection—a selection for a handicap. *J. Theor. Biol.* **53**: 205–14.

Zahavi, A. 1981. Natural selection, sexual selection and the selection of signals. In G.G.E. Scudder and J.L. Reveal (eds) *Evolution today. Proceedings of the Second International Congress of Systematics and Evolutionary Biology*, pp. 133–8. Hurst Institute, Carnegie-Mellon University, Pittsburgh.

Zambrano, M.M., Siegele, D.A., Almiron, M., Tormo, A., and Kolter, R. 1993. Microbial competition: *Escherichia coli* mutants that take over stationary phase cultures. *Science* **259**: 1757–60.

Zanotto, P.M. de A., Gibbs, M.J., Gould, E.A., and Holmes, E.C. 1996*a*. A reassessment of the higher taxonomy of viruses based on RNA polymerases. *J. Virol.* **70**: 6083–96.

Zanotto, P.M. de A., Gould, E.A., Gao, G.F., *et al.* 1996*b*. Population dynamics of flaviviruses revealed by molecular phylogenies. *Proc. Natl. Acad. Sci. USA* **93**: 548–53.

Zawadzki, P. and Cohan, F.M. 1995. The size and continuity of DNA segments integrated in *Bacillus* transformation. *Genetics* **141**: 1231–43.

Zerba, K.E., Ferrell, R.E., and Sing, C.F. 1996. Genotype–environment interaction: apolipoprotein E gene effects and age as an index of time and spatial context in the human. *Genetics* **143**: 463–78.

Zerylnick, C., Torroni, A., Sherman, S.L., and Warren, S.T. 1995. Normal variation at the myotonic dystrophy locus in global human populations. *Am. J. Hum. Genet.* **56**: 123–30.

Zhivotovsky, L.A. and Feldman, M.W. 1995. Microsatellite variability and genetic distances. *Proc. Natl. Acad. Sci. USA* **92**: 11549–52.

Ziegler, E.E., O'Donnell, A.M., Nelson, S.E., and Fomon, S.J. 1976. Body composition of the reference fetus. *Growth* **40**: 329–41.

Ziegler, R.G., Hoover, R.N., Abraham, M.Y., *et al.* 1996. Relative weight, weight change, height, and breast cancer risk in Asian-American women. *J. Natl. Cancer Inst.* **88**: 650–60.

Zinaman, M.J., Clegg, E.D., Brown, C.C., *et al.* 1996. Estimates of human fertility and pregnancy loss. *Fertil. Steril.* **65**: 503–9.

Zinser, H. 1935. *Rats, lice, and history.* Little, Brown, Boston.

Zwaan, B.J., Bijlsma, R., and Hoekstra, R.F. 1995. Direct selection of life span in *Drosophila melanogaster*. *Evolution* **49**: 649–59.

INDEX